14th International Symposium on Laser-Aided Plasma Diagnostics 2009

(LAPD14)

Journal of Physics: Conference Series Volume 227

Castelbrando, Treviso, Italy
21-24 September 2009

Editors:

Leonardo Giudicotti Roberto Pasqualotto

ISBN: 978-1-61738-622-0
ISSN: 1742-6588

Printed from e-media with permission by:

Curran Associates, Inc.
57 Morehouse Lane
Red Hook, NY 12571

Some format issues inherent in the e-media version may also appear in this print version.

Printed by Curran Associates, Inc. (2010)

For permission requests, please contact the Institute of Physics
at the address below.

Institute of Physics
Dirac House, Temple Back
Bristol BS1 6BE UK

Phone: 44 1 17 929 7481
Fax: 44 1 17 920 0979

techtracking@iop.org

Additional copies of this publication are available from:

Curran Associates, Inc.
57 Morehouse Lane
Red Hook, NY 12571 USA
Phone: 845-758-0400
Fax: 845-758-2634
Email: curran@proceedings.com
Web: www.proceedings.com

14th International Symposium on Laser-Aided Plasma Diagnostics 2009

(LAPD14)

Journal of Physics: Conference Series Volume 227

Castelbrando, Treviso, Italy
21-24 September 2009

ISBN: 978-1-61738-622-0
ISSN: 1742-6588

IOP Institute of Physics

TABLE OF CONTENTS

THE AKAZAKI LECTURE

GENERAL LECTURES

TOPICAL LECTURES

SHORT LECTURES

POSTER CONTRIBUTIONS

PREFACE

The Fourteenth International Symposium on Laser-Aided Plasma Diagnostics (LAPD14), was held from 21st to 24th September 2009 in Castelbrando, Treviso, Italy. The series of LAPD symposia was originally started at Kyushu University in 1983, and since then it has been organized each two years alternatively in Japan, Europe and the United States, traveling around the world five times.

Each LAPD Symposium brings together scientists working in different disciplines all related to the diagnostics of any type of plasma by laser or similar techniques. Researchers working on nuclear fusion, industrial process, low temperature plasma chemistry, laser development and material science, are invited to present prominent new diagnostic developments, with the aim of synergetic discussions. The broad spectrum of contributions represents one of the strengths of this symposium, which is an important, unique and fruitful source of cross-fertilization between these fields and a forum of discussions.

The scope of LAPD14 was very broad, including many techniques related to laser probing of plasmas: incoherent and coherent Thomson scattering, polarimetry, interferometry, reflectometry, laser induced fluorescence, laser absorption spectroscopy, laser photodetachment spectroscopy, cavity ringdown spectroscopy, Raman scattering, reflectometry, microwave diagnostics and related laser and hardware developments. LAPD14 was attended by 66 researchers, form 15 different countries who presented a total of 57 papers (13 General, 12 Topical, 10 Short talks and 23 Poster contributions). It is a tradition of LAPD that the first lecture of each meeting, which is more general and aiming to review prominent new developments, is called "the Akazaki lecture" in honor of Prof. Masanori Akazaki of Kyushu University, who was the first Chairman of LAPD from 1983 through 1989. The Akazaki lecture of LAPD14 was given by John Sheffield, who presented his work for the new edition of his worldwide known book "Plasma Scattering of Electromagnetic Radiation".

The site of LAPD14 was chosen for its comfortable environment and the superb location: Castelbrando is an ancient castle whose origin goes back to the Roman age, a fortress built to watch over one of the ancient routes crossing the Alps. Completely restored 10 years ago it is now a high class resort for business meetings and residential holidays, with conference and meeting rooms, restaurants, shops, spa and a beauty farm. The quality of the location and the quietness and beauty of the surroundings made Castelbrando an ideal site for LAPD14 and contributed to the success of the Symposium.

I hope that this volume of Proceedings will serve as a valuable record of the Symposium to all workers in the field. On behalf of the Local Organizing Committee, I would like to thank all participants for their contributions to this Symposium. I also acknowledge all members of the international scientific committee and Prof. Francesco Gnesotto, director of Consorzio RFX, for their support and cooperation.

Leonardo Giudicotti
(Chairman of the Local Organizer Commitee)

The LAPD14 participants in the Castelbrando inner yard

INTERNATIONAL SCIENTIFIC COMMITTEE:

- A. J. H. Donné, FOM-Institute voor Plasmafysica, The Netherlands (Chairman)
- U. Czarnetzki, Ruhr University, Bochum, Germany
- B. Graham, The Queen's University, Belfast, UK
- G. Hebner, Sandia National Laboratories, Albuquerque, USA
- K. Kawahata, National Institute for Fusion Science, Japan
- N. C. Luhmann, Jr., University of California, Davis, USA
- S. Sasaki, Nagoya University, Nagoya, Japan
- H. Soltwisch, Ruhr University, Bochum, Germany
- P. P. Woskov, Massachusetts Institute of Technology, Cambridge, USA

FORMER LAPD CHAIRMEN:

- M. Akazaki (1983 - 1989)
- D. E. Evans (1989 - 1993)
- H. F. Döbele (1993 - 1999)
- K. Muraoka (1999 - 2003)
- A. J. H. Donné (2003 -)

LOCAL ORGANIZERS:

- Leonardo Giudicotti, Consorzio RFX and Padova University (Chairman)
- Roberto Pasqualotto, Consorzio RFX
- Margherita Basso, Consorzio RFX
- Santolo De Benedictis, Institute of Inorganic Methodologies and Plasmas, CNR
- Paolo Innocente, Consorzio RFX
- Alberto Alfier, Consorzio RFX
- Enrico Scek Osman, Consorzio RFX

PREVIOUS LAPD MEETINGS

1983- Fukuoka, Japan: K. Muraoka (Organizer)

1985- Oxford, U.K.: D. Evans (Organizer)

1987- Lake Arrowhead, USA: N. C. Luhmann, Jr. (Organizer)

1989- Fukuoka, Japan: K. Muraoka (Organizer)

1991- Bad Honnef, Germany: F. Döbele (Organizer)

1993- Bar Harbor, USA: P. Woskov (Organizer)

1995- Fukuoka, Japan: K. Muraoka (Organizer)

1997- Doorwerth, Netherlands: A. J. H. Donné (Organizer)

1999- Lake Tahoe, USA: N. C. Luhmann, Jr. (Organizer)

2001- Fukuoka, Japan: K. Muraoka (Organizer)

2003- Les Houches, France: N. Sadeghi (Organizer)

2005- Snowbird, USA: N. C. Luhmann, Jr. (Organizer)

2007- Takayama, Japan: K. Kawahata (Organizer)

2009- Castelbrando, Italy: L. Giudicotti (Organizer)

List of participants

Alfier Alberto	Consorzio RFX, Padova	Italy	alberto.alfier@igi.cnr.it
Alonso Manuel	IST, Lisbon	Portugal	alonso@ipfn.ist.utl.pt
Ando Ayumi	Osaka University	Japan	ando@ppl.eng.osaka-u.ac.jp
Aramaki Mitsutoshi	Nagoya University	Japan	aramaki@nuee.nagoya-u.ac.jp
Bilkova Petra	IPP Prague	Czech Republic	bilkova@ipp.cas.cz
Brombin Matteo	Consorzio RFX, Padova	Italy	matteo.brombin@igi.cnr.it
Brower David	UCLA	USA	brower@ucla.edu
Czarnetzki Uwe	Ruhr University Bochum	Germany	uwe.czarnetzki@ep5.rub.de
De Benedictis Santolo	CNR, IMIP Bari	Italy	santolo.debenedictis@cnr.it
De La Rosa M.Inmaculada	Universidad de Valladolid	Spain	delarosa@opt.uva.es
Den Hartog Daniel	Univ. of Wisconsin Madison	USA	djdenhar@wisc.edu
Dilecce Giorgio	CNR IMIP Bari	Italy	giorgio.dilecce@ba.imip.cnr.it
Donné Tony	FOM Rijnhuizen	The Netherlands	donne@rijnhuizen.nl
Dvorak Pavel	Masaryk University	Czech Republic	pdvorak@physics.muni.cz
Dzierzega Krzysztof	Jagiellonian University	Poland	krzycho@netmail.if.uj.edu.pl
Ehlbeck Joerg	INP-Greifswald	Germany	ehlbeck@inp-greifswald.de
Engeln Richard	Eindhoven University	The Netherlands	r.engeln@tue.nl
Fassina Alessandro	Consorzio RFX, Padova	Italy	fassina@igi.cnr.it
Foley Jill (Elizabeth)	Nova Photonics/PPPL	USA	efoley@pppl.gov
Gil Christophe	CEA, Cadarache	France	christophe.gil@cea.fr
Giudicotti Leonardo	Consorzio RFX, Padova	Italy	leonardo.giudicotti@igi.cnr.it
Glosik Juraj	Charles University Prague	Czech republic	juraj.glosik@mff.cuni.cz
Hatae Takaki	JAEA	Japan	hatae.takaki@jaea.go.jp
Hong Suk-Ho	NFRI	Korea	sukhhong@nfri.re.kr
Innocente Paolo	Consorzio RFX, Padova	Italy	paolo.innocente@igi.cnr.it
Ito Tsuyohito	Osaka University	Japan	tsuyohito@ppl.eng.osaka-u.ac.jp
Kantor Mikhail	Forschungszentrum Juelich	Russia	m.kantor@fz-juelich.de
Kawahata Kazuo	NIFS	Japan	kawahata@lhd.nifs.ac.jp
Kawase Kodo	RIKEN, Nagoya Univ.	Japan	kawase@nuee.nagoya-u.ac.jp
Kempkes Philipp	Ruhr University Bochum	Germany	philipp.kempkes@ep5.rub.de
Knake Nikolas	Ruhr University Bochum	Germany	nikolas.knake@rub.de
Krychowiak Maciej	IPP-Greifswald	Poland	cak@ipp.mpg.de
Lee Jongha	NFRI	Korea	jhlee@nfri.re.kr
Luggenhoelscher Dirk	Ruhr-University Bochum	Germany	dirk.luggenhoelscher@ep5.rub.de
Meo Fernando	DTU, Risø	Denmark	fernando.meo@risoe.dk
Müller Sarah	Ruhr University Bochum	Germany	sarah.mueller@ep5.rub.de
Mukhin Eugene	Ioffe Institute, St. Petersburg	Russia	e.mukhin@mail.ioffe.ru
Nam Yongun	NFRI	Korea	yunam@nfri.re.kr
Nielsen Per	Consorzio RFX, Padova	Italy	aogp@kabelmail.dk
Nishiura Masaki	NIFS	Japan	nishiura@nifs.ac.jp
Park Hyeon K.	POSTECH	Korea	hyeonpark@postech.ac.kr
Pasqualotto Roberto	Consorzio RFX, Padova	Italy	roberto.pasqualotto@igi.cnr.it
Perez Concha	University Valladolid	Spain	concha@opt.uva.es
Riley David	Queens University Belfast	UK	d.riley@qub.ac.uk
Roepcke Juergen	INP-Greifswald	Germany	roepcke@inp-greifswald.de
Roke Sylvie	Max Planck Institut	The Netherlands	roke@mf.mpg.de
Sadeghi Nader	Univ. Joseph Fourier	France	nader.sadeghi@ujf-grenoble.fr
Saito Teruo	University of Fukui	Japan	saitot@fir.u-fukui.ac.jp
Sasaki Koichi	Nagoya University	Japan	sasaki@nuee.nagoya-u.ac.jp
Scannell Rory	CCFE Culham	UK	rory.scannell@ukaea.org.uk
Schweer Bernd	FZ Juelich, IEF-Plasmaphysik	Germany	b.schweer@fz-juelich.de
Sheffield John	University of Tennessee	USA	dacejohn@aol.com
Siciliano Maria Vittoria	INFN Lecce	Italy	sicmaria@le.infn.it
Skiff Fred	University of Iowa	USA	frederick-skiff@uiowa.edu
Smith Roger	University of Washington	USA	ogersmith@gmail.com
Tolstyakov Sergey	Ioffe Institute, St. Petersburg	Russia	sergey.tolstyakov@mail.ioffe.ru
Tsuji-Iio Shunji	Tokyo Tech	Japan	siio@nr.titech.ac.jp
Van der Meiden Hennie	FOM Rijnhuizen	The Netherlands	meiden@rijnhuizen.nl
Van Harskamp Willem-Jan	Eindhoven University	The Netherlands	w.e.n.harskamp@tue.nl
Vasina Petr	Masaryk University	Czech Republic	vasina@physics.muni.cz
Woskov Paul	MIT, Cambridge	USA	ppwoskov@mit.edu
Yokota Yuya	Kyushu University	Japan	yokotay5@asem.kyushu-u.ac.jp

Table of Contributions

The Akazaki Lecture

AK	John Sheffield	Updating "Plasma Scattering of Electromagnetic Radiation"	J. Sheffield

General Lectures

G1	Takaki Hatae	Progress in development of the advanced Thomson scattering diagnostics	T. Hatae, J. Howard, N. Ebizuka, H. Yoshida, M. Nakatsuka, H. Fujita, K. Narihara, I. Yamada, H. Funaba, Y. Hirano, H. Koguchi, S. Kajita and O. Naito
G2	Giorgio Dilecce	LIF diagnostics in volume and surface dielectric barrier discharges at atmospheric pressure	G. Dilecce, P. F. Ambrico, M. Simek and S. De Benedictis
G3	Hyeon K. Park	Collective scattering system for transport study on NSTX	W. Lee, D. R. Smith, E. Mazzucato, H. K. Park, C. W. Domier and N. C. Luhmann, Jr.
G4	Juergen Röpcke	Quantum cascade laser absorption spectroscopy - a new method to study molecular plasma components	J. Röpcke, S. Glitsch, P. Davies, F. Hempel, N. Lang, A. Rousseau, S. Wege and S. Welzel
G5	Paolo Innocente	Scanning beam medium infra-red interferometry for plasma density measurements	P. Innocente, A. Canton, C. Mazzotta and O. Tudisco
G8	Elizabeth Foley	The motional Stark effect with laser-induced fluorescence diagnostic	E. L. Foley and F. M. Levinton
G9	Mitsutoshi Aramaki	Measurement of neutral flow velocity in an ECR plasma using tunable diode laser LIF spectroscopy combined with saturated absorption spectroscopy	M. Aramaki, K. Ogiwara, S. Etoh, S. Yoshimura and M. Y. Tanaka
G10	Sylvie Roke	Sum frequency generation for plasma wall interaction	S. Roke
G11	Fernando Meo	First results and analysis of collective Thomson scattering (CTS) fast ion distribution measurements on ASDEX Upgrade	F. Meo, M. Stejner, M. Salewski, H. Bindslev, T. Eich, V. Furtula, S. B. Korsholm, F. Leuterer, F. Leipold, P. K. Michelsen, D. Moseev, S. K. Nielsen, B. Reiter, J. Stober, D. Wagner, P. Woskov and the ASDEX Upgrade team
G12	Nader Sadeghi	Laser Thomson scattering, Raman scattering and laser- absorption diagnostics of high pressure microdischarges	V. M. Donnelly, S. G. Belostotskiy, D. J. Economou and N. Sadeghi

Topical Lectures

T1	Koichi Sasaki	Diode laser cavity-ringdown absorption spectroscopy for measuring $N_2(A^3\Sigma^+_u)$ density in plasmas	Y. Horikawa, K. Kuriharaand and K. Sasaki
T2	Teruo Saito	Development of terahertz gyrotrons and their application to CTS on LHD	T. Saito, T. Notake, Y. Tatematsu, A. Fujii, S. Ogasawara, La Agusu, T. Idehara, S. Kubo, T. Shimozuma, K. Tanaka, M. Nishiura, K. Kawahata and V. N. Manuilov
T3	Masaki Nishiura	Initial result of collective Thomson scattering using 77 GHz gyrotron for bulk and tail ion diagnostics in the Large Helical Device	M. Nishiura, S. Kubo, K. Tanaka, N. Tamura, T. Shimozuma, T. Mutoh, K. Kawahata, T. Watari, T. Saito, Y. Tatematsu, T. Notake and LHD experiment group
T6	Roger J. Smith	Pulsed polarimetry techniques for fusion plasmas	R. J. Smith
T7	Ayumi Ando	Electron density measurement for plasmas by Terahertz time- domain spectroscopy	A. Ando, H. Kitahara, T. Kurose, K. Kitano, K. Takano, M. Tani, M. Hangyo and S. Hamaguchi
T8	Fred Skiff	Time resolved LIF with a fast-scanning diode laser	F. Skiff and V. Patel
T11	Tsuyohito Ito	Electric field measurements at near-atmospheric pressure by coherent Raman scattering of laser beams	T. Ito, K. Kobayashi, S. Mueller, U. Czarnetzki, and S. Hamaguchi

Short Lectures

S1	Santolo De Benedictis	LIF investigations of O and NO products in air like RF plasma jet	S. De Benedictis, P. F. Ambrico and G. Dilecce
S2	Nikolas Knake	Investigations on the generation of atomic oxygen inside a capacitively coupled atmospheric pressure plasma jet	N. Knake, D. Schröder, J. Winter and V. Schulz-von der Gathen
S3	Paul Woskov	Observations and modeling of the electron cyclotron emission background in the Levitated Dipole Experiment	P. P. Woskov, J. Kesner, D. T. Garnier, M. E. Mauel and S. H. Nogami
S4	Kazuo Kawahata	Recent progress in two-color laser diagnostics	K. Kawahata, T. Akiyama, K. Tanaka, S. Okajima and K. Nakayama
S5	Daniel Den Hartog	Pulse-burst operation of standard Nd:YAG lasers	D. J. Den Hartog, J. R. Ambuel, M. T. Borchardt, J. A. Reusch, P. E. Robl and Y. M. Yang
S7	Maciej Krychowiak	LIF measurements for validation of collisional-radiative modelling of atomic helium in the edge of a fusion plasma	M. Krychowiak, P. Mertens, R. König, B. Schweer, S. Brezinsek, O. Schmitz, M. Brix, U. Samm, R. Wolf and T. Klinger
S9	Jörg Ehlbeck	Laser induced fluorescence measurements on W- and Ba atoms eroded from fluorescent lamp electrodes	J. Ehlbeck, K. Rackow, F. Sigeneger, D. Uhrlandt, K.-D. Weltmann, S. Hadrath and G. Lieder
S10	Juraj Glosik	Application of NIR – CRDS for state selective study of recombination of para and ortho H_3^+ ions with electrons in low temperature plasma	J. Varju, Š. Roučka, T. Kotrík, R. Plašil and J. Glosík

Poster Contributions

P01	Manuel Alonso	Reconstruction activities and first results from the Thomson scattering diagnostic on the TCABR tokamak	M. P. Alonso, A. C. A. Figueiredo, C. A. F. Varandas, F. O. Borges, J. Leonardo, J. I. Elizondo, J. H. F. Sever, O. C. Usuriaga, E. K. Sanada, R. M. O. Galvão, L. A. Berni and M Machida
P02	Rory Scannell	Absolute calibration of LIDAR Thomson scattering systems on large fusion devices	R. Scannell, G. Naylor and M. Walsh
P04	Krzysztof Dzierzega	Thomson scattering from laser induced plasma in air	K Dzierzega, A. Mendys, S. Pellerin, E. Thouin, G. Travaille, B. Bousquet, L. Canioni and B. Pokrzywka
P05	Alessandro Fassina	Edge Thomson scattering in RFX-mod: operation and first measures	A. Alfier, A. Fassina, R. Pasqualotto, S. Dal Bello, V. Cervaro, L. Lotto and M. Tollin
P06	Matteo Brombin	First measurements of the new multichannel far-infrared polarimeter on RFX-mod	M. Brombin, F. Auriemma, A. Canton, L. Giudicotti, P. Innocente and E. Zilli
P07	Christophe Gil	Analysis of the JET FIR interferometer beam phase changes during plasmas and application for fast fringe jump corrections by electronics	C. Gil, A. Boboc, A. Barbuti, P. Pastor and JET EFDA contributors
P08	Yongun Nam	In-vessel optic system design of FIR interferometer/polarimeter system for KSTAR	Y. U. Nam and J. W. Juhn
P09	Yuya Yokota	Reconstruction method of X-mode ultrashort-pulse reflectometry in LHD	Y. Yokota, A. Mase, K. Uchino, Y. Kogi, T. Tokuzawa, K. Kawahata, Y. Nagayama and H. Hojo
P10	Dirk Luggenhölscher	Argon ion velocity distributions in a helicon discharge measured by laser induced fluorescence	D. Luggenhölscher, Y. Celik, Y. K. Pu and U. Czarnetzki
P11	Alberto Alfier	Optimized in-situ window cleaning system by laser blow-off through optical fiber	A. Alfier, R. Pasqualotto, L. Giudicotti, V. Cervaro and L. Franchin
P12	Maria Vittoria Siciliano	Velocity distribution of ions in UV laser induced plasma plumes	A. Lorusso, V. Nassisi, L. Velardi and M. V. Siciliano
P13	M. Inmaculada de la Rosa	Determination of electric field strength and kinetic temperature in the cathode fall region of a hollow cathode discharge	M. I. de la Rosa, C. Pérez, K. Grützmacher, A. B. Gonzalo and J. A. del Val
P14	Concha Pérez	Characterization of hollow cathode fall field strength measured by Doppler-free two-photon optogalvanic spectroscopy via Stark splitting of the 2S level of hydrogen and deuterium	C. Pérez, M. I. de la Rosa and K. Grützmacher
P15	Sarah Müller	Electric field measurements in near-atmospheric pressure nitrogen and air based on a four-wave mixing scheme	S. Müller, T. Ito, K. Kobayashi, D. Luggenhölscher, U. Czarnetzki and S. Hamaguchi
P16	Mikhail Kantor	Advances of Thomson scattering diagnostic on the TEXTOR tokamak	M. Y. Kantor and the TEXTOR team
P17	Jongha Lee	Progress in development of the KSTAR Thomson scattering system	J. Lee and S. T. Oh
P18	Eugene Mukhin	Thomson scattering diagnostics for ITER divertor	E. E. Mukhin, A. S. Kukushkin, S. Yu Tolstyakov, M. M. Kochergin, V. V. Semenov, G. S. Kurskiev, A. G. Razdobarin, K. A. Podushnikova and P. Andrew
P19	Roberto Pasqualotto	Calibration methods for ITER core LIDAR	R. Pasqualotto, L. Giudicotti, A. Alfier and M. J. Walsh

P20	Sergey Tolstyakov	Near-infrared tokamak plasma spectroscopy in support of divertor Thomson scattering diagnostics development for ITER	S. Yu Tolstyakov, E. E. Mukhin, M. M. Kochergin, G. S. Kurskiev, V. V. Semenov, A. G. Razdobarin, K. A. Podushnikova, A.V. Zabuga and Globus-M1 team, V. S. Lisitsa, M. G. Levashova, V. A. Soukhanovskii and NSTX team, M. Beurskens, S. Brezinsek, A. Meigs and JET5 team, P. Andrew
P22	Philipp Kempkes	Electron density measurement in a pulsed-power plasma by FIR laser beam deflection and/or interferometry	P. Kempkes, F. Mackel and H. Soltwisch
P23	Suk-Ho Hong	Developing in-situ ellipsometry for tokamak discharges in KSTAR	S. H. Hong and W. C. Kim

Influence on Ar*($^3P_2$) or He*($^3S_1$) metastable atoms and electron densities of dust formation in argon and helium + 6% acetylene CCP plasmas

N. Sadeghi[1] , I. Stefanovic[2] , B. Sikimic[2] , J. Winter[2])

[1] *Laboratoire de Spectrometrie Physique, University Joseph Fourier and CNRS, Grenoble, France*
E-mail: Nader.Sadeghi@ujf-grenoble.fr
[2] *Institut für Experimentalphysik II, Ruhr-Universität Bochum, 44780 Bochum, Germany*

Dusty plasma models predict important differences in plasma characteristics when nanoparticles are introduced in a rare gas plasma. In the present work, we study the behaviour of metastable atoms and electron densities (N_e) in the discharge and in the afterglow of 0.1 mbar, 8 sccm flow rate Ar or He + 6% C_2H_2, symmetrical CCP discharge (30 cm diameter, 7 cm gap electrodes) at RF power ranging from 10 to 80 W. Density and axial distribution of metastable atoms are deduced by absorption an LIF experiments with an external cavity Diode Laser (DL) and N_e was measured with a 26.5 GHz µ-wane interferometer. To study the kinetics of formation and destruction of these species, the RF power was square wave modulated at 100 Hz. When the plasma is struck in pure argon (Fig. 1) the steady state Ar*($^3P_2$)= 2 10^{16} m^{-3} and N_e=4 10^{15} m^{-3} are reached during the discharge and the decay time in the afterglow is about 3 ms for Ar*. When 0.5 sccm of C_2H_2 is added to the flow, the Ar* density and its decay time are reduced by almost 20 (due to the quenching by C_2H_2) but N_e is almost not changed and its decay becomes faster. A few minutes after starting C_2H_2, nanoparticles are formed. This leads to and almost 50 times enhancement in Ar* and a decrease by an order of magnitude of N_e. The most astonishing is the strong peak in the afterglow of N_e, which indicates an electron production mechanism in the afterglow. When the C_2H_2 flow is stopped, nanoparticles remain trapped between electrodes for several minutes and the behaviours of Ar* and N_e are again modified. Different mechanisms responsible for these important modifications in Ar* and N_e behaviour will be discussed.

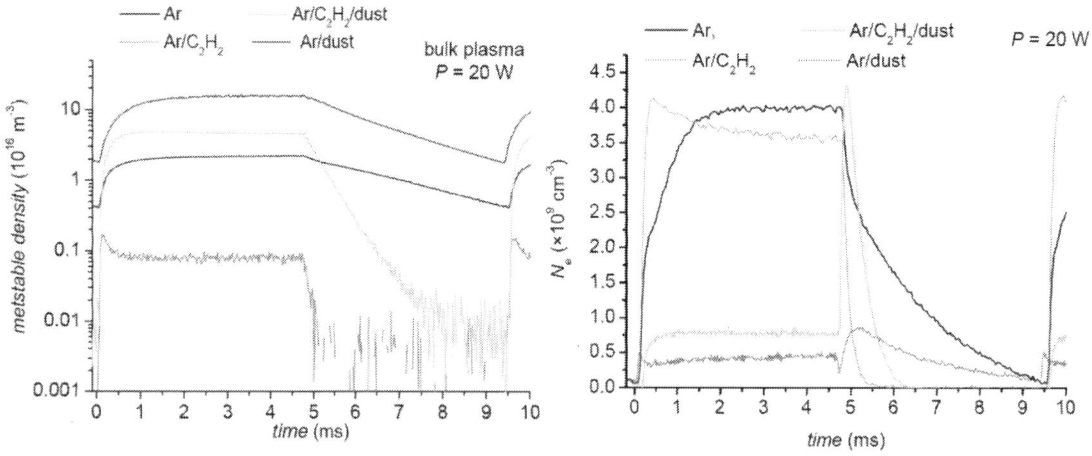

Fig. 1 Variation during the RF pulse (0-5 ms) and in the afterglow of Ar*($^3P_2$) density (left) and electron density (right) at 20 W of RF power in pure argon (black), Ar + C_2H_2 (red), Ar + C_2H_2 with dust (green) and Ar + dust (blue) at the reactor centre.

Terahertz light sources and imaging applications

K. Kawase[1,2], T. Shibuya[1,2], S. Hayashi[2], K. Suizu[1]

[1] *Nagoya University, Furo-cho, Chikusa-ku, Nagoya 464-8603, Japan*
[2] *RIKEN, Hirosawa 2-1, Wako 351-0198, Japan*

Terahertz waves have the property of being able to pass through various objects, such as paper, vinyl, plastics, textiles, ceramics, semiconductors, lipids and powders, just like radio waves. Terahertz waves are also able to image substances, just like X-rays, although unlike Terahertz waves, they do not present risks of irradiation. The property of terahertz waves lies in moderate transmission and moderate absorption. Whereas X-rays sometimes render objects invisible because of their excessive strength of penetration, terahertz waves make things that cannot be seen with X-rays visible, because their strength of penetration is weaker.

Metaphorically speaking, characters written with faint-colored ink on paper cannot be seen under strong light but they become visible when viewed under moderate illumination. The ability of the THz wave to pass through many packaging materials will allow the nondestructive and noninvasive inspections for industries. Our group has been conducting research activities in several directions within the THz field. We have been developing novel widely tunable THz-wave sources using nonlinear crystals, and we suggested a whole range of real-life applications.

Among our research activities for THz sources, we can mention: i) Injection seeded terahertz-wave parametric generators using a microchip laser or a multi-mode laser; ii) Difference frequency generation using a DAST crystal with very high damage threshold; iii) Extremely frequency widened terahertz wave generation in Cherenkov-type radiation using a $LiNbO_3$ slab waveguide; etc.

Among our research activities for applications, we can mention: i) THz CT for the inspection of soot distribution in a ceramic filter ii) Time of flight THz reflection tomography for multi layered thin coatings; iii) THz spectroscopic imaging for the detection of illicit drugs in the envelopes; iv) Laser-THz emission microscope for semiconductor device inspection; iv) Metal mesh sensor for the measurements of molecular bindings or thin film thickness; v) Si rod waveguide sensor for the measurement of liquid; etc.

Although still at the beginning of a long road, the applications of THz waves have started to make visible progress and the results shown here represent a part our contribution to the field. A whole range of other applications for industrial quality testing are underway in many research groups around the world, and are already receiving much interest both from physics investigators and from the those specific communities. Research on improving the nonlinear optical THz-wave sources, performed in our group as well as elsewhere, will probably give the field more chances to find novel applications.

Optical Thomson scatter from a laser-ablated magnesium plume

D. Riley, A. Delserieys, F.Y. Khattak , C.L.S. Lewis

School of Mathematics and Physics, Queen's University of Belfast, Belfast BT7 1NN, UK

We have carried out an optical Thomson scatter study of a KrF laser ablated Mg plume. We have studied the evolution of the electron temperature and density at distances 2mm to 5mm from the target surface. We have observed that the electron density falls more rapidly than the atomic density and believe this is a result of rapid dielectronic recombination. A comparison of the electron density profile and evolution with simple hydrodynamic modelling indicates that there is strong absorption of the laser in the plasma vapour above the target, probably due to photo-ionization. We also conclude that an isothermal model of expansion better fits the data than an isentropic expansion.

Laser-Based Measurement of Magnetic Fluctuation-Induced Particle Flux in a High-Temperature Plasma

D.L. Brower

Department of Physics and Astronomy
University of California-Los Angeles
Los Angeles, California 90095

Particle transport arising from stochastic magnetic fields has long been important in fusion research. Stochastic magnetic fields can arise from global tearing instabilities that often underlie the sawtooth oscillation. Conversely, stochastic fields have also been externally imposed to mitigate the effect of edge localized modes (ELMs) by locally enhancing edge transport in Tokamaks. Radial particle flux due to stochastic magnetic fields is given by $\Gamma_r = <\delta n \delta V_r> \approx V_{//,\alpha} \frac{<\delta n \delta b_r>}{B}$, where α is the particle species, <...> represents a flux surface average, δn are density fluctuations, B and δb_r are the mean and fluctuating magnetic field, and $V_{//}$ is the particle parallel velocity. For the first time, this flux is directly measured in the core of a high-temperature plasma by using a combination of high-speed Faraday rotation and differential-interferometry measurements. The measured electron particle flux can account for the density change on axis during a sawtooth collapse.

Wall characterisation by laser induced desorption, laser induced ablation and laser induced breakdown spectroscopy

B. Schweer, N. Gierse, A. Huber, V. Philipps, M. Zlobinski

Forschungszentrum Jülich GmbH, Association EURATOM-FZ Jülich, Institut für Energietechnik- Plasmaphysik,Trilateral Euregio Cluster, 52425 Jülich, Germany

Laser based methods in combination with spectroscopy are proposed for in-situ characterisation of the first wall in tokamaks. The lasers are used to heat up a defined area at a surface, which then leads to desorption or ablation depending on the applied power and energy densities.

Moderate power densities of about 100 kW/cm^2 for about a few ms are applied for laser induced thermal desorption. Nearly all the gaseous inventory mainly hydrogen isotopes in a surface layer of more than 100 μm thickness is released in a single laser shot, fired during the tokamak discharge. The molecules penetrate into the edge plasma where they are ionized, dissociated and excited. The characteristic line radiation (H_α) is measured absolutely and the time integrated signal can be converted into an inventory using conversion factors which are depending on plasma parameter.

Much higher power densities of 0.5 GW/cm^2 are applied for laser induced ablation spectroscopy. The energy density is above the heat of evaporation or sublimation and the released particles in most cases evaluate a plasma that recombines rapidly. The neutral atoms and clusters penetrate into the edge plasma where they are excited and ionised. The characteristic line intensities of the ablated material are qualitatively measured with dispersive instruments (spectrometer, interference filter) and quantitatively by absolute calibrated detectors (CCD cameras). Therefore for determination of the ablated material the atomic and spectroscopic data of the species must be known.

Also without tokamak discharges the laser induced plasma of the ablated material in front of the surface can be observed spectroscopically. This method is known as laser induced breakdown spectroscopy and applied for material analysis from remote locations (e. g. liquid steel).

Beside the monitoring of surface conditions on plasma facing components in tokamaks all three laser based methods are sensitive diagnostics in laboratory experiments. Because only small areas are exposed by the laser radiation, locally resolved measurements on surface layers are feasible. The released particles can be analysed quantitatively in a quadrupole mass analyser (QMA). For desorbed particles (e.g. hydrogen, hydrocarbon) the QMA can by calibrated absolutely by a known leakage source.

Due to the short existence of the laser induced ablation source (<10 ns) the energy distribution of ablated material can be deduced from time of flight measurements. Absolute measurements of the material fluence are difficult because they need the knowledge of the ionisation efficiency of the QMA for the different masses. In most cases the signals are compared with those from known surfaces.

Here the simultaneous application of laser induced breakdown spectroscopy might support the absolute characterisation of the surface composition.

The talk will give an overview about the principles, constraints and sensitivity limits of the diagnostics for application in a tokamak and laboratory experiment. The properties of the different lasers and detectors are discussed in detail.

Experimental Results from KSTAR Commissioning and Laser Aided Diagnostics on KSTAR*

H. K. Park[1] and KSTAR team[2]
[1] *Pohang University of Science and Technology, Pohang, 790-784, Korea*
[2] *Natioanl Fusion Research Institute, Daejon 305-333, Korea*

Flawless progress of KSTAR during 2008 campaign such as a successful cool down and commissioning of all superconducting systems has been notable. Achieved plasma parameters during this campaign, exceeded the goal of the first plasma (Ip=100kA, pulse length of 100 msec. at 1.5T) in all aspects with the limited voltage and power supply capability has been impressive and provided a good foundation for the success of the 2009 program. Very ambitious goals and schedule for 2010 will be presented. Goal of shaped DND, possibly H-mode plasmas will be pursued with significant facility upgrades of power supplies, in-vessel control coils (IVCC), in-vessel components, heating systems, diagnostics, etc. The long-range program plan is well thought out to provide strong contributions to ITER and the basis for steady-state high performance DEMO. To support advanced physics research on KSTAR for near immediate (2009 and 2010 campaigns) and long term operation, diagnostic preparation is in progress through National Fusion Research Institute (NFRI) staffs and University fusion science programs in collaboration with researcher at NFRI. The essential basic profile diagnostic systems such as Thomson Scattering will be discussed in this paper together with the advanced diagnostic tools such as multi-dimensional imaging tools for unprecedented physics research to be developed for KSTAR will be included.

* Work is supported by Korea Research Foundation

Photo detachment measurements to study the role of negative ions on the excited hydrogen atom density in a magnetized plasma expansion

W.J. van Harskamp, D. C. Schram, M.C.M. van de Sanden, R. Engeln

Department of Applied Physics, Plasma and Materials Group, Eindhoven University of Technology, Eindhoven, The Netherlands

The emission of a weakly magnetized expanding hydrogen plasma jet is dominated by the red H_α emission in the first few centimetres from the exit of the source, but changes to blue at larger distances from the exit due to emission of the higher Balmer lines ($n \geq 4$). Moreover, higher electronic states ($n \geq 4$) of H atoms become stronger populated than the lower states [1]. The reaction pathway proposed for the formation of these highly excited hydrogen atoms is via mutual recombination of positive (H_2^+) and negative ions (H^-). Both species have ro-vibrationally excited hydrogen molecules ($H_2^{r,v}$) as the precursor in the formation process: H_2^+ is formed via charge transfer of H^+ with $H_2^{r,v}$, while H^- is formed by dissociative attachment of electrons with $H_2^{r,v}$. The excited hydrogen molecules have been detected with laser induced fluorescence (LIF) in the vacuum UV [2]. In this contribution we will focus on the role of H^- on the population of the excited hydrogen atoms.

To study the role of negative hydrogen atoms we have used the photo-detachment technique. Laser radiation with a wavelength of 1064 nm, a pulse duration of 8 ns and a beam diameter of 10 mm is directed into the plasma jet perpendicular to the flow direction. The photon energy of the laser light is sufficiently high to remove the extra electron of the H^- ion that is attached to the atom with a binding energy of 0.75 eV. If we assume that the population into the excited H-atom states is partly due to mutual ion-ion recombination, the destruction of H^- ions, through the laser-induced photo-detachment process, will disturb the plasma chemistry and thus the emission originating from the excited states. The change in emission of the Balmer-α line is recorded at the center of the crossing of the plasma jet and the laser beam.

The emission of the Balmer-α line is measured to be constant during steady state operation. When pulsed laser radiation is directed into the plasma jet, the emission of the Balmer-α line decreases temporally. This decrease is up to 50 % of the steady-state intensity. It will be argued that the laser induces a decrease in the H^- ion density, which results in a decease in population of excited hydrogen atoms via the mutual ion-ion recombination process (as is mentioned above).

[1] J.J.A. van den Dungen, O. Gabriel, H.S.M.M. Elhamali, D.C. Schram, R. Engeln, *IEEE Transactions on Plasma Science* **36**, 1028 (2008)
[2] O. Gabriel, et al., *Phys. Rev. E* **78**, 016407 (2008)

Forward coherent Thomson scattering for ion temperature measurements on Magnum-PSI: a feasibility study

H.J. van der Meiden

FOM-Institute for Plasma Physics Rijnhuizen, Association EURATOM-FOM, P.O.Box 1207, 3430 BE Nieuwegein, The Netherland.*
** partners in the Trilateral Euregio Cluster*

In this presentation, forward coherent Thomson scattering (CTS) is proposed to measure the scattering spectrum reflecting the ion velocity distribution of low temperature plasmas, the so-called ion feature. CTS is a direct non-intrusive diagnostic method to provide the ion temperature and for instance the axial velocity of a plasma jet. These parameters are required to determine the total particle flux produced by a cascaded arc plasma source. CTS is feasible for the linear plasma generators Pilot-PSI and Magnum-PSI because high densities can be achieved at low temperatures. The design considerations of the CTS diagnostic are described in this presentation. Calculations show that CTS is possible at laser wavelengths of about 1000 nm, so with a Nd:YAG or a ruby laser. At this wavelength, a scattering angle of 20-30 degrees is sufficiently small to achieve a scattering parameter $1<\alpha<2$. The narrow spectrum will be sampled at a spectral resolution of 0.005 nm. The measured ion feature provides the ratio of the electron and ion temperature T_e/T_i and combined with the electron temperature determined by the standard operating TS incoherent system, T_i can be accurately calculated. The estimated statistical error of T_i, based on Poisson statistics and uncertainty in α ($d\alpha$) due to required *f*-number for a viewing system, is expected to be below 15 % at $n_i = 1.0 \times 10^{21}$ m^{-3} and $T_i = 2.5$ eV, for a single spatial element of 2.5 mm at a laser energy of 1.5 J at 1064 nm.

Conceptual design of High resolution Thomson scattering system on COMPASS tokamak

Petra Bílková[1], Milan Aftanas[1], Petr Böhm[1], Vladimír Weinzettl[1], David Šesták[1], Radek Melich[1], Jan Stöckel[1], Rory Scannell[2] and Mike Walsh[2]

[1]*Institute of Plasma Physics AS CR, v.v.i., Association EURATOM IPP.CR, Czech Republic, bilkova@ipp.cas.cz*
[2] *EURATOM/UKAEA Fusion Association, Culham Science Centre, Abingdon, Oxfordshire OX143DB, United Kingdom*

A new High resolution Thomson scattering (HRTS) diagnostic system is being designed for the COMPASS tokamak (R_0=0.56m, a=0.2m, B_T=0.8-2.1T, k=1.6), which was reinstalled at Institute of Plasma Physics Prague, Czech Republic. The first plasma has been achieved in December 2008.

The HRTS on COMPASS is aimed to measure spatial profiles of both electron temperature and density in the core and the edge regions serving as an essential diagnostic frame for a fulfilment of the COMPASS physical program focused to pedestal studies.

The contribution describes a conceptual design of Thomson scattering diagnostic. The HRTS system will reach requested parameters - time resolution at least 30 Hz all over the plasma discharge and spatial resolution of 1 cm for core plasma and 3 mm for the pedestal region. First, a laser part of the system is introduced including a laser beam path as well as a timing scheme of laser operation. Next, we describe a detection system consisting of collection optics, optical fiber setup taking into account a possibility of multiplexing technique, polychromators based around APD technology and the data acquisition system. Finally, calculations of both the collection optics and detection etendue and estimates of detection throughput resulting in a number of detected photoelectrons are presented.

Thermostatic control of optical isolator for long-term stability of magneto-optic sensing with orthogonally polarized two-frequency laser

S. Tsuji-Iio, Y. Maruyama, T. Miyazaki, H. Tsutsui, R. Shimada

Tokyo Institute of Technology, 2-12-1-N1-11 Ookayama, Meguro-ku, Tokyo 152-8550, Japan

Magnetic sensors with low drifts are required for long-pulsed fusion devices. Magnetic-optic polarimeters can avoid the problems of zero-point drifts of long integration with commonly used pick-up loops. Although high angle resolution of the Faraday rotation is easily achieved with the use of a pair of photoelastic modulators (PEMs), the resolution depends on the polarization angle with respect to the optical axes of PEMs. A Zeeman laser which emits orthogonally polarized two beams with slightly different frequencies facilitates heterodyne measurement of the Faraday rotation with high accuracy. When optical fibers are, however, employed for free optical path from the light source to Faraday elements, reflected lights from a fiber coupler degrades the lasing and hence the stability of the beat frequency. An optical isolator for orthogonally polarized two-frequency lasers can suppress the optical feedback. Its performance is maximized around 25 deg. Celsius since it utilises a paramagnetic TGG (terbium-gallium-garnet) rod for compact isolator with permanent magnet. By placing the optical isolator in a thermostatically controlled chamber, the Faraday angle resolution of the magnetic-optic polarimeter with polarization maintaining optical fiber is improved up to around 0.02 deg. for 1000-s measurement.

All papers published in this volume of *Journal of Physics: Conference Series* have been peer reviewed through processes administered by the proceedings Editors. Reviews were conducted by expert referees to the professional and scientific standards expected of a proceedings journal published by IOP Publishing.

Updating Plasma Scattering of Electromagnetic Radiation

J Sheffield

Institute for Secure and Sustainable Environment, 311, Conference Center Building, 600 Henley Street, Knoxville, TN 37996, USA.

E-mail: jsheffi1@utk.edu

Abstract. The monograph *Plasma Scattering of Electromagnetic Radiation* was published by Academic Press in 1975. A Russian edition, Atomidzat, came out in 1978. An updated version is being prepared by D. Froula, S. Glenzer. N Luhmann, and J. Sheffield for publication in 2010 by Elsevier. The new version will discuss the broader applications of Thomson scattering, which include the full range of plasmas used in research and industry. The expansion of the field has been made possible by the growing number of powerful radiation sources (from X-rays to microwaves), detectors, and innovative techniques. When the book was published, the highest temperatures in laboratory plasmas were around 2 keV for the electrons. Compare this to today's 25 keV where the relativistic effects are dramatic. The application to low temperature plasmas with T_e in the range of 1 – 30+ eV, important in industry, has grown. Important capabilities have been developed in the areas of energetic particle, micro-instability, and high energy density plasma measurements. For the future, we look forward to the use of scattering as a diagnostic on the large new fusion facilities—NIF, LMJ, and ITER.

1. Introduction

An updated version of *Plasma Scattering of Electromagnetic Radiation* [1] is being prepared for publication in 2010. This paper reviews advances that have been made since 1973, when the original manuscript was completed. It is recommended that readers, who can't wait until 2010, look at the numerous excellent books and review articles that have been written in the meantime: for example, Hutchinson [2], Bretz [3], DeSilva [4], Luhmann et al. [5], Donné et al. [6].

Application of Thomson scattering in two extremes of electron temperature stand out: very high temperatures, up to 25 keV, achieved in fusion research in which relativistic effects are important; and low temperatures, 1 to 30+ eV, and minute plasmas in industrial applications. Measurements of microinstabilities have been important to validating theories of transport in toroidal devices. Measurements in warm dense matter have burgeoned; Glenzer and Redmer [7] discuss the use of X-ray lasers. In all cases, measurement capabilities have been enhanced by the growing range of radiation sources, detectors and innovative techniques.

Consequently, Thomson scattering will play an important role in the diagnosis of the major new fusion facilities that are built or under construction—the National Ignition Facility (NIF), Laser MégaJoule (LMJ), and International Thermonuclear Experimental Reactor (ITER).

2. Sources, Detectors, and Innovative Techniques

Key developments in Thomson scattering techniques are discussed in Luhmann et al. [5], Donné et al. [6], and Muraoka [8]. The earliest Thomson scattering measurements in the laboratory were made one point at a time in the plasma. The television Thomson scattering system (TVTS), developed at PPPL, and applied on PLT and TFTR Johnson et al. [9], allowed the collection of scattered signals from many points along the input beam, for example,74 points with 1 cm resolution. This approach made it easier to obtain 2-dimensional plots of n_e and T_e. In general, compared to 1975, a far wider range of efficient detectors exist today. In the wavelength range of 115 to 1040 nm, the quantum efficiency of various photocathodes runs from 15 to 30%. In the infrared region, the quantum efficiency varies from 0.15 to 1.0 and is typically 0.5. The use of micromachining and thin-film lithography fabrication with new materials and antenna coupling schemes resulted in sensitive high speed microbolometers. These have found widespread application in the millimeter wave and THz region.

A second important advance was the use of a high-repetition rate YAG laser at Garching, Roehr et al. [10], which permitted data to be obtained at 100 Hz or more. This laser plays an important role in measurements on industrial plasmas. Today, a wide range of sources is available, see table 1.

Table 1. Some representative sources: used by 1973 in italics.

Source	λ_i	Pulse duration	Peak power	Energy
X-ray	3×10^4 eV	100 ps	5×10^{15} photons on target	
Argon ion	488 nm	CW	~100 W	
Nd-YAG 2xf	532 nm	~ 10 ns	~ 50 MW	0.5 J
Ruby	694.3 nm	15 ns	1.7 GW	~25 J
Alexandrite	750 nm	350 ps		1-2 J
Nd-YAG	1.064 μm	~ 20 ns	> 100 MW	>2 J
CO_2	10.6 μm	CW	25-60 W	
CH_3OD	57.2 μm	CW	1.6 W	
CH_3OH	118:8 μm	CW	1.6 W	
HCN	337 μm	~ 20 nsec	1 kW peak	20 mJ
D_2O	385 μm	1 μs	2.5 MW	~ 2.5 J
Gyrotron	0.87 mm	CW	80 W	
$^{13}CH_3F$	1.22 mm	~ 1 μs	4 kW	0.5 mJ
Gyrotron	1.76 - 5.0 mm	CW- long pulse	1 MW peak	

A further advance was the light detection and ranging (LIDAR) system was jointly developed by the University of Stuttgart and JET, Salzmann et al. [11]. It uses a very short laser pulse 0.3 ns duration (length 90 mm) and collects backscattered light; permitting measurements to be made along a chord through the plasma.

A triple grating spectrograph with a mask used to filter out the central wavelength radiation, was used by Kono and Nakatani [12]. van der Mullen et al. [13] applied the technique in measurements on a magnetic multipole plasma with electron temperature in the range 1.5 – 5 eV, and density $10^{18} - 5 \cdot 10^{18}$ m^{-3}.

3. Relativistic effects

High temperature and relativistic effects on the scattered spectra were analyzed from the early 1960s, but their true impact only came in as fusion plasmas reached successively higher temperatures, starting in the 1970s. The spectral profile of the scattered radiation is given for the case in which E_i is perpendicular to the scattering plane and only the component in the direction of E_i is measured, by

$$S(\omega_s,\theta,2\alpha)=\left(\frac{\omega_s}{\omega_i}\right)^2\int\left|1-\frac{(1-\cos\theta)\beta_e^2}{(1-\beta_i)(1-\beta_s)}\right|^2(1-\beta^2)f(\beta)\delta(\vec{k}\cdot\vec{v}-\omega)d^3\beta \qquad (1)$$

where ω_s and ω_i are the scattered and incident frequency, respectively, θ is the scattering angle, and β= v/c. At high temperatures the relativistic Maxwellian must be used for $f(\beta)$

The correction to the non-collective scattered spectrum, to first order in $\Delta\lambda/\lambda$, is given in [1]:

$$1-3.5\frac{\Delta\lambda}{\lambda_i}+\frac{c^2\Delta\lambda^2}{4a^2\lambda_i^3\sin^2(\theta/2)},\ \text{where } a=(2kT_e/m_e)^{1/2} \qquad (2)$$

This formula yields T_e and n_e accurate to about 5% up to 10 keV for 90° and 50° scattering.

Matoba et al. [14] computed the fully relativistic spectrum, and also gave a correction expression valid to second order in $\Delta\lambda/\lambda$: This formula yields T_e and n_e accurate to about 5% up to 50 keV for 90° and 50° scattering. Zhuravlev and Petrov [15] derived an analytic expression for the relativistic case ignoring the variation in the so-called depolarization term. Selden [16] found good agreement between Matoba et al. and Zhuravlev and Petrov of 0.1% for 90° scattering at 20 keV and 1.0% at 100 keV. Naito et al. [17] derived an analytic formula accurate to 0.1% at 100 keV. Beausang and Prunty [18] derived formulae relevant to LIDAR scattering when $\theta = \pi$, see Figure 1. Differences in the various calculations mainly affected the density measurement. The $(\omega_s/\omega_i)^2$ effect can be seen in Figure 2.

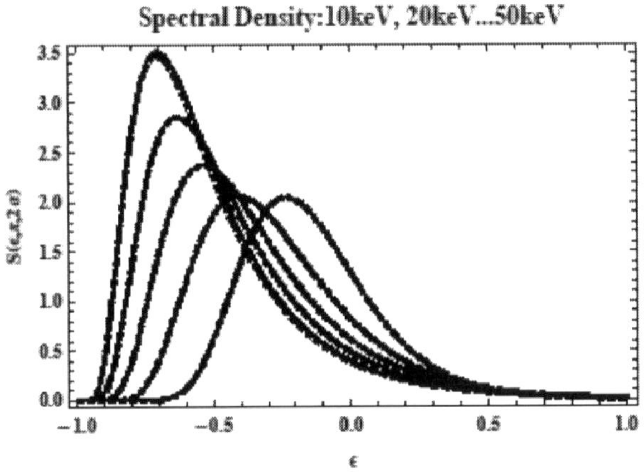

Figure 1. $S(\varepsilon,\pi,2\alpha)$ versus ε for LIDAR scattering at T_e 10, 20, 30 ,40 ,50 keV, courtesy Beausang [18].

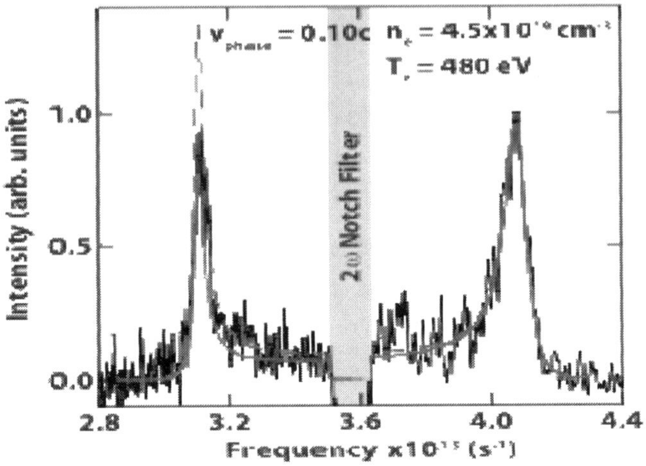

Figure 2. The term $(\omega_s/\omega_i)^2$ is important in collective scattering as shown for the case of a laser produced gas-jet plasma, courtesy D. Froula (private communication).

4. Industrial and other low temperature plasmas

Plasmas have a wide range of applications outside fusion energy research, for example in fluorescent lighting, plasma displays, semiconductor processing, making coatings, cleaning surfaces, destroying noxious chemicals, and plasma cloaking. Thomson scattering is an attractive diagnostic option because it can make local measurements without perturbing the plasma, as can be the case for Langmuir probes.

Key characteristics of such plasmas are a low temperature in the 1 – 30+ eV range, and electron densities in the range of 10^{16} to 10^{24} m^{-3}, often with $n_n >> n_e$. As discussed in the reviews of Muraoka et al [19]), Warner and Hieftje 20] and van de Sande [21] such parameters present a challenge for using Thomson scattering as a diagnostic, because of the low photon count rate at the lower densities, the narrow scattered line width, large Rayleigh scattering, and stray light and background plasma light. In addition for etching, chemically-reactive gases are commonly used.

The industrial plasmas are usually operated in steady-state; consequently, the approach that has been used with pulsed sources is to: integrate the signals over as many as a 10^3 pulses; measure the Rayleigh component separately and subtract it; and have two or three monochromators in series to reduce the stray light.

Measurements have been made on: impulse breakdown plasma in atmospheric air, Uchino et al. [22]; electron cyclotron resonance (ECR) sources, Bowden et al. [23]; radio frequency inductively coupled sources, Hori et al[24]; magnetic neutral loop discharges, Sakoda et al. [25]; capacitatively coupled radio frequency sources, Wesseling and Kronast [26]; micro-discharge plasmas, Noguchi et al. [27]; neon-mercury positive column, Bakker and Kroesen [28]; atmospheric argon plasma, Zaidi et al. [29]; tin vapor discharge Kieft et al. [30]; microwave plasma torch, van der Mullen et al. [31]; and a magnetic multipole, Maurmann et al. [32].

In the system used by Bowden at al. [23], an ECR plasma was operated in argon at 1 mTorr (3.5 x 10^{19} particles/m^3). The electron density and temperature determined from the fitted Gaussian spectrum ranged from 2.1 - 6.2 x 10^{17} m^{-3} and 2.4 - 3.1 eV, respectively. In one experiment, they operated a YAG laser at 532nm and at 1 kHz repetition rate with 1 mJ pulses of duration 7 ns, and beam divergence of 0.5 mrad. To improve the signal they used spherical mirrors to reflect the laser light up to 26 times through the scattering volume. A difficulty with this approach was the relative increase in background light because of the longer duration of the measurement. The lower limit on density measurement capability was estimated as 5 x 10^{16} m^{-3}. A more satisfactory approach was to operate the laser at 10 Hz repetition rate with 0.5 J pulses of 10 ns duration and a beam divergence of 0.5 mrad. The estimated lower limit on density capability was 1 x 10^{16} m^{-3}.

Hori et al. (24) demonstrated the ability to measure non-Maxwellian electron energy distributions in an inductively-coupled plasma, see Fig. 3.

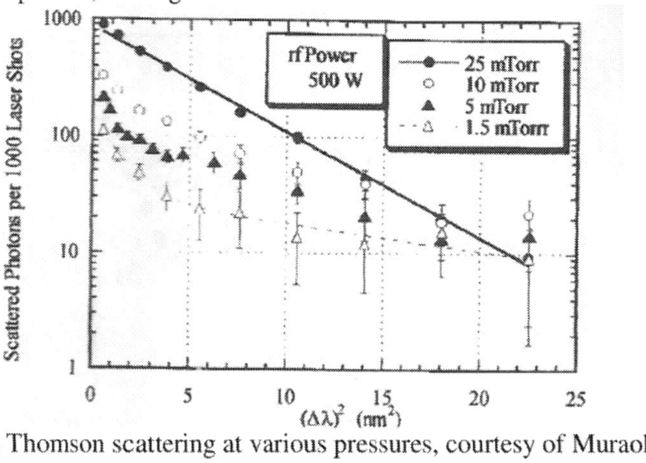

Figure 3. Thomson scattering at various pressures, courtesy of Muraoka.

Sakoda et al. [25] measured the electron density (n_e ~ 0.7 - 5.5 x 10^{17} m^{-3}) and temperature (0.9 - 2 eV) profiles in a neutral loop discharge.

A tin vapor discharge is one of the sources under development for extreme ultraviolet lithography. During the pre-pinch phase, the electron temperature and density are, respectively, in the range of 5 – 30+ eV, and 10^{23} –10^{24} m^{-3}. Measurements on such plasmas have been made by Kieft et al. [30].

Noguchi et al. [27] and Hassaballa et al. [32] made measurements in small scale (0.1 - 1 mm) plasmas of a micro-discharge, as used in a plasma display. A key to their success was the suppression of strong laser stray light with the aid of a triple-grating spectrometer. The plasma parameters were $n_e \sim 8 \times 10^{18}$ m^{-3} and $T_e \sim 0.5 - 1.4$ eV.

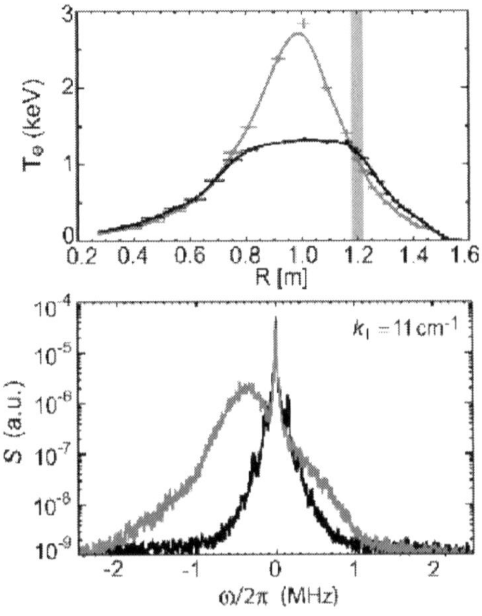

Figure 4. Temperature profiles (top) and spectral density of fluctuations (bottom) at 0.3 (red line) and 0.43 s (black line). The shaded area (top) is the measurement location where L_{Te} is 15 and 50 cm, respectively. Negative frequencies (bottom) correspond to wave propagation in the electron diamagnetic direction, courtesy Mazzucato.

5. Micro-instability measurements

Tynan et al. [33] have reviewed the progress in experimental drift turbulence studies, because these modes are believed to be responsible for enhanced transport in many toroidal confinement devices. Ion and electron pressure gradients drive ion and electron diamagnetic currents. However, in the absence of dissipation of the electron motion parallel to the magnetic field, any density fluctuations will be in phase with the potential fluctuations resulting in no net flow. But, in the presence of dissipation, owing to collisions or to trapped electrons, for example, a phase shift will occur and this leads to a flux. At sufficiently high pressure the fluctuations can perturb the magnetic field and lead to additional loss mechanisms.

Early studies in a tokamak by Mazzucato [34] showed scattered spectra that were consistent with drift instabilities. However, at the time, the theory of instability growth and saturation was not developed sufficiently to clearly identify whether these modes were responsible for the observed anomalous transport. Since then, there has been growing evidence that in the case of toroidal plasmas, such as the tokamak and stellarator, the ion temperature gradient instability (ITG), trapped electron modes (TEM), and electron temperature gradient instability (ETG) cause anomalous transport in the plasma core.

The modes evolve, owing to the changes in the background plasma and through 3-wave coupling, to give a broad wavenumber spectrum. At some point, the turbulent energy cascades to smaller scales and modes where it may be damped by viscosity or Landau damping. Eventually, a new equilibrium balance is reached at a balance between the destabilizing and stabilizing forces.

Substantial progress, reviewed by Hammett [35] has been made both theoretically and computationally using gyrokinetic theory coupled to the Particle-in-Cell (PIC) approach.

Experimental diagnostics measure conditions under the constraints of viewing sight-line and time and space. Because of these limitations, synthetic data, allowing for the experimental realities is constructed from the code calculations to compare with the experimental data.

An important contribution was made on Tore Supra using CO_2 laser light scattering, Hennequin [36] with the measurement of wavenumber fluctuations in the range 3 cm^{-1} $< k_\perp <$ 26 cm^{-1}. The measurement showed how the spectrum and decorrelation times scaled with normalized poloidal gyroradius. Among earlier work was the study of microturbulence on the KT-5 tokamak also using CO_2 light scattering, Chang et al. [37].

Mazzucato et al. [38] made localized measurements, enabled by the toroidal curvature of field lines, at 280 GHz of fluctuations on NSTX. The results are consistent with the behavior of ETG modes, see Fig. 4. Rhodes at al. [39] employed the backscattering of 94 GHz waves on DIII-D using the second harmonic resonance as a beam dump, to study instabilities in the range 0 to 40 cm^{-1}.

6. Energetic ions, high energy density plasmas, and burning plasma measurements

6.1. Energetic ions

Energetic ions, with energy $E_f > T_i$, derive from two main sources: heating by neutral beams and radio-frequency waves (RF); and from fusion. In the case of RF heating, the ions are accelerated from a background distribution to give a high energy tail. In the case of beams and fusion the ions start at an energy E_{f0}, and then slow down by collisions, initially with electrons and then with other ions. In general, $v_i < v_f < v_e$ for thermal ions, fast ions and electrons respectively. Suggestions for using scattering of electromagnetic radiation to measure the alpha particle distribution function were made by Hutchinson et al. [40] for CO_2 laser light, and by Woskov [41] for millimeter waves.

To date, the experimental tests have involved measurements of ions resulting from neutral beam injection. A major problem with using 10.6 μm light is the requirement for the very small scattering angle < 1° needed to obtain a sufficiently high value of $\alpha = 1/k\lambda_D$. Nevertheless, a system has been installed on JT-60U with an improved CO_2 laser with high energy (~17J), high repetition rate (15 Hz), and improved S/N ratio, Kondoh et al. [42].

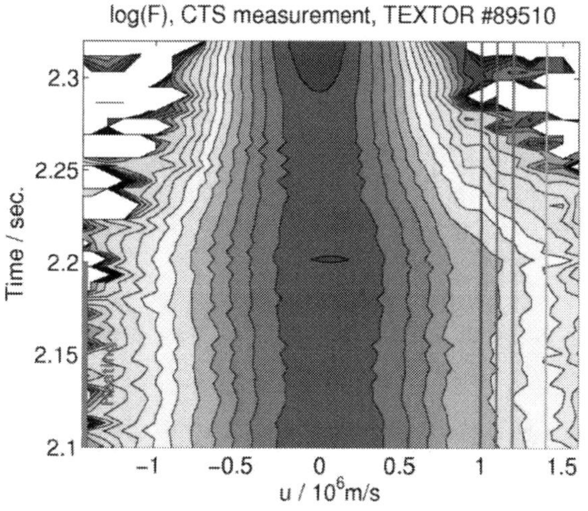

Figure 5. Contour plot of the logarithm of the measured ion velocity distribution. The auxiliary heating was turned off at t = 2.2 s, with permission from H. Bindslev, S.K. Nielsen, L. Porte, J.A. Hoeksma, S.B. Korsholm, F. Meo, P.K. Michelsen, J.W. Oosterbeek, E.L Tskadse, P. Woskow, and the TEXTOR team, Phys. Rev. Lett. **97** 205005, 2006. Copyright 2006 by the American Physical Society. hhtp://link.aps.org/abstract/PRL/v97/p205005

A successful approach has been to use longer wavelength radiation in the millimeter range, where there are essentially no limitations on scattering angle: the ion feature always dominates, thereby allowing the scattering geometry to be selected to meet other objectives. In this frequency range, the challenge is the relatively strong ECE raising the background noise. The first gyrotron-based fast ion measurements, coming from neutral beams, were made on JET, Bindslev et al. [43] Subsequently, a similar system was deployed to measure beam ions on TEXTOR, Bindslev et al. [44] The system used a 200 kW, 110 GHz gyrotron source of 0.2 s duration at a scattering scattering angle of 150° to 170°. The gyrotron was operated with a train of up to100, 2 ms duration pulses with an adjustable time delay between pulses to temporally track fast ion evolution during the plasma shot, see Fig. 5.

Measurements have also been made on ASDEX-U, Korsholm et al. [45]. A 77 GHz system to measure both the bulk and tail ion distribution has been designed for the LHD stellarator by Nishiura et al. [46].

6.2. High energy density plasmas

The development of very high power lasers for inertial fusion made it possible to create high energy density plasmas in the laboratory. Ultraviolet probe lasers may be used for densities less than 10^{21} m^{-3}. For higher densities X-ray lasers (He-alpha, Ly-alpha and K-alpha radiation) have been used (Fig. 6.) for both non-collective and collective scattering measurements, Boehly et al. [47] and Glenzer and Redmer [7].

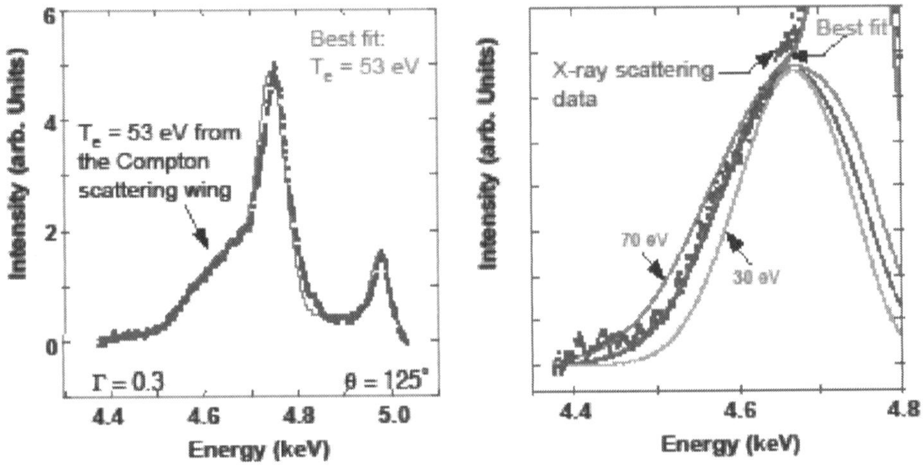

Figure 6. Scattering spectrum (blue dots) from isochorically heated beryllium. Elastic and inelastic componenets are observed from both the titanium He-alpha and Ly-alpha probe X-rays (left). The spectrum is well fit with synthetic scattering spectra (red line). These results characterize the solid density plasma regime with an error bar of about 10%, courtesy Glenzer [48].

6.3. Burning plasmas

Thomson scattering systems are being considered to measure core (LIDAR), edge, and divertor electron temperature and density in ITER, Walsh et al. [49]. Relativistic effects must be included because T_e up to 40 keV is anticipated.

In fact, finite velocity effects should also be taken into account for very energetic fusion ion products, and energetic beam ions, such as the 1 MeV deuteron beams (v_{fk0}/c = 0.03) proposed for ITER. To first order in ω/ω_I, the correction factor is given approximately by

$$\left(1+2\frac{\omega}{\omega_i}\right) \cong \left[1+4\sin\left(\theta/2\right)\right]\frac{v_{fk0}}{c} \qquad (3)$$

Building upon the experience measuring beam ions, Bindslev et al. [50] have proposed to use a 60 GHz gyrotron, which is below the fundamental ECE resonance, to measure the fusion alphas in ITER The measurement would be made in the presence of the beam 1 MeV deuterons. For both types of diagnostic, a major challenge will be in developing components that can handle the harsh ITER environment.

The National Ignition Facility (NIF) will include both the capability for high-energy and ultra-short laser produced X-ray probing of the plasmas, Glenzer and Redmer [7].

Acknowledgements

I appreciate the input from my coauthors, Dustin Froula, Siegfried Glenzer, and Neville Luhmann, and from the many people who responded to our request for suggestions for improving our book. I am also grateful for the support of Patricia Osborn of Elsevier.

References

[1] Sheffield J 1975 *Plasma Scattering of Electromagnetic Radiation* Academic Press New York,.
[1] Hutchinson I H 1987 *Principles of Plasma Diagnostics*, Cambridge University Press Ch. 7.

[2] Bretz N 1997 Rev. Sci. Instrum **68** 2927.
[3] DeSilva A W 2000 Contrib. Plasma Phys. **40** 1-2 23-35.
[4] Luhmann Jr N C et al 2008 Fusion Sci and Tech **53** 335.
[5] Donné A J H et al 2008 Fusion Sci. and Tech. **53** 397.
[6] Glenzer S H and Redmer R to be published in Rev. Mod. Phys.
[7] Muraoka K 2006 Problems of Atomic Science and Technology No 6. Series: Plasma Physics (12) p.236-240.
[8] Johnson D et al 1985 Rev. Sci. Instrum. **56** 1015.
[9] Roehr H et al 1982 Nuclear Fusion **22** 1099.
[10] Salzmann H et al 1987 Nuclear Fusion **27** 1925.
[11] Kono A and Nakatani K 2000 Rev. Sci. Instrum. **71** 2716.
[12] van der Mullen J et al 2004 Spectrochimica Acta Part B **59** 929.
[13] Matoba T et al 1979 Jap. J. Appl. Phys. **18** 1127.
[14] Zhuravlev V A and Petrov G D 1979 Sov J. Plasma Phys. **5** 3.
[15] Selden A C, 1980 Phys. Letts. **79**A 405 and Culham Report CLM-R220 1982.
[16] Naito O et al 1993 Phys. Fluids B **5** 4256.
[17] Beausang K V and Prunty S L, 2008 Plasma Phys. Control. Fusion **50** 095001.
[18] Muraoka K et al 1998 Plasma Phys. Control. Fusion **40** 1221.
[19] Warner K and G M Hieftje 2002 Spectrochimica Acta Part B **57** 201
[20] Van de Sande M 2002 Technische Universiteit Eindhoven Proefschrift ISBN 90-386-1919-7.
[21] Uchino K et al 1982 Japanese J. Appl. Phys **21** L696.
[22] Bowden M D et al 1999 Plasma Sources Sci. Technol. **8** 203.
[23] Hori T et al 1998 J. Appl. Phys. **83** 1909.
[24] Sakoda J et al 1997 Japan, J. Appl. Phys. **36** L67.
[25] Wesseling H J and B Kronast 1996 J. Phys. D. Appl. Phys. **29** 1035.
[26] Noguchi Y et al. 2002 J. Appl. Phys. **91** 613.
[27] Bakker L P and G M W Kroesen 2001 J. Appl. Phys. **90** 3720.
[28] Kieft ER et al 2005 Rev. Sci. Instrum. **76** 093503.
[29] Zaidi S H et al 2001 AIAA-2001-0415, 39th AIAA Aerospace Meeting and Exhibit, Reno, NV.
[30] Maurmann S et al 2004 J. Phys. D: Appl. Phys. **37** 2677.
[31] Hassaballa S et al 2004 IEEE Trans. on Plas. Sci. **32** 127.
[32] Tynan G R et al 2009 to be published in Plas. Phys. Control. Fusion.
[33] Mazzucato E 1978 Phys. Fluids **21** 1063.
[34] Hammett G W 2009 to be published in Phys. Plasmas.
[35] P Hennequin et al 2006 C.R. Phys. **7** 670.
[36] Chang L et al 1990 Chinese Phys. Lett. **7** 16.
[37] Mazzucato E et al 2008 Phys. Rev. Lett. **101** 075001.
[38] Rhodes T L et al 2006 Rev. Sci. Instrum. **77** 10E922.
[39] Hutchinson D P et al 1985 Rev. Sci. Instrum. **56** 1075.
[40] Woskov P 1987 Rep. PFC/RR-87-16 MIT Cambridge.
[41] Kondoh T et al 2007 Plasma and Fusion Research: Regular Articles vol **2** S1111.
[42] Bindslev H et al 1999 Phys. Rev. Lett. **83** 3206.
[43] Bindslev H et al 2007 Plasma Phys. Control Fusion **49** B551-B562.
[44] Korsholm S et al 2006 Rev. Sci. Instrum. **77** 10E514.
[45] Nishiura M et al 2008 Rev. Sci. Instrum. **79** 10E731.
[46] Boehly T et al 1997 Optics Comm. **133** 495.
[47] Glenzer S H et al 2003 Phys. Rev lett. **90** 17.
[48] Walsh M J et al 2009 ITER Topic: IT/P6-25.
[49] Bindslev H et al 2004 Rev Sci. Instrum. **75** 3598.

Progress in Development of the Advanced Thomson Scattering Diagnostics

T Hatae[1], J Howard[2], N Ebizuka[3], H Yoshida[4], M Nakatsuka[4], H Fujita[4], K Narihara[5], I Yamada[5], H Funaba[5], Y Hirano[6], H Koguchi[6], S Kajita[4] and O Naito[1]

[1] Japan Atomic Energy Agency, 801-1 Mukoyama, Naka-shi, Ibaraki, 311-0193, Japan
[2] Australian National University, Canberra, ACT 0200, Australia
[3] Nagoya University, Furo-cho, Chikusa-ku, Nagoya 464-8603, Japan
[4] Osaka University, Yamadaoka, Suita-shi, Osaka, 565-0871, Japan
[5] National Institute for Fusion Science, Oroshi-cho, Toki-shi, Gifu, 509-5292, Japan
[6] National Institute of Advanced Industrial Science and Technology, Tsukuba-shi, Ibaraki 305-8568 Japan

E-mail: hatae.takaki@jaea.go.jp

Abstract. We have been studied the advanced Thomson scattering diagnostics from viewpoints of new concepts, laser technology and spectrum analysis. This paper summarizes results of development on technologies for advanced Thomson scattering diagnostics.

1. Introduction

Incoherent Thomson scattering [1] is one of the essential diagnostics for measurement of local electron temperature (T_e) and density (n_e), and is widely used in fusion plasma experiment. Thomson scattering is established diagnostic technique; nevertheless, further improvement on the measurement capability has been required for physical understanding of the fusion plasma. For example, a low temperature measurement (<100 eV) is needed to study a divertor plasma, while a high temperature measurement (~30 keV) is also needed to study a reactor-class core plasma. On the other hand, high repetitive measurement is necessary to observe transient phenomena such as formation/collapse of a transport barrier. And a necessity of profile measurement with high spatial resolution has risen. To realize such requirements effectively, development of new technology is crucial. We have been studied the advanced Thomson scattering diagnostics to meet such requirements from viewpoints of new concepts, laser technology and spectrum analysis. This paper summarizes results of development on technologies for advanced Thomson scattering diagnostics.

2. New concepts for Thomson scattering

2.1. Polarization interferometer based on Fourier transform spectroscopy

In the Thomson scattering diagnostics, grating spectrometers or filter polychromators have been utilized to analyze the scattered spectrum thus far. Although these are widely used, there are some disadvantages: the throughput is relatively small in the case of grating spectrometer, the transmissivity of rear placed spectral channels decreases when the interference-filter-type polychromator has many wavelength channels. The main advantage over polarization interferometers based on the Fourier transform spectroscopy is amenability to imaging, simplicity of design and cost. There may be some transmission advantages because of the use of a small number of high transparency components. Furthermore, this is suitable for measurements with restricted plasma background light. Polarization interferometers for the Thomson scattering diagnostics have developed to evaluate the validity.

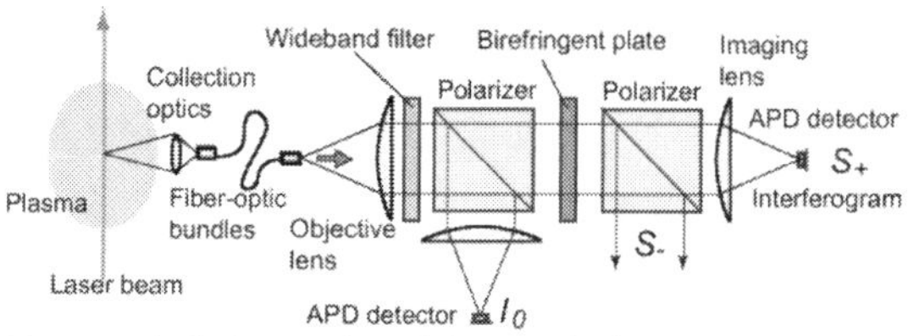

Figure 1. Schematic of the dual channel polarization interferometer. [4]

A method based on measurement of the optical coherence of scattered radiation at a fixed optical delay has been proposed for Thomson scattering [2]. A dual channel polarization interferometer utilizing a fixed-thickness birefringent plate has been developed [3]. The polarization interferometer is composed of a birefringent plate of fixed optical delay sandwiched between polarizers as shown in Figure 1. By suitably choosing the optical delay, these independent outputs provide sufficient information to determine both the electron temperature and density. The scattered signals at the complementary output ports are $S_{\pm}=0.5I_0\{1\pm\zeta_T(\phi_0)\}$, where I_0 is the spectrally integrated light intensity, $\zeta_T(\phi_0)$ is the Thomson spectrum fringe visibility at monochromatic birefringent phase delay ϕ_0. The value of I_0 is in proportion to n_e, and value of $\zeta_T(\phi_0)$ is function of T_e.

Since proof-of-principle tests is carried out in TPE-RX reversed-filed-pinch (RFP) machine using the existing YAG laser Thomson scattering system, parameters for design of a dual channel polarization interferometer are fixed as follows: $T_e \leq 1$ keV, $n_e \geq 1\times10^{19}$ m^{-3}, scattering angle 90°, YAG laser wavelength 1064 nm. A YAG laser Thomson scattering system has been already installed in TPE-RX, and this system has a conventional filter polychromator (4 spectral channels). The electron temperature of one spatial point at the plasma center was measured using the filter polychromator and the polarization interferometer alternately. To compare at different temperature ranges, data during PPCD (high temperature data) and before PPCD phase (low temperature) are measured. As a result, similar T_e values within data variation and error bar ranges were successfully measured by both interferometer and polychromator [4].

Measurement of one point on the interferogram using the dual channel polarization interferometer is insufficient to evaluate the electron temperature with the wider range. We considered that measurement of the whole or the main part of interferogram is essential for the wider electron temperature range. When the whole or the main part of interferogram is measured, it is possible to reconstruct the Thomson scattered spectrum accurately. We have been developing a prototype multichannel polarization interferometer using the Wollaston prism to evaluate the availability for the Thomson scattering diagnostics as shown in Figure 2 [4]. The light from the optical fiber bundles is collimated by a lens, and is introduced to an interferometer part. The interferometer part is simple structure that the Wollaston prism is sandwiched between two polarizers. An interferogram which is localized at a plane inside the Wallaston prism is imaged onto a photo cathode of an image intensified CCD camera by an imaging lens. Interferograms of various monochromatic light sources were measured by the multichannel polarization interferometer. In this test, a green He-Ne laser (λ = 543.5 nm), a red He-Ne laser (λ = 632.8 nm) and a near-infrared LED (λ = 850 nm) were used as light sources. These spectra were reconstructed from interferograms by FFT. Since 15-periods of the fringe can be observed for 632.8 nm He-Ne laser, effective wavelength resolution becomes ~60 nm in this

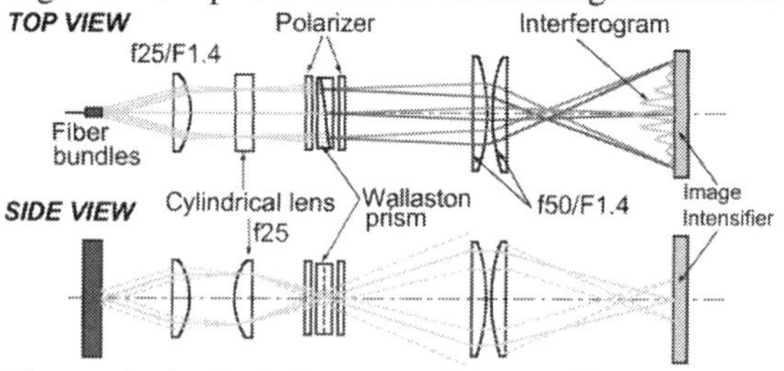

Figure 2. Optical diagram of the multichannel polarization interferometer using a Wollaston prism. [4]

system. It is possible to measure the electron temperature of more than 100 eV (scattering angle 90°) in this system. To measure low temperature of less than 100 eV, it is necessary to increase the fringe frequency.

Polarization interferometers based on the Fourier transform spectroscopy for the Thomson scattering diagnostics have been examined to evaluate the validity. The validity of dual channel polarization interferometer has been proved by the plasma measurement. It seems that both dual channel and multichannel interferometers are useful instruments to observe the Thomson scattering spectrum. Based on these results, we have investigated imaging interferometers for analysis of Thomson scattering spectra [5]. We have also considered an application of Solc filter to Thomson scattering [5]. A Solc filter is a multiwaveplate polarization interferometer whose passband can be tailored according to the number of plates and their respective thicknesses and orientations. It is possible that this method realize the multichannel polychromator by the simple optics.

2.2. Multipass Thomson scattering using SBS phase conjugate mirror

In the conventional Thomson scattering system, a focused laser beam is injected into the plasma in order to generate scattered light, and the beam is then dumped by a beam dumper after passing through the plasma. Since the beam transmission loss in the plasma is negligible, the energy of the dumped beam is almost the same as that of the input beam. The large laser energy which passed through the plasma is abandoned without recycling. It is natural to consider that the beam that has passed through the plasma can be utilized again for measurement. In the case of using a flat mirror to turn the beam, since the beam has a positive divergence angle, the returned beam spreads after the beam is reflected by the mirror. The significant stray light will be generated when the spread beam touches the vacuum vessel and laser port, leading to the difficulty in Thomson scattering measurement. In the case of using a concave mirror, although it is possible to decrease the generation of stray light, precise beam alignment is required. We have focused on a phase conjugate mirror based on stimulated Brillouin scattering (SBS) [6] to address a multipass Thomson scattering. The SBS is one of the most common nonlinear optical processes for achieving a phase conjugate wave with the inclusion of a small Brillouin frequency shift. The part of the nonlinear optical medium that causes phase conjugation is called a "phase conjugate mirror (SBS-PCM)" in order to distinguish it from a conventional mirror. We confirmed 95% of the reflectivity at 145 W (2.9 J × 50Hz) of average power using JT-60U Thomson scattering laser [7,8]. In the high average-power operation, the SBS-PCM worked stably without laser breakdown of the SBS medium.

A double-pass Thomson scattering measurement using SBS-PCM was carried out in JT-60U [7,8]. A laser beam passing through the plasma is reflected by the SBS-PCM in place of a beam dumper, and the reflected beam is automatically returned along the same path as the incident one by the phase conjugate effect, and passed through the plasma again. If a conventional mirror is used for this purpose, very precise beam alignment is frequently required to reduce the amount of stray light. However, such alignment is completely unnecessary in the case of the SBS-PCM. After an initial simple positioning of the SBS-PCM, alignment-free operation can be realized by the phase conjugate effect. Two ohmic plasma discharges with similar electron densities and temperatures were performed, and the electron temperature and density were measured with and without the double-pass scattering method. As a result, the intensity of the scattered light became 1.6 times larger and the relative errors of electron

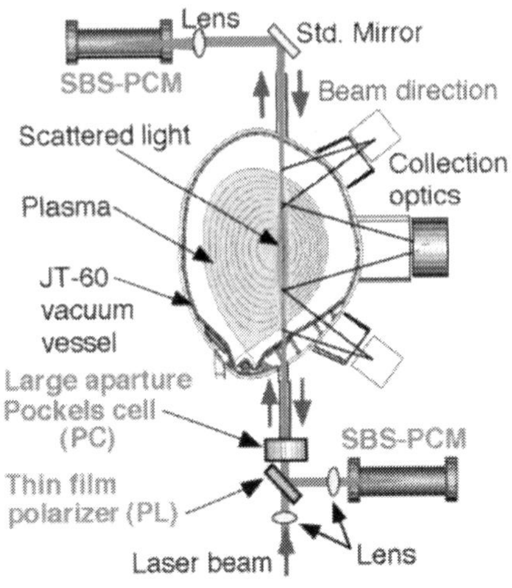

Figure 3. Optical layout for multipass Thomson scattering using SBS-PCMs [8]

temperature and density were reduced to 2/3 in comparison with single-pass scattering.

In the case of double-pass scattering, scattered light can be increased by up to a factor of 2, as mentioned above. However, most of the input energy remains in the double-passed beam. We have proposed the multi-pass Thomson scattering method that utilizes the input beam energy more effectively and that generates over twice the amount of scattered light [7,8]. Since the multi-pass scattering method is an extension of the double-pass method, the components of the multi-pass scattering system are similar to those used in the double-pass system, except that another SBS-PCM with a lens, a thin film polarizer, and a large aperture Pockels cell are added. Multi-pass scattering generates over twice the amount of scattered light by confining the laser beam between a pair of SBS-PCMs. The laser beam can be confined between a pair of SBS-PCMs by control of the laser beam's polarization. As mentioned, alignment-free operation can be realized by the SBS-PCM after an initial adjustment. In the case that the optical loss is under 10%, the scattering light is amplified up to about 5 times larger than that of single pass when the laser beam passes the plasma after 10 times. When an appropriate optical delay is installed between a pair of SBS-PCMs, temporal separation of each pulse is possible. Consequently, very high repetitive measurement is available by burst laser pulses. Advantage of this method is that one laser system is enough to produce burst pulses. We have been preparing the multipass Thomson scattering system for the proof-of-principle tests in LHD.

3. Laser technology for Thomson scattering

The laser system is a key component of Thomson scattering diagnostics. To improve the spatial resolution, an increase of laser energy is necessary. Producing shorter laser pulses is crucial for improvement of spatial resolution in the case of the LIDAR. To observe the transient phenomena by Thomson scattering, high repetition laser is required.

In the solid-state laser, development of high output energy and high repetition laser system is not easy, and the performance is limited by the thermal effects of laser rod in the high power amplifier (thermal lens effect, birefringence) and parasitic oscillation. To solve these issues, the SBS-PCM was applied to a conventional YAG laser system in JT-60U [7-10]. Before introducing the SBS-PCM, the laser system is composed of an oscillator and a series of four amplifiers. This system produced a Gaussian-like beam with an energy of 1.5 J, a pulse length of ~30 ns, and a divergence of < 0.5 cm·mrad (half angle) at a 30 Hz repetition rate (average power is 45 W). The average power is limited by issues above. The optical configuration was changed from single-pass single amplifier stage to double-pass twin amplifier stages employing SBS-PCMs. The optical layout of the high average power laser system is shown in Figure 4. The performance of the laser was drastically enhanced by the improvements and finally achieved 373 W (= 173 W + 200 W) of average power (7.46 J × 50 Hz). The average power exceeded eight times that of the original laser system. This laser system was in routine operation for Thomson scattering diagnostics in the JT-60U.

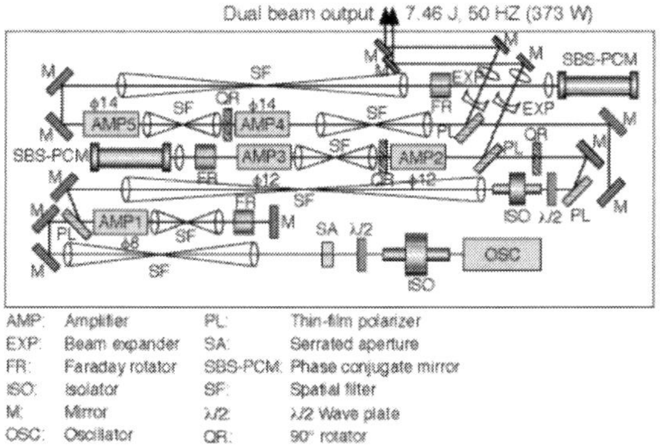

Figure 4. High-average-power YAG laser system employing SBS-PCMs in JT-60 [8]

As another application of SBS-PCM, we have studied SBS pulse compressor (SBS-PC). To obtain pico-second pulses, the mode-locking scheme which requires complicate optical configuration has been used so far. We focus on SBS-PC for application to the LIDAR. The SBS-PC is similar to the SBS-PCM, and uses a very long SBS cell (or plural long SBS cells). The optical layout is simple compared to the mode-locking, and the reflected pulse is compressed automatically by the long SBS cell. In core LIDAR system in ITER, YAG laser of the 1064-nm wavelength is considered to improve the measurement error very high T_e up to 40 keV. Required pulse width for the core LIDAR in ITER is 300 ps, we set the target at less than 300 ps. Since a compression ratio corresponds to a phonon lifetime, a FC-40 liquid having a short phonon life time of 240 ps was selected to obtain the target pulse width. We fabricated two long SBS cells as SBS-PC. The cell had 1.5 m in a length, 4 cm in a diameter and AR-coated windows at both ends, and the FC-40 was filled. One cell was used as an SBS generator, another cell was used as an SBS amplifier. As a result, a 13-ns YAG laser pulse was temporally compressed to a 160 ps phase conjugated pulse at 1064-nm wavelength as shown in Figure 5 [11]. The maximum reflectivity was over 80% without any damage. It seems that this technique is applicable to the LIDAR.

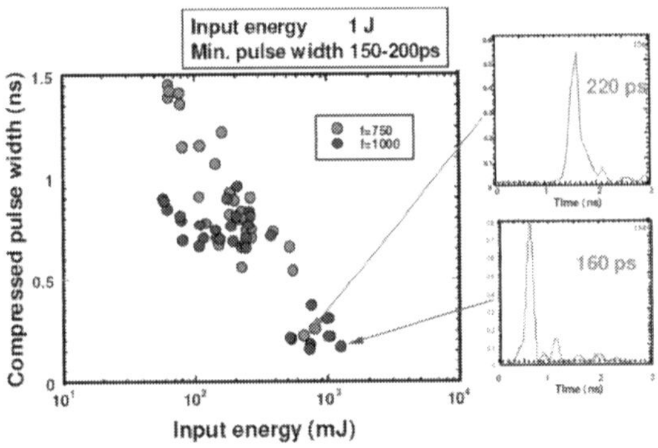

Figure 5. Result of SBS pulse compression. A 13 ns YAG laser pulse was compressed to 160 ps phase conjugate pulse [11]

To increase the laser gain, development of new laser materials is a solution. Recently, ceramic material which is equivalent to the laser performance of the conventional laser crystal has been developed. Furthermore, it became that ions which was difficult to dope in the conventional laser crystals was possible to dope in the ceramics. We studied an application of Cr^{3+} and Nd^{3+} co-doped YAG (Cr, Nd:YAG) ceramics [12] to high power laser. In the Cr, Nd:YAG ceramics, absorbed energy by Cr^{3+} is transferred to Nd^{3+}, the laser efficiency is improved as a result. We produced a Cr, Nd:YAG ceramics rod that the length and the diameter are 100mm and 8mm, Cr and Nd concentrations are 0.05% and 0.8%, respectively. The amplification test was carried out to compare the performance between the conventional Nd:YAG crystal (Nd concentration is 1.0%) and Cr, Nd:YAG ceramics using JT-60U laser system. When pumping energy and repetition rate were 93 J and 50 Hz, respectively, the extracted energy of the ceramic rod was 1.08 J which corresponded to 1.5 times larger extracted energy of the crystal (0.71 J). Therefore, the Cr, Nd:YAG ceramics is promising material to improve the laser efficiency easily.

4. Spectrum analysis

To evaluate high T_e, consideration of relativistic effect is indispensable. Fully relativistic formulae for Thomson scattering diagnostics have been developed. A compact analytic formula for fully relativistic Thomson scattering spectrum including depolarization has been presented [13]. By rational approximation, an analytic formula with high accuracy (relative error < 0.1% at 100 keV) is obtained, which is applicable to a wide range of plasmas. The applicability of Thomson scattering measurement is expanded to a class of non-Maxwellian plasmas which can be described by generalized Lorentzian (kappa) distributions. A fully relativistic scattering spectrum has been presented in an analytic form which allows the possibility of a more robust estimation of the bulk electron temperature in the existence of high energy component [14]. A simple formula for reconstructing fully relativistic electron momentum distribution functions from Thomson scattering data has been also presented [15]. Although the formula assumes the isotropy in momentum space, it may be useful for obtaining detailed information on the electron distribution function in fusion plasmas. The first formula has been used to evaluate the electron temperature in JT-60U. It seems that these formulae are useful for Thomson scattering diagnostics in various plasma experiments.

To observe the Thomson scattering spectrum in the wide T_e range, investigation for the necessary number of wavelength channels and optimum band width of these channels in a spectrometer is significant to evaluate T_e with high accuracy. This examination also contributes to the minimization of the development cost for the spectrometer. The necessary number of wavelength channels for a spectrometer has been investigated in Thomson scattering diagnostics [16]. The measurement errors (σ) are evaluated for a realistic case and their lower limits ($\overline{\sigma}$) are given in explicit forms. For a given number of wavelength channels, optimal channel allocation in wavelength space is sought by minimizing the average relative errors. As a scale of the goodness of measurement, the ratio $\langle \overline{\sigma}/\sigma \rangle$

averaged over the electron temperature of interest is introduced. This quantity ranges from 0 to 1 and tends to 1 as the lowest achievable error is approached. If $\langle \bar{\sigma}/\sigma \rangle \geq 0.9$ is sufficient, then for a typical ruby laser system measuring 400-700 nm, seven channels is adequate number for covering the electron temperature of 0.01-10 keV and five channel for 0.1-10 keV. Designs for polychromators of the ITER edge Thomson scattering system have been investigated utilizing this scheme [17,18].

5. Conclusions

We have developed new technologies for an advanced Thomson scattering diagnostics to meet various requirements for understanding physical phenomena in fusion plasmas. These technologies for the advanced Thomson scattering diagnostics may contribute future Thomson scattering diagnostics. For example, a polarization interferometer is applicable for T_e measurement with wide range and imaging measurement. A multipass Thomson scattering and high average power laser employing SBS-PCM may improve the S/N ratio and repetition rate for the measurement, and allows measurement with high spatial resolution. An SBS-PC is a promising technology for LIDAR to generate short laser pulse. A Cr,Nd:YAG will be used for high-efficiency and high average power laser system. Fully relativistic formulae will contribute to the spectrum analysis with wide T_e range. A guideline to optimize wavelength channels will be useful for a design of spectrometer.

Acknowledgments

This work has been supported in part by Grant-in-Aid for Scientific Researches on Priority Area "Advanced diagnostics for burning plasma" from The Ministry of Education, Culture, Sports, Science and Technology (No. 17044009, 18035016, 20026011), and has been performed with the support and under the auspices of the NIFS Collaboration Research Program (NIFS08KCHH018), the Institute of Laser Engineering - Osaka university Collaborative Research Program and JT-60 Collaborative Research Program.

References

[1] Sheffield J 1975 *Plasma Scattering of Electromagnetic Radiation* (New York: Academic) pp 191-210
[2] Howard J 2006 *Plasma Phys. Control. Fusion* **48** 777
[3] Hatae T, Howard J, Hirano Y, Naito O, Nakatsuka M and Yoshida H 2007 *Plasma Fusion Res.* **2** S1026
[4] Hatae T, Howard J, Ebizuka N, Hirano Y, Koguchi H, Kitamura S, Sakuma T and Hamano T 2009 *J. Plasma Fusion Res. SERIES* **8** 680
[5] Howard J and Hatae T 2008 *Rev. Sci. Instrum.* **79** 10E704
[6] Yoshida H, Kmetik V, Fujita H, Nakatsuka M, Yamanaka T and Yoshida K 1997 *Appl. Optics* **36** 3739
[7] Hatae T, Nakatsuka M and Yoshida H 2004 *J. Plasma Fusion Res.* **80** 870
[8] Hatae T, Naito O, Nakatsuka M and Yoshida H 2006 *Rev. Sci. Instrum.* **77** 10E508
[9] Yoshida H, Nakatsuka M, Hatae T, Kitamura S and Kashiwabara T 2003 *Jpn. J. Appl. Phys.* **42** 439
[10] Yoshida H, Nakatsuka M, Hatae T, Kitamura S, Sakuma T and Hamano T 2004 *Jpn. J. Appl. Phys.* **43** L1038
[11] Yoshida H, Hatae T, Fujita H, Nakatsuka M and Kitamura S 2009 *Opt. Express.* **17** 13654
[12] Ikesue A, Kamata K and Yoshida K 1995 *J. Am. Ceram. Soc.* **78** 2545
[13] Naito O, Yoshida H and Matoba T 1993 *Phys. Fluids* B **5** 4256 (Erratum: 1994 *Phys. Plasmas* **1** 806)
[14] Naito O, Yoshida H, Hatae T, Nagashima A and Matoba T 1996 *Phys. Plasmas* **3** 1474
[15] Naito O, Yoshida H, Hatae T and Nagashima A 1997 *Phys. Plasmas* **4** 1171
[16] Naito O, Yoshida H, Kitamura S, Sakuma T and Onose Y 1999 *Rev. Sci. Instrum.* **70** 3780
[17] Kajita S, Hatae T and Kusama Y 2008 *Rev. Sci. Instrum.* **79** 10E726
[18] Kajita S, Hatae T and Naito O 2009 *Fusion Eng. Des.* **84** 1713

LIF diagnostics in volume and surface dielectric barrier discharges at atmospheric pressure

G Dilecce[1], **P F Ambrico**[1], **M Simek**[2] **and S De Benedictis**[1]

[1]Istituto di Metodologie Inorganiche e dei Plasmi CNR, sede di Bari - Via Orabona 4, 70125 Bari - ITALY

[2]Institute of Plasma Physics, Za Slovankou 3, 18200 Prague, Czech Republic

E-mail: giorgio.dilecce@ba.imip.cnr.it

Abstract. In this paper we review our recent work based on Laser Induced Fluorescence (LIF) diagnostics applied to Dielectric Barrier Discharges (DBD) at atmospheric pressures. Emphasis is given to the main issues that have to be faced in this application through the detailed description of the most difficult case, that of CH LIF detection. The application to kinetic studies of LIF diagnostics is then illustrated through the application of LIF to the CN radical and of Optical-Optical Double Resonance (OODR) LIF to the measurement of $N_2(A^3\Sigma_u^+)$ metastable.

1. Introduction

Atmospheric pressure dielectric barrier discharges are at present a well established issue in the plasma chemistry community, but, in spite of the large amount of publications, the accurate application of optical methods to the diagnostics of kinetic processes is still in its infancy. This can be ascribed to both historical and technical reasons. The technical reason is that the collision regime at atmospheric pressure (ATP), and the reduced discharge dimensions (the inter-electrode gap in particular), impose severe limits and difficulties to the application of optical diagnostics. This is particularly true for Laser Induced Fluorescence (LIF). On the other hand application of LIF is necessary when ground state transient species must be looked for. Optical Emission Spectroscopy (OES), infact, is rarely applicable as a monitor of ground states species. In our recent studies on OES in dielectric barrier discharges, we have found a number of examples in which observable emissions are due to excitation of electronic states by energy transfers with metastable molecules (NO-γ System, nitrogen Herman Infrared System [1, 2] dissociative excitation (CH 4300 Å System [3]), recombination reactions (CN violet System [4]). In such a way that a relationship between emission features and the corresponding ground-state molecules cannot be established. Absorption techniques would be eligible for ground state molecules measurement, and are free of collision quenching problems, but dielectric barrier discharges are highly transient systems, whose study requires time-resolved measurements. To date, absorption techniques, do not have the necessary time resolution. In this talk we will present our recent work on the application of LIF to dielectric barrier discharges at ATP, starting with the most difficult case we have encountered, that of LIF detection of CH radical [3]. We will illustrate it first as a benchmark of all the difficulties that can be encountered in LIF application to ATP DBDs. We will then describe our most complex and successful achievement, the detection of nitrogen triplet metastable $N_2(A^3\Sigma_u^+)$ by Optical-Optical Double Resonance (OODR) LIF, a case in which it has been possible to obtain the absolute calibration of the LIF measurement by means of a method based on optical emission [1]. Such technique has also been applied to a surface dielectric barrier discharge [2, 5], a device that is particularly hostile for any optical diagnostic method. Finally we will address the detection of CN radical, and its use in combination with optical emission spectroscopy [4], that clearly shows how much LIF, with its good time resolution, is useful for kinetic studies.

2. CH LIF detection

The LIF measurements of CH density at atmospheric pressure has been addressed in combustion research. Almost all possible transitions involving the three electronic states $A^2\Delta$, $B^2\Sigma^-$ and $C^2\Sigma^+$ have been employed to this end (see, for example [6], [7], [8] and references therein). In spite of the difficulties due to the low emission yield of the excited states, particularly increased by the pre-dissociative character of B and C states, densities as low as $10^{11} - 10^{12}$ cm^{-3} have been detected in atmospheric pressure flames [8], and even planar-LIF has been demonstrated in [9].We have attempted CH radical detection by LIF in He-CH_4 and N_2-CH_4 mixtures. We have been able to detect a measurable LIF signal in He-CH_4 only, finding a low signal strictly restricted to the active discharge time region [3]. LIF detection of CH in a Dielectric Barrier Discharge (DBD) at atmospheric pressure is infact an archetype of all the difficulties that can be encountered in the application of this spectroscopic technique. These difficulties are relevant to the discharge kinetics, the discharge device structure, the CH molecular structure, the kinetics of CH electronic states at atmospheric pressure. We examine these issues in detail. Low average CH densities are expected, due to its high reactivity and to the pulsed and spatially inhomogeneous nature of the discharge. Filamentary discharges are a sum of short and tiny microdischarges that fill a small part of the volume for a short time. CH, produced by electron impact in the microdischarges, does not fill the entire discharge volume, due to its high reactivity/short lifetime, resulting in low average densities. The LIF setup must then be regulated for maximum sensitivity.

A drawing of the discharge apparatus is reported in figure 1. Interelectrode spacings are 5 mm, in He-based mixtures, and 1.5 mm in N_2 based mixtures. The small dimensions, of the order of the laser beam ones, and the fluorescing nature of the materials employed, are a big concern in a LIF experiment, due to the resulting strong diffused light that is superimposed

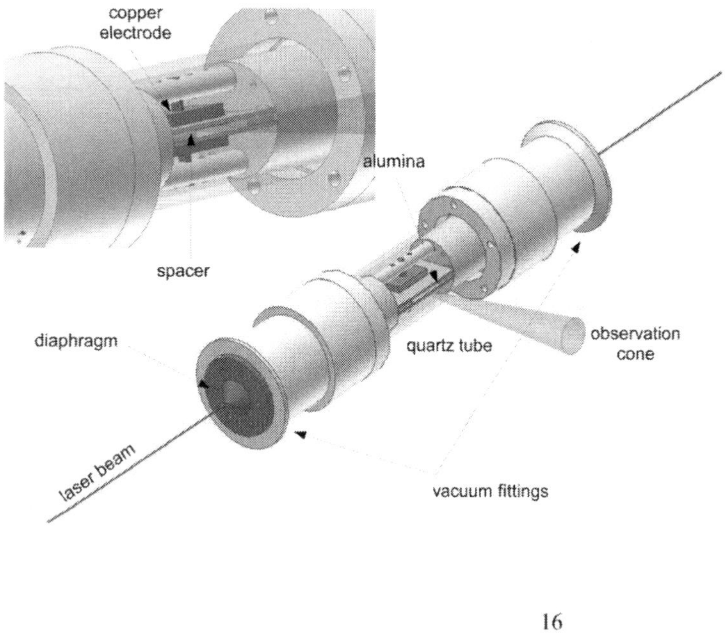

Figure 1. Drawing of the discharge apparatus. Gas inlet, pumping, In and out optical windows are grouped in assemblies (not shown) connected to the two vacuum fittings.

Journal of Physics: Conference Series **227** (2010) 012003 doi:10.1088/1742-6596/227/1/012003

to the true LIF signal. Strong direct laser scattered light calls for a LIF excitation-detection scheme with different bands, allowing spectral separation of the LIF signal from the scattered laser. With the requirement of maximum sensitivity, however, another type of indirect laser scattered light comes into play. The hardware materials of the discharge apparatus irradiated by spatial 'wings' of the laser beam produce a broadband fluorescence, at wavelengths larger than the laser one, that cannot be completely filtered out, unless the LIF detection wavelength is lower than the laser one, that, as we will see in the next, is not the actual case. The spatial wings of the laser beam are produced by diffusion from all the beam manipulation optics and from the discharge vessel windows. Careful alignment and spatial cleaning of the laser beam can minimize, but not remove completely this indirect scattering. The cleaning of the laser beam is obtained by diaphragms made of black-anodized aluminium, two of which are placed inside the discharge vessel (see figure 1).

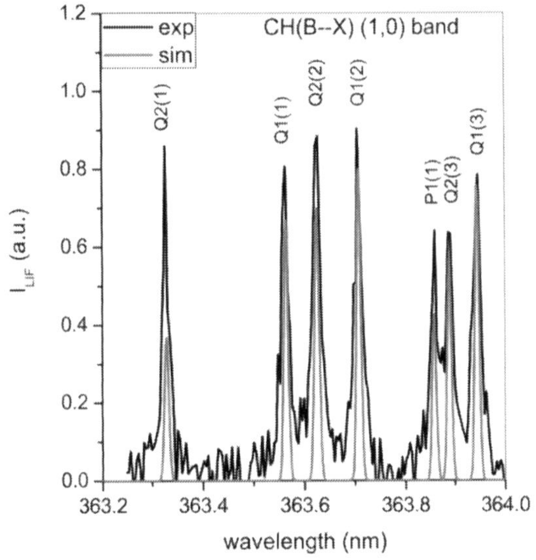

Figure 2. LIF excitation spectrum showing the region chosen for laser absorption. The simulated spectrum is shown for wavelength identification only. Simulations are calculated by LIFBASE [10].

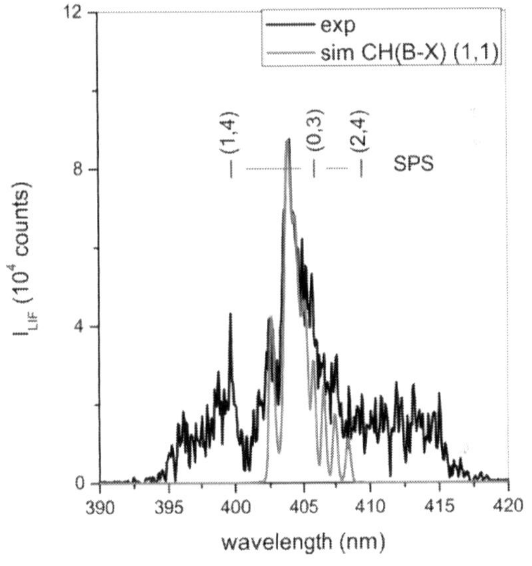

Figure 3. Fluorescence spectrum measured by ICCD. The position of N_2 Second Positive System (SPS) emission bands superimposed to the LIF band is marked (simulation by LIFBASE).

All available electronic transitions, CH($A^2\Delta \rightarrow X^2\Pi$), CH($B^2\Sigma^- \rightarrow X^2\Pi$) and CH($C^2\Sigma^+ \rightarrow X^2\Pi$) are strongly diagonal, limiting the choice of an efficient excitation-detection scheme in which the detected fluorescence band is different from that chosen for laser absorption. This is of fundamental importance since, as we have seen, the amount of direct and indirect laser scattered light is quite large. We have chosen the following scheme:

$$CH(X^2\Pi, v=0) + h\nu_L \rightarrow CH(B^2\Sigma^-, v=1) \rightarrow CH(X^2\Pi, v=1) + h\nu_F \qquad (1)$$

that gives the best combination of diagonal/off-diagonal absorption/emission coefficients. In particular the $Q_1(2)$ rotational line is pumped with λ_L=363.798 nm, and the observation is made with the monochromator centered at about 405 nm. Direct laser scattered light is suppressed by additional use of a bandpass interference filter. Part of the indirect discharge materials fluorescence, that is broadband, instead, enters the spectral observation window. Finally, nitrogen Second Positive System emission bands fall within the fluorescence observation

window. These represent a tolerable background in He-based mixtures, where air impurities are present, but, in N_2 based mixtures such a background becomes overwhelming, preventing any LIF signal detection. In figure 2 and figure 3 LIF excitation and fluorescence spectra are shown respectively, measured in a He+0.14%CH_4, showing the correct attribution of the measured signal to CH LIF and the low useful signal extracted from the SPS background.
The collision quenching of CH(B) state by N_2 and By He is not known. It has been measured only for methane air combustion mixtures. It could be relatively slow if we imagine an analogy with CH(A) state quenching whose rate constant is $2.7 \times 10^{-13} cm^3 s^{-1}$ by N_2 [11], and much lower by He [3]. Nevertheless CH(B, v=1) state is predissociative, with predissociation onset at the rotational level N'=7. Even if we excite a lower rotational level by laser absorption, fast rotational energy transfers (RET) quickly bring excitation beyond the predissociation limit. In [7] calculations keeping RETs into account gave a fluorescence quantum yield for N'=2, due to predissociation only, of 4.13×10^{-3} at T=1200 K, and 3.75×10^{-3} at T=1900 K. Due to the temperature and mixture dependence of collision processes, extrapolation to our case is difficult, but it is reasonable to imagine, in our conditions, a fluorescence quantum yield lower than 10^{-2}.

3. OODR-LIF on $N_2(A^3\Sigma_u^+)$

The classical single-photon LIF scheme for the detection of N_2(A) involves the FPS $N_2(B^3\Pi_g \rightarrow A^3\Sigma_u^+)$ transitions, both for excitation and detection [12]. The quantum yield for fluorescence from the B-state is very unfavorable at ATP (order of 10^{-4}), and we already noticed in [13] that at pressures above 1 Torr the LIF signal became difficult to detect. We then proposed in [14] an Optical-Optical Double Resonance (OODR) LIF scheme in which the metastable is pumped to C-state by absorption of two photons resonant with FPS and SPS transitions:

$$\begin{aligned} N_2(A^3\Sigma_u^+, v'' = 0) + h\nu_{L1} \rightarrow N_2(B^3\Pi_g, v' = 3) + h\nu_{L2} \rightarrow \\ \rightarrow N_2(C^3\Pi_u, v) \rightarrow N_2(B^3\Pi_g, v') + h\nu_F \end{aligned} \quad (2)$$

Such a scheme takes advantage of the two orders of magnitudes larger quantum yield of C-state at ATP. Following the photons absorption, fluorescence is observed from the same excited v-level (direct fluorescence) and from all the vibrational levels lower than v. The latter is the consequence of vibrational relaxation and we will call it collision fluorescence. We have used the direct fluorescence the measurement of metastable density in a volume [1] and surface discharge [2, 5]. An interesting by-product has been the possibility to measure, from the direct fluorescence, the electronic quenching rate constants, and from collision fluorescence the vibrational relaxation rate constants, of the whole $N_2(C, v)$ manifold [15, 16]. From the technical point of view, the OODR-LIF solution requires in principle two lasers that must be synchronized within one ns by appropriate electronics. This was the solution adopted in our laboratory. There are nevertheless some possibilities to use a single laser setup. Already in [14], we used a single dye laser setup in which, by accidental misalignment of the cavity, in addition to the principal beam at λ_1, a spurious mode was produced at about $\lambda_2 = \lambda_1 + 30$ nm. By a BBO crystal it was then possible to produce the mixed wavelength $1/\lambda_{mix} = 1/\lambda_1 + 1/\lambda_2$. Tuning the dye for $\lambda_1 \sim 685$ nm we could produce a second beam, naturally synchronized to the first one, at $\lambda_{mix} \sim 350$ nm, implementing the scheme of formula (2). More reproducible arrangements were proposed in [14]. One by a Ti-sapphire laser, whose fundamental wavelenght can excite the FPS (1,0) at about 891 nm, with its third harmonic that would fall in the spectral region of the SPS (3,1) band. Another more versatile solution, could be the use of a tunable laser, dye or OPO, pumped by a Nd-YLF laser, whose fundamental wavelenght, at 1051 nm, is suitable for absorption from the FPS (0,0) band. The choice of the SPS band for the second photon absorption would be then free. The modeling of the OODR process at atmospheric pressure is an hopeless task, so that the possibility to deduce absolute densities of the metastable relies upon a calibration strategy. We adopted in [1] the method proposed in [17] by which the metastable density is proportional to the ratio of SPS and NO-γ emission intensities and to the NO density in the post-discharge. We then selected a condition, the post-discharge of a diffuse APTD in pure N_2, in which the LIF sampled volume is the same homogeneous volume from which the emissions come out (note that this is not true in a filamentary discharge). The calibrated OODR-LIF could then be applied to different conditions (discharge and filamentary regime). By applying this strategy we found a metastable density of the order of 10^{13} cm^{-3} in the pure nitrogen APTD, a value in close agreement with the calculations of [18]. Addition of small oxygen concentrations to the discharge was observed to induce a transition to the filamentary regime, starting from 250 ppm of oxygen, that appears not to be caused by a marked decrease of the metastable density in the discharge, but rather by a considerable increase of its quenching rate. Such an increase, due to collision quenching by O_2 and O, strongly reduces the survival of the metastable between two discharge events in the voltage cycle, and reduces its ability to reach the surface as well. These observation are consistent with the idea that the APTD regime is sustained by a space charge memory effect due to the nitrogen triplet metastable, as outlined before. We also apply OODR-LIF measurements of the nitrogen triplet metastable to a Masuda-type surface discharge [2]. Surface discharges develop in a thin (order of 100 μm) layer above the surface. The laser diagnostic cannot then, for geometrical reasons, probe the discharge layer. We then measured the N_2(A) density at 1 mm from the discharge, in a so-called space afterglow. In addition, the calibration method described above could not be applied since it is based on the measurement of emissions that come instead from the discharge, i.e. from a volume different from that probed by the laser. We then applied the OODR-LIF calibrated in the volume discharge as above, finding a density of about 10^{11} cm^{-3} in the afterglow. This afterglow density was traced-back in the discharge, by estimations based on NO-γ emissions, to about 10^{12} cm^{-3}, one order of magnitude lower than in the volume discharge. Such a lower density might be explained by larger diffusion losses due to the vicinity of the production zone (the discharge) to the surface. However investigations under modulated regime ($T_{ON}/T_{OFF} = (1-10)ms/20ms$) [5] revealed that the upper limit of the metastable density in the space afterglow is about $10^{12}cm^{-3}$. The density value decreased by increasing the discharge duration T_{ON} with respect to the diffusion time (about 3 ms) of the metastable necessary to reach a steady-state density in the space afterglow. That is likely due to the production of a larger density of quenching products (N, N_2(X,v)). This fact, combined with some difference in the material of the discharge electrodes and of the driving ac frequency, justifies the lower value measured in [2] by running the DBD under continuous regime. An important aspect is that in the space afterglow a stable cloud of metastables is established, which is not modulated by the discharge voltage cycles.

4. LIF on CN($X^2\Sigma^+$)

We addressed in [4] the issue of the CN violet system emission in a N_2-CH_4 DBD. This issue was particularly interesting since the emission revealed a largely supra-thermal vibrational excitation of the CN($B^2\Sigma^+$) emitting state, that looked surprising at high pressure. By the way we had already done the same observation in a N_2-C_6H_6 DBD [19]. Such kind of vibrational distributions had been largely studied in hydrocarbon low-pressure flames (i.e. flowing afterglows at pressures from 1 to few tens of Torr) in the early 60's, but no trace can be found in the literature of investigations carried out at atmospheric pressure. In [20] active nitrogen flames with a variety of 'fuels' (halogenated hydrocarbons, cyanogen derivatives, CH_4 and C_2H_2) were examined, finding out three different kinds of CN(B) vibrational excitations. In particular the distribution obtained under 'blue flame' conditions showed observable emission up to v=15, with a large population of the v=5-7 levels group. The 'blue flame' distribution was attributed to the reaction:

$$N^* + CN(X^2\Sigma^+) \rightarrow CN(B^2\Sigma^+, A^2\Pi) + N_2(X^1\Sigma) \tag{3}$$

In [21] N* was argued to be the $N_2(A^3\Sigma_u^+)$ metastable,while in [22] and [23] vibrationally excited nitrogen N_2(X, $v \geq 12$) was proposed for this role. In order to investigate the occurrence of reaction (3), and its precursor, we implemented a LIF measurement of CN($X^2\Sigma^+$) according to the following scheme:

$$CN(X^2\Sigma^+, v') + h\nu_L \rightarrow CN(B^2\Sigma^+, v) \rightarrow CN(X^2\Sigma, v'') + h\nu_F \tag{4}$$

with $v - v' = 0$ and $v - v'' = -1$. In addition we performed OODR-LIF measurements on $N_2(A^3\Sigma_u^+)$. Both measurements were not absolute, but only a tracing of the relative time evolution of CN and N_2(A) that we used for comparison to the corresponding time evolution of the CN violet system emission. No correlation was found between these three quantities that could justify the effectiveness of reaction (3) that was then excluded as a source of CN(B) excitation. Note that in this assessment exercise, the time resolution of LIF diagnostics was of fundamental importance. LIF measurements were even more useful since, by inspection of fluorescence spectra it was possible to verify that, even at atmospheric pressure, the vibrational relaxation in the CN(B, v) manifold was not fast. The vibrational distribution observed from violet system emission were then quasi-nascent, and could be used for the identification of the excitation reaction. We found then finally, and with a significant help from LIF diagnostics, that two processes were active in the emission on CN violet system, namely the recombination $C + N + M \rightarrow CN(B) + M$ [24] and the reaction $N + CH \rightarrow CN(B) + H$ [20].

5. Acknowledgements

This work has been made in the frame of CNR-AVCR 2007-2009 agreement

6. References

[1] Dilecce G, Ambrico P F and De Benedictis S 2007 *Plasma Sources Sci. Technol.* **16** 511 – 522
[2] Ambrico P F, Simek M, Dilecce G and De Benedictis S 2008 *Plasma Chem. Plasma Process.* **28** 299–316
[3] Dilecce G, Ambrico P F and De Benedictis S *J. Phys. D: Appl. Phys.* in press
[4] Dilecce G, Ambrico P F, Scarduelli G, Tosi P and De Benedictis S 2009 *Plasma Sources Sci. Technol.* **18** 015010
[5] Ambrico P F, Simek M, Dilecce G and De Benedictis S 2009 *Appl. Phys. Lett.* **94** 231503
[6] Hirano A, Ipponmatsu M and Tsujishita M 1992 *Opt. Lett.* **17** 303–304
[7] Luque J, Klein-Douwel R J H, Jeffries J B and Crosley D R 2000 *Appl. Phys. B* **71** 85 – 94
[8] Moreau C, Desgroux E T P, Pauwels J, Chapput A and Barj M 2003 *Appl. Phys. B* **76** 597 – 602
[9] Tsujishita M, Ipponmatsu M and Hirano A 1993 *Jpn. J. Appl. Phys.* **32** 5564–5569
[10] Luque J and Crosley D 1999 Int. rep. p99-009 Tech. rep. SRI
[11] Tamura M, Berg P A, Harrington J E, Luque J, Jeffries J B, Smith G P and Crosley D R 1998 *Combust. Flame* **114** 502–514
[12] De Benedictis S and Dilecce G 1997 *J. Chem. Phys.* **107** 6219–6229
[13] De Benedictis S and Dilecce G 1993 *CHem. Phys.* **178** 547–560
[14] Dilecce G, Ambrico P F and De Benedictis S 2005 *Plasma Sources Sci. Technol.* **14** 561 – 565
[15] Dilecce G, Ambrico P F and De Benedictis S 2006 *Chem. Phys. Lett.* **431** 241 – 246
[16] Dilecce G, Ambrico P F and De Benedictis S 2007 *Chem. Phys. Lett.* **444** 39 – 43
[17] Simek M 2003 *Plasma Sources Sci. Technol.* **12** 421–431
[18] Massines F, Gherardi N, Naudé N and Segur P 2005 *Plasma Phys. Control. Fusion* **47** B577–B588
[19] Franceschi P, Guella G, Scarduelli G, Tosi P, Dilecce G and De Benedictis S 2007 *Plasma Process. Polym.* **4** 548 – 555
[20] Iwai T, Savadatti M I and Broida H P 1967 *J. Chem. Phys.* **47** 3861–3874
[21] Boden J and Trush A 1968 *Proc. R. Soc. A* **305** 93
[22] Taieb G and Legay F 1970 *Can. J. Phys.* **48** 1956–1965
[23] Provencher G M and McKenney D J 1972 *Can. J. Chem.* **50** 2527–2536
[24] Washida N, Kley D, Becker K H and Groth W 1975 *J. Chem. Phys.* **63** 4230–4241

Collective scattering system for transport study on NSTX

W Lee[1], D R Smith[2], E Mazzucato[2], H K Park[1], C W Domier[3], and N C Luhmann Jr[3]

[1] POSTECH, Pohang 790-784, South Korea
[2] Princeton Plasma Physics Laboratory, Princeton, NJ 08543, USA
[3] University of California at Davis, Davis, CA 95616, USA

E-mail: woochanglee@postech.ac.kr

Abstract. The problem of degraded energy confinement due to excessive transport across the magnetic field remains one of the critical issues for magnetically confined plasma. Radial transport of ion energy is relatively well understood, but that of the electrons has remained as anomalous, greatly exceeding the neoclassical prediction. It has been suggested that the anomalous electron thermal transport is explained by an electron gyro-scale turbulence driven by the electron temperature gradient (ETG) instability. A 280-GHz collective Thomson scattering system has been employed to address the electron gyro-scale fluctuations in National Spherical Torus Experiment (NSTX) plasmas. The spatial resolution of the targeting wavenumber is greatly affected by the configuration of the magnetic field since the radial fluctuation is perpendicular to the local magnetic field line. The effect of the toroidal field curvature and magnetic shear on the spatial resolution of the scattering system is investigated numerically. An absolute power calibration was performed to determine the power response of the heterodyne detection system. These spatial resolution studies and absolute power calibration were applied to estimate the normalized density fluctuations from the measured scattering signals in NSTX plasmas.

1. Introduction

A sufficient energy confinement time is critical to achieve a burning plasma in magnetically confined devices and the confinement time primarily depends on radial transport of the particle and energy. Although the ion thermal and particle transports can be suppressed to the neoclassical level, the electron thermal transport has always been anomalous in the tokamak plasmas. Recent studies have suggested the electron gyroscale turbulence driven by the electron temperature gradient (ETG) instability serves as the underlying mechanism for the electron thermal transport. Experimental studies in various devices have been performed to investigate the electron thermal transport driven by the electron temperature gradient (ETG) instability [1, 2, 3, 4]. Experiments on NSTX are significantly extending the studies [5, 6, 7]. The NSTX plasma, which has a large electron gyro-radius, is well suited for electron thermal transport studies.

The five-channel, 1 mm wavelength (280 GHz) collective scattering system has been installed on NSTX to measure the electron gyroscale fluctuations [8]. The five-channel detection system was designed to simultaneously measure the frequency spectra of density fluctuations at five discrete wavenumbers. The range of wavenumbers extends up to 20 cm^{-1} which corresponds to $k_\perp \rho_e \leq 0.7$, where $k_\perp$ is the perpendicular component of the fluctuations with respect to the local magnetic field and ρ_e is the electron gyro-radius. Probe and scattered beams are nearly on the midplane, so that the detected fluctuations are primarily radial or perpendicular to the magnetic surface. Steerable launching and receiving optics have been used to position the scattering volume ranging from the magnetic axis to the outer edge of the plasma.

Plasma density fluctuation levels were estimated from the scattering signals measured in H-mode and high electron temperature L-mode NSTX plasmas. The estimation of the scattered power (or fluctuation level) was carried out based on absolute power calibration of the detection system and numerical calculation of the scattering length of each channel.

2. Basic principle of the collective scattering experiment

The scattering process is based on the three-wave coupling among the incident and scattered electromagnetic waves and a plasma density fluctuation wave. In the scattering process, energy and the momentum have to be conserved as $\omega_s = \omega_i + \omega$ and $\boldsymbol{k}_s = \boldsymbol{k}_i + \boldsymbol{k}$, where the subscripts s and i refer to the scattered and the incident wave, respectively. Since the frequencies of the probe and scattered beams are significantly higher than that of the plasma waves, the momentum matching condition yields the Bragg relationship:

$$\boldsymbol{k} = 2\boldsymbol{k}_i \sin(\theta_s/2), \tag{1}$$

where θ_s is the scattering angle.

A local density fluctuation interacts with the probe electromagnetic wave and generates scattered radiation. The time-average scattered power in the frequency range of $\omega \rightarrow \omega + d\omega$ and in the solid angle $d\Omega$ centered on $\boldsymbol{R}$ is [8, 9]

$$P_s(\boldsymbol{R}, \omega) d\Omega d\omega = \frac{P_i r_e^2}{A} d\Omega \frac{d\omega}{2\pi} \overline{|n_e(\boldsymbol{k}, \omega)|^2}, \tag{2}$$

where P_i is the probe beam power, $r_e = e^2/4\pi\epsilon_0 m_e c^2 = 2.82 \times 10^{-13}$ cm is the classical electron radius, A is the cross sectional area of the probe beam, and $\overline{|n_e(\boldsymbol{k}, \omega)|^2}$ is a Fourier analyzed time-averaged density fluctuation. The scattered beam is considered as a Gaussian beam, so that the solid angle $d\Omega = \lambda_i^2/A$ is derived through the far-field approximation, where λ_i is the probe beam wavelength. Assuming a properly designed optical system, the receiver system collects all the scattered power within the solid angle, so that the above scattered power becomes

$$P_s(\boldsymbol{R}, \omega) d\omega = \frac{P_i r_e^2 \lambda_i^2}{A^2} \frac{d\omega}{2\pi} \overline{|n_e(\boldsymbol{k}, \omega)|^2} = P_i r_e^2 \lambda_i^2 L_v^2 \frac{d\omega}{2\pi} \frac{\overline{|n_e(\boldsymbol{k}, \omega)|^2}}{V^2}, \tag{3}$$

where L_v is the scattering volume length along the probe beam and V is the scattering volume. The NSTX scattering system has a probe beam with $P_i \sim 100$ mW and $\lambda_i \sim 0.1$ cm, so that the time-averaged density fluctuation in the frequency range $\omega \rightarrow \omega + d\omega$, can be obtained from the measured scattered power in Eq. 3 as

$$\delta n_e^2(\boldsymbol{k}, \omega)\ [/\text{cm}^3] \equiv \frac{d\omega}{2\pi} \frac{\overline{|n_e(\boldsymbol{k}, \omega)|^2}}{V^2} \sim 10^{25} \times \frac{P_s(\boldsymbol{R}, \omega) d\omega}{L_v^2}\ [\text{mW}]. \tag{4}$$

3. Absolute power calibration

The power calibration of the heterodyne detection system was carried out using a millimeter wave source of a known output power together with calibrated attenuators [10]. Raw data of the calibration are shown in Fig. 1. The power responses of the five-channel detection system

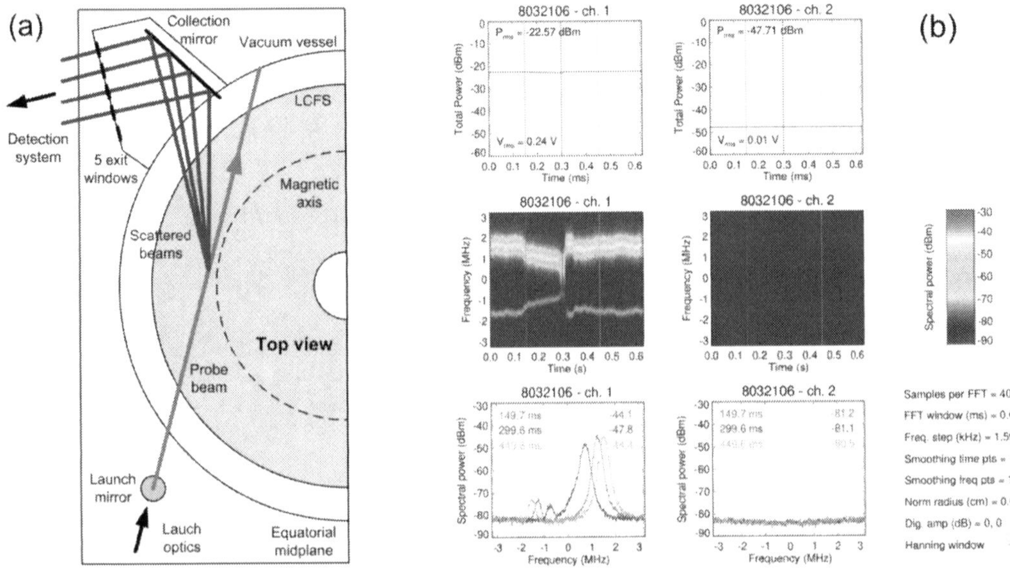

Figure 1. (a) A layout of the NSTX collective scattering system and (b) raw data of the absolute power calibration of the detection system. The first row is the total output power, and second and third rows are Fourier analyzed spectral powers. An input power of -49 dBm directly came to channel 1 of the detection system and the output power was -23 dBm, so that the power response of channel 1 is 26 dB. The noise level in the other channels was ~ -48 dBm, which corresponds to a minimum detectable input power of ~ -74 dBm. In 4096 points frequency space, the noise level is ~ -80 dBm, which corresponds to a minimum detectable input power of ~ -110 dBm.

were 26, 36, 30, 33, and 30 dB, respectively. The minimum detectable power (or noise level) is ~ -75 dBm (or $\sim 3 \times 10^{-8}$ mW), which corresponds to the normalized density fluctuation level, $\delta n_e/n_e$, of $\sim 6 \times 10^{-6}$ in the assumption of $n_e \sim 10^{13}$ /cm^3 and $L_v \sim 10$ cm. In the Fourier space with 4096 sample points, the noise level is ~ -110 dBm (or $\sim 10^{-11}$ mW) and the minimum detectable $\delta n_e/n_e$ is $\sim 1 \times 10^{-7}$ in the same assumption.

4. Scattering length on the detected scattered power

As shown in Eq. 4, it is essential to properly take into account the scattering length in the estimation of the fluctuation levels. The scattering length decreases as the scattering angle increases since the scattering volume is basically an overlap region of the probe beam and scattered beam. The exact scattering length is determined by the magnetic field configuration since the plasma turbulence is anisotropic in general with $k_\perp \gg k_\parallel$ where the subscripts $\perp$ and $\parallel$ represent the perpendicular and the parallel components to the magnetic field, respectively. The variation of the magnetic field direction within the scattering volume can modify the detecting efficiency of the receiver due to the varying momentum matching condition shown in Eq. 1.

The scattering length in nonuniform magnetic fields was numerically calculated in references [11, 12, 10]. The scattering lengths were found to be primarily affected by the toroidal curvature. Figure 2 shows the calculated scattering lengths for fluctuation wavenumbers ($k_\perp = 4$, 8, 12, 16, and 20 cm^{-1}) at several toroidal toroidal curvature radii (R = 110, 120, 130, and 140 cm). The scattering length decreases as the fluctuation wavenumber increases and toroidal curvature radius decreases.

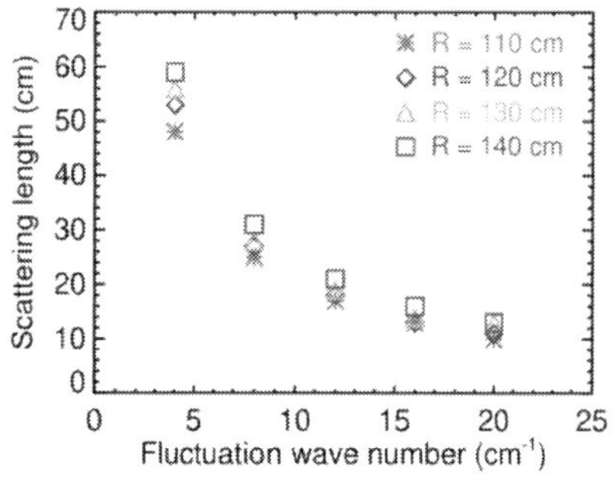

Figure 2. Scattering lengths for several fluctuation wavenumbers. The scattering length decreases as the wavenumber increases and the toroidal curvature radius decreases.

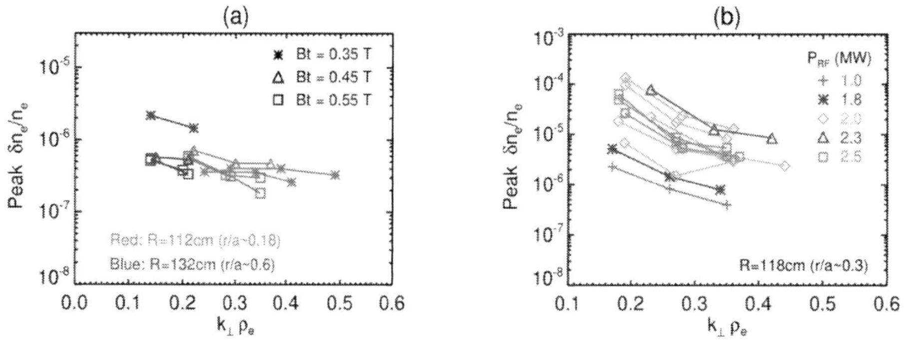

Figure 3. Peak normalized density fluctuations, $\delta n_e(\boldsymbol{k},\omega)/n_e$, in (a) H-mode plasmas ($B_t = 0.35 - 0.55$ T, $I_p = 0.7$ MA, and $P_{NB} = 4$ MW) and (b) high-T_e L-mode plasmas ($B_t = 0.55$ T, $I_p = 0.6$ MA, and $P_{RF} = 1.0 - 2.5$ MW). Note that the $\delta n_e(\boldsymbol{k},\omega)/n_e$ is a Fourier spectrum of $\delta n_e(\boldsymbol{k},t)/n_e$.

5. Estimation of the fluctuation levels

The previous calibration result and scattering length calculation were applied to the NSTX plasmas to estimate fluctuation amplitudes from the measured scattering signals.

Figure 3 shows the peak value of the normalized density fluctuations, $\delta n_e(\boldsymbol{k},\omega)/n_e$, measured in both (a) H-mode plasmas and (b) high-T_e L-mode plasmas. Note that the $\delta n_e(\boldsymbol{k},\omega)/n_e$ is a Fourier spectrum of $\delta n_e(\boldsymbol{k},t)/n_e$. The H-mode plasmas were heated by 4-MW NB power at a plasma current of 0.7 MA and three toroidal magnetic fields of 0.35, 0.45, and 0.55 T. The red points were measured at the plasma core of $R \sim 113$ cm (or $r/a \sim 0.16$) and the blue points were measured at the intermediate region of $R \sim 132$ cm (or $r/a \sim 0.6$). The L-mode plasmas were heated by high harmonic fast wave (HHFW) of power from 1.0 to 2.5 MW at a plasma current of 0.6 MA and toroidal magnetic field of 0.55 T. The HHFW power preferentially heats electrons rather than ions, so that the electron temperature of the plasmas can rise above 4 keV. All scattering measurements were performed in the plasma core region of $R \sim 118$ cm (or $r/a \sim 0.3$). The normalized density fluctuations were $10^{-7} \sim 10^{-5}$ in H-mode plasmas and $10^{-7} \sim 10^{-4}$ in high-T_e L-mode plasmas. The normalized fluctuation levels tend to slightly decrease as the normalized wavenumber, $k_\perp \rho_e$, increases since the scattering length is smaller at higher wavenumber.

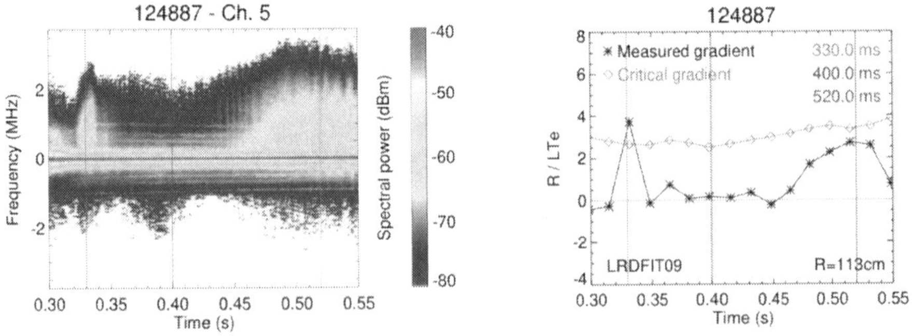

Figure 4. Density fluctuation signal in H-mode plasma (left) and its normalized electron temperature gradient(right). Fluctuation appears when the normalized gradient near the critical value.

For exact interpretation of the scattered power, other factors must be considered such as the Faraday rotation of the linearly polarized probe and scattered beams. Specifically, a vertically polarized probe beam enters into the vacuum vessel and its polarization is changed by the magnetic field. The polarization of the scattered beam is also changed until it reaches the detector which is horizontally polarized. Here, it is simply assumed that a half of the scattered power enters to the detector considering the detector which has preference in polarization.

6. Effect of T_e gradient on fluctuations

Recent studies [13, 14, 15, 16] have predicted that the electron gyro-scale fluctuations are associated with the electron temperature gradient (ETG) instabilities, and that is experimentally observed in NSTX plasmas [5]. Fluctuations appear when the electron temperature gradient is above or near a critical value which is calculated for the onset of ETG modes (see Fig 4). In addition, it is known that the electron gyro-scale fluctuations are suppressed in a strongly reversed (negative) magnetic shear [6, 16] and $\boldsymbol{E} \times \boldsymbol{B}$ flow shear [7]. In this paper, the effect of the electron temperature gradient on the fluctuations was investigated in the previous H-mode and high-T_e L-mode plasmas. Here, the normalized electron temperature gradient, R/L_{T_e}, is given as [15]

$$R/L_{T_e} = -R/T_e(dT_e/dr) \tag{5}$$

where R is the major radius.

Figure 5 shows the results. In H-mode plasmas, fluctuations weakly correlate with the normalized electron temperature gradient in the range of $2 \sim 8$. On the other hand, in high-T_e L-mode plasmas, fluctuations strongly correlate with the electron temperature gradient in the range of $6 \sim 10$ for all wavenumbers. This result is additional evidence that the electron gyroscale density fluctuations in NSTX are induced by the ETG instabilities.

7. Conclusion

Electron gyro-scale density fluctuations were observed in NSTX plasmas by using the 1-mm wavelength scattering system. Fluctuations were measured at the core region ($r/a \sim 0.16 - 0.3$) in both H-mode and L-mode plasmas and at the intermediate region ($r/a \sim 0.6$) in H-mode plasmas. The fluctuations were in the normalized wavenumber, $k_\perp \rho_e$, from 0.14 to 0.5, which is the wavenumber range of the ETG instability.

These fluctuation levels were estimated from the scattering signal based on the power calibration of the detection system and calculation of the scattering length. The peak normalized

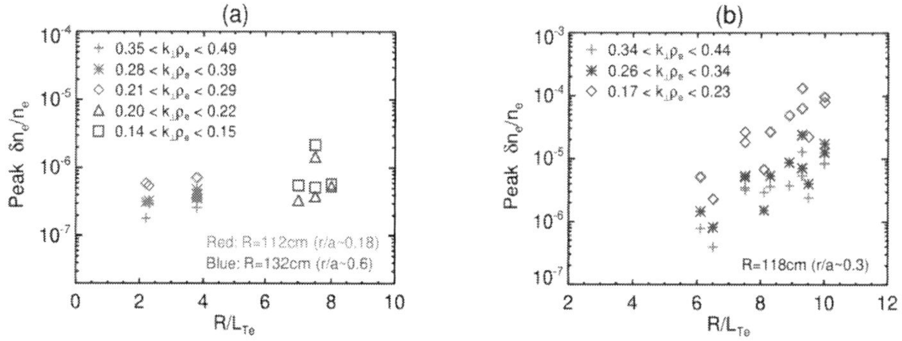

Figure 5. Effect of normalized electron temperature gradient, R/L_{T_e}, on the density fluctuation in (a) H-mode and (b) high-T_e L-mode plasmas. In high-T_e L-mode plasma, fluctuations strongly correlate with the electron temperature gradient.

density fluctuations, $\delta n_e(\boldsymbol{k},\omega)/n_e$, were $10^{-7} \sim 10^{-5}$ in H-mode plasmas and $10^{-7} \sim 10^{-4}$ in high-T_e L-mode plasmas. Fluctuations measured in high-T_e L-mode plasmas strongly correlate with the electron temperature gradient.

Acknowledgments
The authors would like to thank the NSTX team for support and hard work. This work was supported by the U.S. Department of Energy under Contract Nos. DE-AC02-09CH11466 and DE-FG02-99ER54518, Department of Physics at POSTECH, and Korea National Fusion Research Institute.

References
[1] Rhodes T L, Peebles W A, Nguyen X, VanZeeland M A, deGrassie J S, Doyle E J, Wang G and Zeng L 2006 *Rev. Sci. Instrum.* **77** 10E922
[2] Horton W, Hoang G T, Bourdelle C, Hu B, Garbet X, Ottaviani M and Golas L 2004 *Phys. Plasmas* **11** 2600
[3] Gusakov E Z, Gurchenko A D, Altukhov A B, Bulanin V V, Esipov L A, Kantor M Y, Kouprienko D V, Lashkul S I, Petrov A V and Stepanov A Y 2006 *Plasma Phys. Control. Fusion* **48** B443
[4] Gurchenko A, Gusakov E, Altukhov A, Stepanov A, Esipov L, Kantor M, Kouprienko D and Dyachenko V 2007 *Nucl. Fusion* **47** 245
[5] Mazzucato E, Smith D, Bell R, Kaye S, Hosea J, LeBlanc B, JRWilson, Ryan R, Domier C, N C Luhmann, Jr, Yuh H, Lee W and Park H 2008 *Phys. Rev. Lett.* **101** 075001
[6] Yuh H Y, Levinton F M, Bell R E, Hosea J C, Kaye S M, LeBlanc B P, Mazzucato E, Peterson J L, Smith D R, Candy J, Waltz R E, Domier C W, N C Luhmann, Jr, Lee W and Park H K 2009 *Phys. Plasmas* **16** 056120
[7] Smith D R, Kaye S M, Lee W, Mazzucato E, Park H K, Bell R E, Domier C W, LeBlanc B P, Levinton F M, N C Luhmann, Jr, Menard J E and Yuh H 2009 *Phy. Rev. Lett.* **102** 225005
[8] Smith D R, Mazzucato E, Lee W, Park H, Domier C W and N C Luhmann, Jr 2008 *Rev. Sci. Instrum.* **79** 123501
[9] Sheffield J 1975 *Plasma Scattering of Electromagnetic Radiation* (New York: Academic)
[10] Lee W, Park H K, Cho M H, Namkung W, Smith D R, Domier C W and N C Luhmann, Jr 2008 *Rev. Sci. Instrum.* **79** 10E723
[11] Mazzucato E 2003 *Phys. Plasmas* **10** 753
[12] Mazzucato E 2006 *Plasma Phys. Control. Fusion* **48** 1749
[13] Horton W, Zhu P, Hoang G T, Aniel T, Ottaviani M and Garbet X 2000 *Phys. Plasmas* **7** 1494
[14] Dorland W, Jenko F, Kotschenreuther M and Rogers B N 2000 *Phys. Rev. Lett.* **85** 5579
[15] Jenko F, Dorland W and Hammett G W 2001 *Phys. Plasmas* **8** 4096
[16] Jenko F and Dorland W 2002 *Phys. Rev. Lett.* **89** 225001

Journal of Physics: Conference Series **227** (2010) 012005 doi:10.1088/1742-6596/227/1/012005

Quantum Cascade Laser Absorption Spectroscopy – a New Method to Study Molecular Plasma Components

J Röpcke[1], S Glitsch[1], P Davies[2], F Hempel[1], N Lang[1], A Rousseau[3], S Wege[4] and S Welzel[1a]
[1]INP Greifswald, F.-Hausdorff-Str. 2, 17489 Greifswald, Germany
[2]University of Cambridge, Lensfield Road, Cambridge CB2 1EW, United Kingdom
[3]LPP, Ecole Polytechnique, CNRS, 91128 Palaiseau Cedex, France
[4]Plasway, Hauptstr. 7a, 01328 Dresden, Germany

E-mail: roepcke@inp-greifswald.de

Abstract The recent development of quantum cascade lasers (QCLs) offers an attractive new option for the monitoring and control of industrial plasma processes and for trace-gas analysis as well as for highly time-resolved studies on the kinetics of plasma processes. The contribution reviews selected examples of the application of QCLs for infrared absorption studies in basic research and for plasma monitoring and control in industry.

1. Introduction

Plasmas with molecular feed gases are used in many applications such as thin film deposition, semiconductor processing, surface activation and cleaning, and materials and waste treatment. Within the last decade this has stimulated the adaptation of infrared spectroscopic techniques because most of the used compounds and their decomposition products are infrared active. The availability of quantum cascade lasers (QCL) opens up new attractive possibilities for the application of Infrared Laser Absorption Spectroscopy (IRLAS) for in situ plasma process monitoring and control. QCLs emit in the mid infra red spectral range (MIR) at room temperature, i.e. without the need for cryogenic cooling. Because of this it is possible to design very compact MIR sources which are characterized by single mode operation and high radiance power (mW). Sensitivities in the range of ppb with time resolutions of better than ms are feasible. The aim of the present contribution is to review recent achievements using QCLs for plasma diagnostics and monitoring in research and industry. This review will include (i) time resolved studies on pulsed DC discharges using QCL, (ii) application of QCLs and PIR-fibres for monitoring and control of industrial surface treatment and (iii) in situ monitoring of plasma etch processes with a QCL arrangement in semiconductor industrial environment.

2. Chirp-based spectroscopic studies of NO containing discharges

2.1. Highly time resolved QCLAS in pulsed plasmas

The broad range of applications where the chemistry is exploited in N_2-O_2 containing discharges, such as pollution abatement or atmospheric pressure plasmas used for thin film deposition or

[a] present address: Eindhoven University of Technology, P.O. Box 513, 5600 MB, Eindhoven, The Netherlands

biomedical applications, has stimulated extensive studies of these plasmas. Different diagnostic techniques (e.g., laser induced fluorescence, mass spectrometry, IRLAS) have been applied [1 - 3]. The important role of surface reactions on the molecule formation has also been recognised [4]. Additionally, the filamentary structure of discharges at atmospheric pressure usually complicates the experimental approach. Thus, low pressure plasmas, being typically more homogeneous in nature, are preferably used for fundamental studies. Consequently, the time resolved in situ detection of ground state NO densities in pulsed plasmas may provide an improved understanding of the main chemical processes and a discrimination of gas phase and surface reactions.

Since pulsed QCLs exhibit an inherent frequency sweep (also known as frequency down chirp) of typically 0.005 cm^{-1}/ns (150 MHz/ns) an entire absorption spectrum may be acquired within a few tens of nanoseconds and thus enable highly time resolved plasma diagnostics. Recently, this approach has been used to study the time decay of NO in situ in single discharge pulses [5]. Considering the increase of the gas temperature in the DC discharge the NO concentration was found to remain almost unchanged during a 1 ms plasma pulse. However, the application of successive pulses suggested a conversion of NO. Selecting an adequate absorption line provides a sensitive, non-invasive temperature probe if the experiments can be carried out under static and flowing gas conditions.

These initial experiments have been extended to long plasma pulses of 100 ms to study the suggested chemical removal of NO after several short plasma pulses, which would correspond to the millisecond time scale. The experimental setup is almost essentially the same as employed in ref. [5] and thus only the few changes are described here. A replaced power supply enabled the long plasma pulse experiments to be performed. In order to facilitate a chemical modelling in the future the gas mixture consisted of 1 % NO in pure N_2 (instead of an Ar/N_2 mixture). The sensitivity of the measurement system could be increased by means of an optical double pass configuration yielding about 100 cm absorption path length through the cylindrical discharge tube. A new thermoelectrically cooled detector (neoplas control GmbH) was also employed. The detector element (VIGO) was slightly more sensitive than before at the expense of an increased rise time. Finally, this led to less pronounced rapid passage structures in the recorded spectra (figure 1). Since the absolute NO number densities were retrieved from calibrated line strength values of unresolved R 5.5 transitions at 1897.35 cm^{-1} the change of the detector performance is automatically included (see ref. [5] for details of the calibration procedure). The shift of QCL emission to the present spectral position was accompanied by a higher laser output power to achieve a reasonable signal-to-noise ratio in the double pass configuration.

2.2. Results

NO absorption signals have been recorded under flowing and static gas conditions at constant pressure (266 Pa) during the entire plasma pulse and the off-phase (figures 2 and 3). The time resolution, which could be in principle 90 ns (i.e. the QCL pulse length), was adapted to the detected variations of the absorption signal. Examples are shown in figure 1. The experiments have been carried out at two discharge currents, i.e. 19 mA and 35 mA (figure 2). The lower current case corresponds to the minimum input power required to obtain a stable discharge.

The integrated NO absorption coefficient (product of number density and line strength) as a function of time (figures 2 and 3) can clearly be divided into 3 parts during the plasma on-phase. The early stage plasma pulse ($t \leq 1$ ms) is characterised by a gradually decreasing absorption coefficient. The drop of typically not more than 15 % is comparable to the mainly temperature induced decrease of the NO line strength which was also observed for short plasma pulses. After ~ 1 ms the decrease of the NO signal is more pronounced until the integrated absorption coefficient levels off after 25 ms at ~ 20 % of its initial value (figure 2, 35 mA). This plateau is caused by the detection limit of the configuration (3×10^{13} cm^{-3}). During the off-phase an increase of the NO signal can be observed. After about 5 s the initial value is reached which agrees with the residence time of the system under flowing conditions. It is also clear from figure 2 that the onset of the rapid decrease (5 ms) and the detection limit (80 ms) are observed later if the discharge current is reduced to 19 mA. Applying a 35 mA discharge pulse yields a difference between the temporal evolution of the NO absorption signal under static and flowing gas conditions (figure 3) while almost no difference is detected for 19 mA. Since the additional depletion of NO under flowing conditions is caused by an increase of the gas

Journal of Physics: Conference Series **227** (2010) 012005 doi:10.1088/1742-6596/227/1/012005

temperature at constant pressure the observed difference between the two gas regimes can be used to estimate the (relative) temperature increase during the discharge pulse [5].

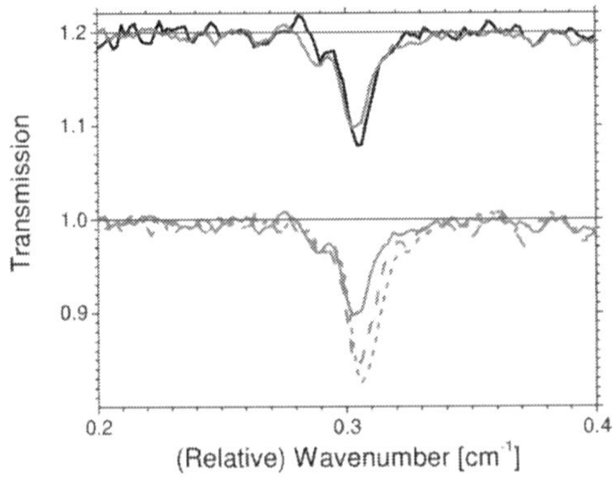

Figure 1. Sample transmission spectra of NO at 1897.35 cm^{-1} under flowing (grey) and static (black) gas conditions. The absorption features were recorded 0.1 ms (dotted), 1 ms (dashed) and 10 ms (solid) after the ignition of the discharge using 35 mA. (The comparison between the two gas conditions has been shifted for clarity.)

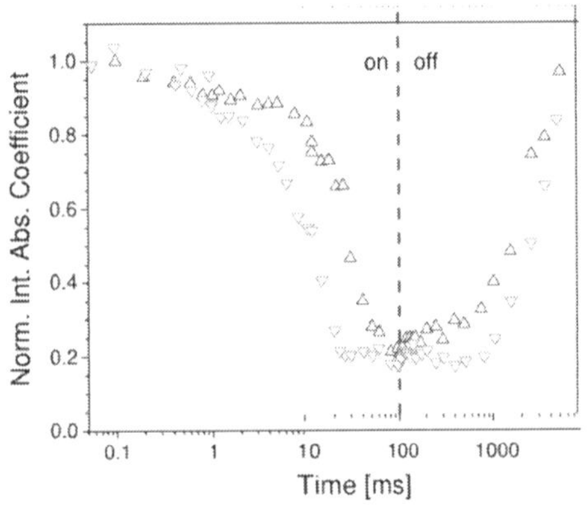

Figure 2. Time dependence of the integrated NO absorption coefficient during 100 ms plasma pulses under flowing gas conditions at 19 mA (triangle up) and 35 mA (triangle down). The plateau at ~ 0.2 corresponds the detection limit of the system.

Figure 4 shows the temporal evolution of the absolute gas temperature deduced from figure 3 based on the difference of the integrated NO absorption coefficients [5]. During the first milliseconds the gas temperature changes only slightly. After ~ 5 ms a considerable increase from 320 K to almost 500 K can be observed. Nevertheless, this temperature increase and the corresponding decrease of the line strength cannot explain the decrease of the NO signal below the detection limit. Thus, chemical processes have to be considered.

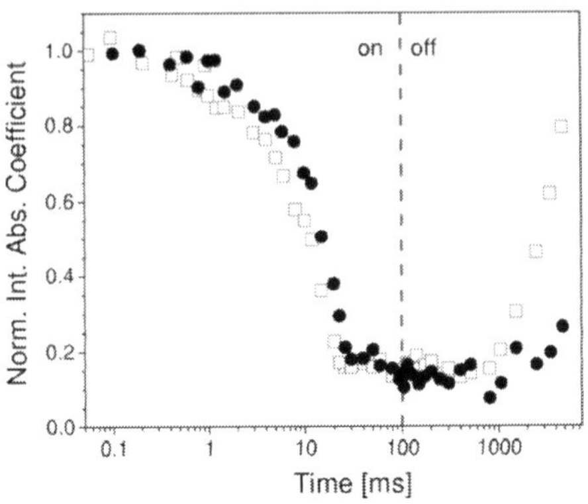

Figure 3. Comparison of the time dependent integrated NO absorption coefficient under static (full black) and flowing (open grey) gas conditions observed for a 35 mA plasma pulse.

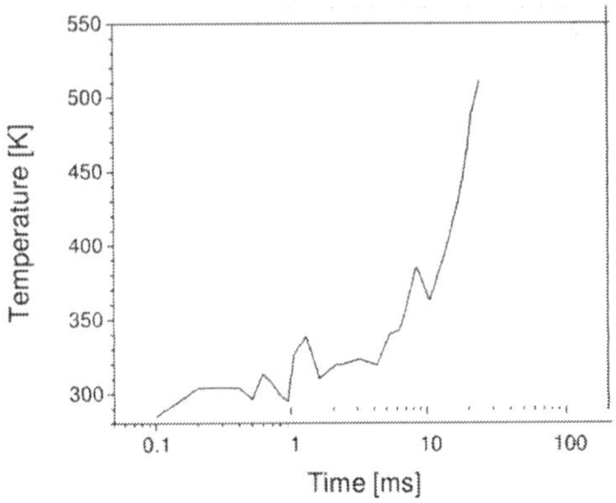

Figure 4. Estimated temporal evolution of the absolute gas temperature for a 35 mA plasma pulse inferred from the difference of the integrated NO absorption coefficients. Data close to the detection limit (t > 20 ms) were considered.

A calculation using a simplified model based on the measured injected power density [5] yields a much more pronounced temperature increase which appears unrealistic. A straightforward downscaling of the injected power density leads to the conclusion that less than 50 % of the input power are used for neutral gas heating in the present case of pure N_2 as background gas. The increase after 10 ms may be explained by vibrational-translational exchange. The identification of the main reactions requires a chemical modelling work which is now in progress.

3. Time resolved studies on industrial pulsed DC discharges containing Boron

3.1 General considerations

Non-equilibrium low-pressure molecular plasmas containing boron are valuable tools in plasma enhanced chemical vapor deposition (PECVD) processes as well as in etching systems to deposit or remove thin films and to treat surfaces for improved properties. Various precursor molecules like BCl_3, BF_3 or B_2H_6 are used to provide active boron-containing species in the plasmas. In particular Boron Trichloride (BCl_3) is a source gas of considerable interest in plasma research and industry since it is used for the surface modification of steel components in mechanical engineering and automotive industries as well as for plasma etching of semiconductor devices. In contrast to the widespread application of gas discharges containing BCl_3 in industry, not much is known about the internal discharge parameters and phenomena, including processes of precursor fragmentation, plasma heating and chemical conversion processes. Even though IRLAS is well suited for this purpose, it is rarely applied for investigations in gas discharges containing BCl_3. Due to the high intensity and its good match with CO_2 laser modes, in particular the ν_3 vibrational band of BCl_3 around 958 cm^{-1} was used for investigations in BCl_3 [6,7] with and without rf discharge in the lab and for plasma process monitoring [8].

3.2 Experimental

For the present work QCL absorption spectroscopy (QCLAS) was applied for the investigation of the discharge chemistry in a industrial pulsed-dc-discharge-reactor (Type CVD4, ELTRO GmbH, Germany). The plasma reactor was a bell furnace style (∅ 80 cm, h: 150 cm) with walls heatable to temperatures up to 900 °C. The substrate holder acts together with the electro conductive steel substrates as the cathode of the discharge while the bottom is grounded together with all other metal parts and forms the anode. The plasma is ignited as a thin layer around the substrates, as typically for pulsed dc-discharges. The process gases ($BCl_3 + H_2 + Ar$) were pumped through a port at the other side of the plate. The furnace and the plasma processing were controlled by an industrial computer system. First measurements have been performed using the Q-MACS Process 2 (QCL Measurement and Control System). The optical design consists of a pulsed QCL as the infrared source with an infrared-fibre to allow a flexible and enclosed laser beam handling and remote positioning of the laser source and control unit. The optics are simplified to the greatest possible extent to simplify the alignment and enhance the robustness of the system. Figure 5 shows the scheme of the experimental arrangement.

3.3 Results

The data acquisition unit of Q-MACS Process was designed for the requirements of the TDLWintel software package [9,10]. This program can be used to control QCLs and to analyze the recorded spectra in real time and to display results of measured concentrations. By using this rapid scan software with real-time line shape fitting routines a time resolution of up to 1 ms can be provided, allowing the study of kinetic processes of infrared active compounds in plasmas or gases. To increase the signal to noise ratio the measurement can be averaged for several seconds, being a sufficient temporal resolution for most of the industrial plasma process monitoring applications.

The lack of absorption line data for several molecular source gases used in industrial plasmas is a major problem for the application of infrared absorption spectroscopic techniques for process analysis and control. Knowledge of exact wave number positions and line strength data of absorption features is an important precondition for quantitative measurements by absorption spectroscopy. Therefore this spectral information has to be acquired before measurements at industrial reactors can be performed. For the determination of the line data from the recorded spectra first the theoretical absorption line width was calculated from the pressure-, Doppler- and instrumental broadening. Then an initial guess of the unknown line positions and strengths was created and varied until the residuals of a least square fit, using this data, in comparison to the recorded spectra were below a given value. Following this, a final calculation of the line strength from resulting shapes was performed. Then a data file containing the line parameter was produced, which is compatible to the HITRAN database format [11].The data set consists of 289 lines. This is a much smaller number than the 1600 lines per cm^{-1} expected from literature [6] but sufficient to reproduce the actual spectra. The BCl_3 data file was used for measurements with QCLs (Figure 6). If pulsed QCLs are tuned by a small DC current ramp provided by TDLWintel, they show typically a smaller scanning range and bigger instrumental broadening, compared to TDLs. Nevertheless, it was possible to fit absorptions of BCl_3 and to verify test gas concentrations using QCLs. Therefore, spectral fits of BCl_3 spectra using the present data can be used for on-line process control. The data set gives sufficient flexibility for the monitoring of industrial plasma processes with changes of the total pressure or the gas temperature.

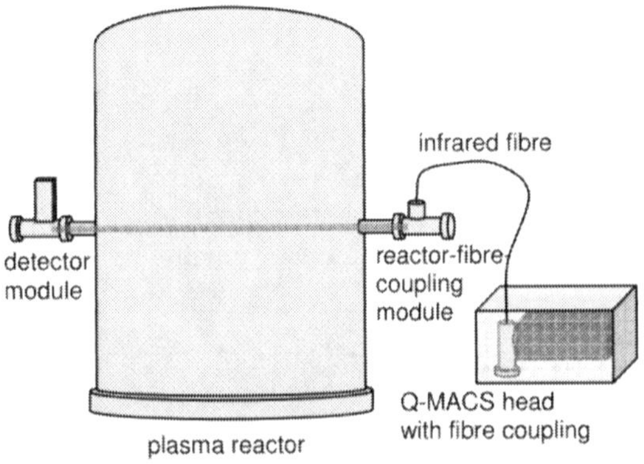

Figure 5. Experimental arrangement of the Q-MACS Process 2 with IR-fibre coupling to access the industrial plasma reactor.

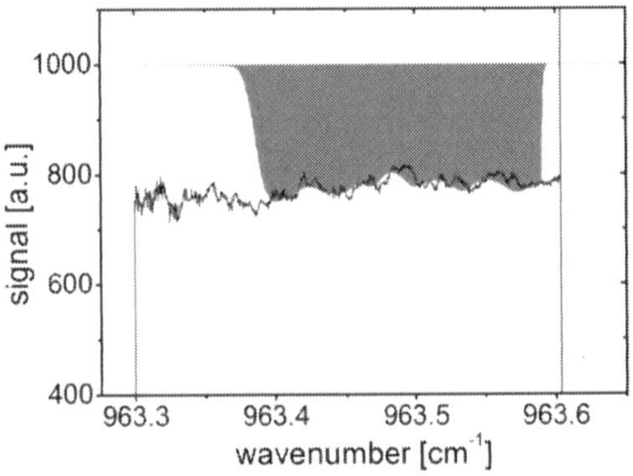

Figure 6. Test-fit of a BCl_3 spectrum measured with QCLAS.

Figure 7 shows an example of a plasma process monitoring with various contents of BCl_3 in the gas mixture and for various plasma duty cycles. Even though fluctuations caused by long-term changes of the irradiated laser intensity were observed, these results prove that it is possible to monitor industrial plasma processes using the Q-MACS Process 2. Particularly, it is possible to monitor the development of the concentration of the precursor gas BCl_3 in the pulsed DC plasma reactor while driving process plasmas to deposit Boron based hard coatings.

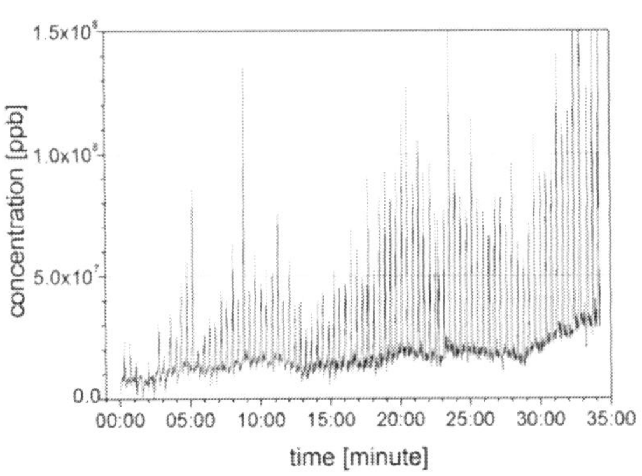

Figure 7. Change of the BCl_3 signal in the plasma process

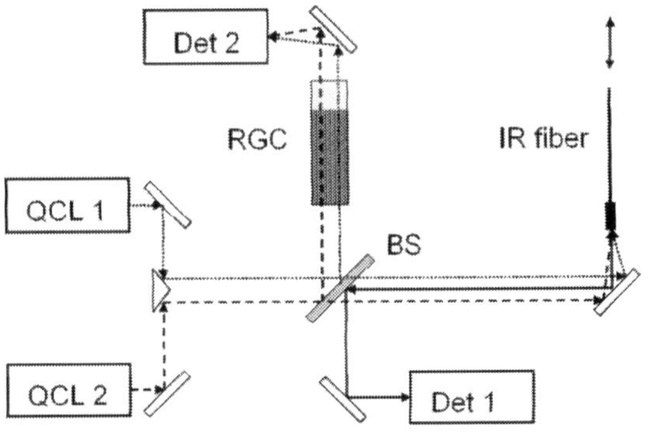

Figure 8. Schematic of the optical board of the QCLAS arrangement with two QCLs, measurement detector (Det 1), reference detector (Det 2), reference gas cell (RGC), beamsplitter (BS) and the IR fiber. The IR radiation is coupled back to the IR fiber with an external mirror in the reactor [15].

4. In situ monitoring and control of plasma etch processes in semiconductor industry

4.1. QCLAS for MFC control

In plasma etch processes the gas composition in the plasma bulk is a key parameter determining process stability and process results. It depends on various factors like (i) feed gas flows and pump speed, (ii) gas adsorption and desorption at the reactor walls, (iii) chemical reactions in the plasma and at the surfaces and also (iv) on total gas pressure and temperature. Up to now, typically optical emission spectroscopy, self excited electron resonance spectroscopy and mass spectrometry have been applied for online monitoring of plasma etch processes. The drawback of all these methods is the lack of direct and quantitative access to etch relevant properties, like the concentration of precursors or of molecular etch products in the bulk plasma.

With the compact and modular measurement device Q-MACS [12,13], it is possible to use the potential of the QCL for plasma process control. First applications of an especially designed Q-MACS for in situ process monitoring of the industrial production of DRAMs have been published recently [14]. It was shown, that important etch products, like the SiF_4 molecule, can be monitored in situ during etch processes for deep trench structures in silicon in real-time and under conditions of high

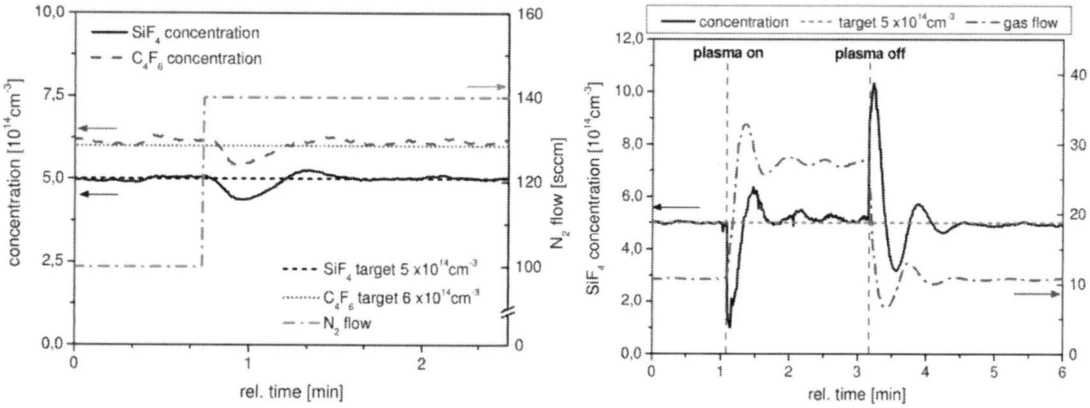

Figure 9. Left figure – Controlling of MFCs by QCLAS for constant SiF_4 and C_4F_6 concentrations at target values during changing the admixture of N_2 in the gas phase at constant pressure of 30Pa. Right figure – Controlling the SiF_4 concentration at the target value of $5\times10^{14}cm^{-3}$ when the SiF_4/N_2 plasma is switched on and off. The total pressure was kept constant at 20Pa [15].

volume production. With this, such concentrations can be used for process control purposes, like feeding them back to the process hardware in a control loop. This has successfully been shown for the control of mass flow controllers (MFC) [15]. The feed gas flow can be adjusted at the MFC on the basis of the measured concentration in the plasma reactor via QCLAS in order to keep the target concentration constant, whatever the process is.

Additionally, it could be demonstrated, that process control with the help of QCLAS is not only applicable for one species but also for two molecular species simultaneously. For this purposes the optical board of the Q-MACS was redesigned having integrated a second QCL as depicted schematically in figure 8. The IR radiation of both QCLs is combined by a beam splitter and transmitted via an optical fiber to the reactor. Using a mirror inside the plasma chamber the IR radiation is reflected back to the detector (Det 1) on the optical board. The radiation of both lasers is measured with the same detector, because they are multiplexed. In addition, in a reference path the spectral emission stability of the QCLs is monitored using a gas cell filled with C_2H_4 and controlled via line locking. For this proof of principle a planar microwave (MW) plasma reactor has been used as plasma source. Details of the plasma reactor can be found elsewhere [16].

The QCLs were selected for the detection of the SiF_4 and C_4F_6 molecule, which are important precursors in semiconductor industry. The QCLAS measurements of the precursor C_4F_6 have been based on FTIR studies of the absorption behaviour of the molecule [15]. For the measurement of C_4F_6 one QCL was selected to emit at $973cm^{-1}$, whereas the other QCL detects SiF_4 at $1028cm^{-1}$.

4.2. Results

To demonstrate the potential of QCLAS for process control, the in situ measured molecular concentrations were fed back via a proportional-integral-derivative (PID) control loop to the MFCs. The aim was to control the gas feed into the MW plasma reactor via concentration measurements in the chamber. Therewith, the gas concentration is the control variable, the gas flow is the actuating variable respectively. If the actual concentration deviates from a given target concentration, a special developed control loop changes the flow at the MFC, until the difference vanishes. The results for the measurements in the gas phase are shown in the left graph of figure 9. In a first step a gas composition of SiF_4, C_4F_6 and N_2 was monitored keeping the total pressure constant at 30Pa. With the help of the PID control loop the concentrations of SiF_4 and C_4F_6 were kept constant independently but simultaneously even in case of changing the admixture of N_2. Based on these gas phase experiments the control of species concentrations was also applied for a SiF_4/N_2 MW plasma. The right side of figure 9 shows (i) how effective the plasma leads to a fragmentation of the SiF_4 precursor and (ii) how the concentration of the SiF_4 is kept constant by controlling its flow. In this example the PID control loop enforced more than a doubling of the SiF_4 flow when the plasma is switched on in order to keep the SiF_4 concentration constant in the reactor. In the plasma off-phase the target concentration of the SiF_4 precursor of $5\times10^{14}cm^{-3}$ is again automatically achieved after approximate 1min. This system response is directly connected to the time needed for measurable changes in the concentrations in the reactor since the flow was changed at the MFC. Therefore it strongly depends on properties like reactor geometry, flowing conditions, pump speed and the MFC itself. Compared to typical process times for etching deep trench structures in silicon for DRAM production of more than 10min [14], the MFC control via QCLAS in the present configuration would be adaptable.

Acknowledgments

K.-D. Weltmann is gratefully acknowledged for his continuous encouragement and general support. The authors are especially thankful to D. Dahl, H. Zimmermann, F. Weichbrodt and U. Macherius for their permanent support. The stimulating climate in working with O. Guaitella and C. Lazzaroni is especially acknowledged. The authors give sincere thanks to S. Strämke and C. Harward for valuable long term cooperation.

References

[1] De Benedictis S, Dilecce G, Simek M 1997 *J. Phys. D: Appl. Phys.* **30** 2887
[2] Castillo M, Herrero V J, Mendez I, Tanarro I 2004 *Plasma Sources Sci. Technol.* **13** 343
[3] Rousseau A, Dantier A, Gatilova L V, Ionikh Y, Röpcke J, Tolmachev J A 2005 *Plasma Sources Sci. Technol.* **14** 70
[4] Zijlmans R A B 2008 *Ph.D. Thesis* Eindhoven University of Technology
[5] Welzel S, Gatilova L, Röpcke J and Rousseau A 2007 *Plasma Sources Sci. Technol.* **16** 822
[6] Richton R E and Farrow L A 1982 *J. Chem. Phys.* **76** (11) 5256
[7] Farrow L A 1985 *J. Chem. Phys.* **82** (8) 3625
[8] Hempel F, Artyushenko V, Weichbrodt F and Röpcke J 2009 *J. Phys.: Conf. Series* **175** 012003
[9] Zahniser M S, Nelson D D and Kolb C E 2002 in Applied Combustion Diagnostics, Kohse-Hoinghaus K and Jeffries J (eds), Taylor and Francis, New York 648
[10] Nelson D D, Shorter J H, McManus J B and Zahniser M S 2002 *Appl. Phys. B* **75** 343
[11] Rothman L S, Jacquemart D, Barbe A, Benner D C, Birk M, Brown L R, Carleer M R, Chackerian C, Chance K, Coudert L H, Dana V, Devi V M, Flaud J M, Gamache R R, Goldman A, Hartmann J M, Jucks K W, Maki A G, Mandin J Y, Massie S T, Orphal J, Perrin A, Rinsland C P, Smith M A H, Tennyson J, Tolchenov R N, Toth R A, Vander Auwera J, Varanasi P, Wagner G 2005 *J. Quant. Spectrosc. Radiat. Transfer* **96 (2)**: 139
[12] Röpcke J, Welzel S, Lang N, Hempel F, Gatilova L, Guaitella O, Rousseau A and Davies P B 2008 *Appl. Phys. B* **92** 335
[13] Stancu G D, Lang N, Röpcke J, Reinicke M, Steinbach A and Wege S 2007 *Chem. Vap. Deposition* **13** 351
[14] Lang N, Röpcke J, Steinbach A and Wege S 2009 *IEEE Trans. Plasma Sci.* in print.
[15] Lang N, Röpcke J, Wege S and Steinbach A 2009 *Eur. Phys. J. Appl. Phys.* in print.
[16] Ohl A 1998 *J. Phys. IV France* **8** 7

Scanning beam medium infra-red interferometry for plasma density measurements

P Innocente[1], A Canton[1], C Mazzotta[2] and O Tudisco[2]
[1]Consorzio RFX, Euratom-ENEA Association, Corso Stati Uniti, 4 35127 Padova, Italy
[2]Centro Ricerche Energia, Ass. EURATOM-ENEA-CNR, CP 65 Frascati, Roma, Italy
E-mail: paolo.innocente@igi.cnr.it

Abstract. In magnetically confined fusion experiments interferometry is commonly used to measure plasma electron density. This because interferometry is a very reliable technique, not affected by calibration problem. The main drawbacks of interferometers are the integral characteristic of the measurement and the fringe loosing problem. Scanning beam interferometry can always overcome the former problem and in some cases the second as well. The advantages of the scanning beam technique can better exploited in two colours vibration compensated medium infra-red (MIR) interferometers. Indeed the short wavelength of the probing beams provide small beam sizes allowing plasma density measurement along many non-overlapping paths, furthermore the vibration compensation system relaxes the requirement of massive interferometer structures embracing the fusion device, actually allowing to install back-reflecting mirror directly attached to the experimental machine.

As an example of application of this technique the two-color medium-infra-red-compensated scanning beam interferometer installed on the Frascati Tokamak Upgrade (FTU) experiment is presented. We present also a preliminary design of a scanning interferometer for the new proposed Fusion Advanced Studies Torus (FAST) experiment. The interferometer uses retroreflectors installed inside the vessel to back reflect the scanning beams; this will allow realizing scanning beams measuring from the plasma edge. This feature in principle can solve the fundamental problem of interferometers: the fringe loosing problem.

1. Introduction

Interferometry technique is widely used in magnetically confined nuclear fusion experiments [1, 2, 3, 4, 5] to measure plasma electron density. Interferometers implementing this technique differ in many expects: probing wavelength (ranging from visible light to microwave), detection method (Homodyne or Heterodyne), optical scheme (Michelson or Mach-Zehnder). Though previous differences all interferometers share most of the advantage and disadvantages. On the advantage side can be inscribed an easy data interpretation, no-calibration requirement and a high reliability while on disadvantage side have to enroll the integral nature of measurement and the fringe loosing problem [6]. Data interpretation is easy because, with a proper choice of the source wavelength, probing beam phase shift is proportional to the line integral of the electron density along beam path, furthermore since the ratio between the electron density line integral and phase shift depends only from fundamental constants and source wavelength, the measurement does not require any calibration. Finally interferometers have proven to be very reliably thanks to the availability of good laser sources and proper optical and mechanical setup.

The main disadvantage of interferometry is its integral nature: it does not provide a local measurements of density but only an average value. This feature in some cases can indeed be an advantage, for example when interferometer is used for feedback density control line averaging provides a direct evaluation of average particles content avoiding local fluctuation. The integral nature is a real disadvantage in plasma transport studies where a good evaluation of density profiles is required. In this case a high number of channels is required to Abel invert integral measurements, but the required number of channels is so high to be impractical in terms of costs and maintenance effort.

The second disadvantage is related to the differential and periodic nature of the phase shift measurements: to get density phase shift a reference baseline has to be obtained measuring interference phase before or after plasma discharge when density is zero, furthermore interference phase can be measured only within $[0, \pi]$ while phase shift can be much larger involving many fringes F (with fringes number $F = 2n\pi$) [6]. For this reason to correctly measure phase shift, F has to be computed starting from (or ending up to) the time of the reference baseline. To get F the interference phase must be fast sampled in order to avoid phase variations larger than π between two consecutive samples. If for any reason F is lost, a new reference baseline is needed. Not loosing fringes number is not a problem for short time low density measurements, but can be a hard task in presence of fast transient phenomena (like pellet injection, disruptions) or for long times. To solve this problem two non-conventional interferometers setup have been proposed: dispersion interferometer [7] and differential interferometer [8] but no one up to now is used in medium-large size fusion experiments.

Scanning beam interferometry applied to fusion experiments is a way to overcome previous interferometry disadvantages. In scanning beam interferometer one or more probing beam fast scan plasma cross section, providing in this way a high number of equivalent channels. In all setup scanning beam interferometry provides a fringe loosing immune (except for the starting value) measurement of the integral density profile, sharing in this way the advantage of differential interferometers but with a better signal to noise ratio. Furthermore in some cases vacuum accesses allow to build scanning interferometers with scans spanning from outside to inside plasma cross section, in these cases zero density baseline is performed on each scan allowing an easy recover of incidental fringe loosing. This provides fringes loosing immune interferometers.

In this paper we first present the scanning beam interferometer developed for the Frascati Tokamak Upgrade (FTU) experiment, a medium size tokamak device [9, 10] ($R = 0.935\,m$, $a = 0.3\,m$) targeted to high magnetic field operation studies ($B_T = 4 \div 8\,T$, $I_p < 1.6\,MA$). The FTU scanning solution is similar to that of a previously developed FIR (Far-Infra-Red) interferometer [12] but we implemented it on a two colors vibration compensated MIR (Medium-Infra-Red) interferometer to extend the scanning beam advantages by improving spatial and temporal resolution. Good spatial resolution is obtained by the small size of the probing beams ($\approx 1cm$) such that two parallel scanning beams probing most of the plasma cross section provides more than 30 parallel lines of sight. Good time resolution is obtained using high frequency mechanical scanning devices available for small size beams. Moreover, the short wavelengths can be easily frequency shifted at high speed by Bragg cells for the heterodyne detection, providing a high measurement bandwidth. This allows to correctly measure the highest densities presently reached in FTU (about $7 \div 8 \cdot 10^{20}\,m^{-3}$) and the fastest density changes observed during pellet injection with time slopes as fast as $10^{25}\,m^{-2}s^{-1}$.

Thanks to the high number of equivalent channels, FTU interferometer is able to solve the most important disadvantage of the interferometers, but the vacuum vessel ports do not allow realizing scanning fans going externally to the plasma cross section. This is possible for the interferometer of the new proposed Fusion Advanced Studies Torus (FAST) experiment. In this new device it will be possible to realize scans crossing together and going from inside to outside plasma cross section. That will improve density profile reconstruction and will allow to realize an interferometer immune to fringe jumps, which is particularly important on a long pulse device as it will be the FAST experiment.

In Sec. 2 we present the main features of the FTU scanning interferometer. In Sec. 3 we present optical setup of the proposed FAST scanning interferometer. Finally in Sec. 4 we summarize main features of scanning interferometry.

2. FTU interferometer

2.1. Diagnostic setup

A detail description of the FTU interferometer can be found elsewhere [13], in this section we summarize its main features.

The optical scheme of the FTU interferometer is a double pass Mach–Zehnder type. To scan plasma cross section it uses two probing beams and two tilting mirrors (Scanner) placed at the focus of an Off-axis Parabolic Mirrors (OPM), see Fig. 1. The OPM convert the Scanner angular beam deflection to a parallel scan, sampling plasma section by parallel channels. On each fan a Roof reflector (RR) send the beam back displaced of 2 *cm* in the toroidal direction. RR are mounted directly on the FTU mechanical structure.

Considering the beam diameter inside the plasma cross section, the beam movement during a sampling interval and requiring non-overlapping measurements, the FTU scanning interferometer corresponds to a traditional system with more than 30 channels. Furthermore allowing partial beam overlapping and considering the MHz bandwidth of the interferometer more than a two times larger number of lines of sight can be sampled on each scan.

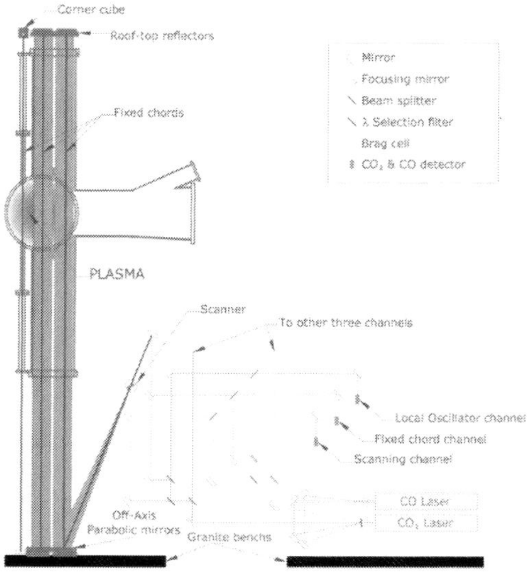

Figure 1. Simplified optical scheme of one scanning chord, one fixed chord and local oscillator channel.

The instrument uses a CO_2 laser ($\lambda = 10.6\ \mu m$) as the main light source, and a CO laser ($\lambda = 5.4\ \mu m$) as the compensation one. Heterodyne detection at 40 and 30 MHz is obtained by frequency shifting the reference beams with two acousto-optic modulators (Bragg cells). The CO_2 and CO lasers beams are sent to two Bragg Cells that split the beams and frequency shift the deflected beams by 40 MHz and 30 MHz. To reduce optical components, the same detector is used for both wavelengths: the CO_2 signal is separated from the CO signal electronically thanks to the different modulation frequency.

A crucial part of the scanning interferometer is the beam deflector, to such purpose FTU interferometer uses two 8 kHz Counter Rotating Scanners (CRS) produced by GSI General Scanning whose 7.8x5.5 mm mirror oscillates at 8 kHz with a maximum amplitude of 15°. We preferred a Mechanical Deflector because it simplifies optical scheme though it isn't the fastest solution [14].

2.2. Results

Scanning interferometer measurements provide good evaluation of density profile evolution in many experimental configurations but the specific features of the MIR interferometry (high time resolution, high diffraction limit) are highlighted in discharges with pellet injection. Indeed pellet injection produces fast, large and localized density variations [15], these plasma conditions are often hard to account for the long wavelength low frequency modulation FIR interferometers.

Opposite MIR interferometer perform very well as can see in Fig. 2 which shows the time evolution of the line electron density along some chords of the scanning interferometer. The scanning interferometer can follow the fast density rise for all the five injected pellets.

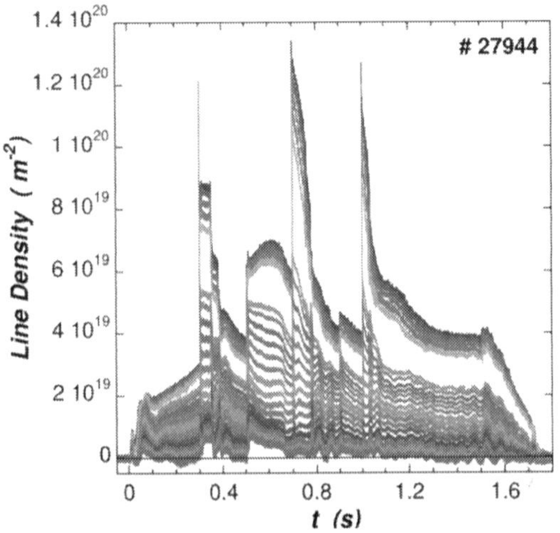

Figure 2. Line density measures on a 500 kA discharge where 3 big pellets were injected from equatorial port (t=0.3, 0.7 and 1.0 s) and two small pellet were injected from vertical port (t=0.5 and 0.9 s).

2.3. Roof mirror real time tilt feed-back

The main problem of the FTU scanning interferometer was related to the RR tilts: being directly attached to the FTU vessel it suffers large displacements and tilts during discharge operation, mainly due to the strong force applied from the high toroidal field. Thanks to the two-color compensation system the RR displacement do not affect density measurements but tilts along scanning direction does it. Indeed, at the first order of approximation, the RR two-mirrors 90° set-up compensates RR rotations along its longitudinal axis in such a way that they does not produce any deflection or path length variation on the reflected beams. Opposite tilts of the RR along scanning direction produce two effects: reduce interference amplitude changing the direction of the reflected beams, introduce a measurement error when the CO_2 and CO beams are not exactly superimposed.

The measurement error is proportional to the RR tilt and is due to the different path length variations of CO_2 and CO beams, the largest residual errors of the FTU interferometer were due to RR tilts.

To reduce error an off-line procedure has been implemented based on the measurement of ratio between residual error and RR tilt after each discharge. This is possible because the compensated interferometer can measured both the line integral density and the path length variation, then tilt amplitude can be estimated evaluating path length variation on a scan. The measured error-tilt ratio can be applied to correct the tilt error. Such computation was routinely performed and it allowed to reduce the residual error to the lowest values.

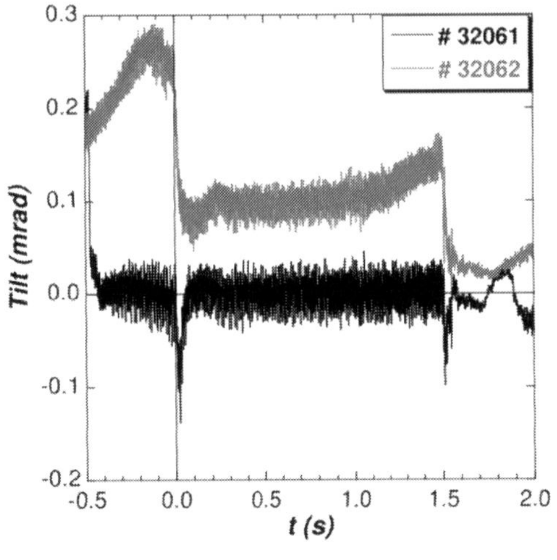

Figure 3. Comparison of RR tilt angle on two plasma discharges with (# 32061) and without (# 32062) tilt feedback control.

To improve the situation and to solve also the reduction of interference amplitude problem we developed a feed-back system to cancel in real time RR tilt angle. We installed on the RR a piezo translator by which we can remotely control vertical angle. To compute the tilt angle with a 10^{-3} *mrad* resolution we sample the same interferometric signals that are also stored to compute after pulse line densities. Sampling is synchronized to beam scans by the zero crossing of the CRSs angular position signals. In this way RR tilts are computed every $4\,ms$, an usual proportional–integral–derivative controller (PID controller) has been implemented as the feedback mechanism. In Fig. 3 is shown a comparison between tilts measured on a scanning channel when feedback control is on or off, the comparison is performed on two identical plasma discharges. Feedback action starts at -0.5 s and in a very short time is able to cancel slow time variation RR tilt, some residual tilt is only present at beginning and at the end of the discharge when fast and large tilt variations are present.

3. FAST Interferometer

The FAST [16] experiment has been proposed as an European ITER [17] satellite facility with the aim of preparing ITER operation scenarios and helping early DEMO [18] design and R&D. FAST should also serve as test bed for the development of compatible ITER and DEMO diagnostics.

Following the good results of FTU interferometer a MIR scanning interferometer sharing most of its characteristics has been also proposed for FAST. Based on FAST reference design the proposed scanning interferometer tries to take advantage of vacuum vessel assembly features and to overcome some limitation of ports setup. In particular two good opportunities are the large free volume available behind the limiter on vacuum vessel bottom side and the large horizontal port. On the other side a strong limitation is the low portion of plasma that could be sampled by two ports passing line of sights. To improve the reliability of the system the interferometer has been divided in two independent modules: a vertical interferometer and an horizontal one. The vertical interferometer uses the top vertical port to measure the density on two fans and one fixed line of sight. The horizontal interferometer enter through the horizontal port to sample the plasma on three horizontal fans.

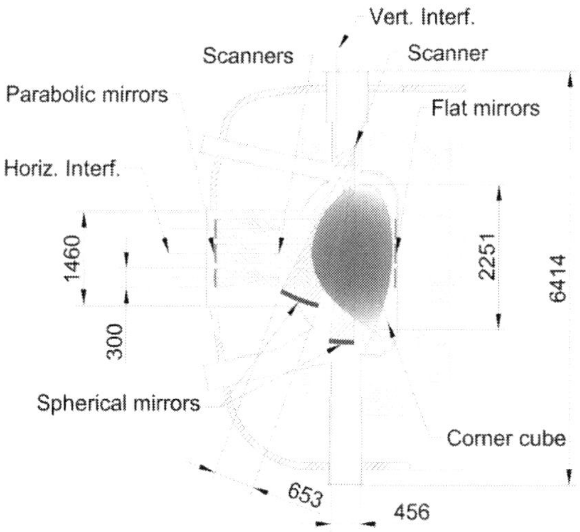

Figure 4. FAST interferometer: conceptual optical scheme of measuring channels

To send back the beams to the scanners two different solutions for retro-reflectors have been proposed: spherical mirrors for the vertical fans and flat mirrors for horizontal fans. Spherical mirrors reduces the overall dimensions required to realize a scan, flat mirrors have been selected for the horizontal interferometer due to the small size available between limiter surface and vacuum vessel, to reduce mirrors damage they are installed $10\,mm$ behind limiter shadow and only a small slice of the mirrors, having a toroidal length of $20\,mm$, is exposed to the plasma. The drawback of the spherical and flat mirrors retro-reflector compared to roof mirrors is the possible interferometer misalignment caused by toroidal mirrors tilt. To overcome this problem we will apply the mirror position feedback technique both to the scan direction and to the perpendicular direction. The scan direction will be controlled at sub-micron precision by the same interferometric method used on FTU, while the perpendicular direction will be controlled by quadrant detectors measuring returned beam positions. To ensure a line density measurement also in case of mirrors feedback failure, a vertical channels using a corner cube as retro reflector has been foreseen.

One of the nice feature of the designed FAST interferometer is that the edge vertical fan scans the plasma cross section starting from outside LCFS where plasma density reasonably will be zero. This, beside providing a very good edge profile measurements, will allow to recover also fringe count in case of interferometer failure.

4. Summary

MIR scanning interferometry can be effectively applied to fusion device experiments to provide good reliable measurements of electron density profile time evolution. Providing a high number of equivalent measurements channels with minimum experimental effort, the scanning technique overcome the limitations of traditional interferometry. This is confirmed from the operation of the FTU scanning interferometer.

Scanning technique extends the good features of interferometry but to get the best it has to be optimized to the experimental device peculiarities. An example of this optimization are the different optical schemes adopted for the FTU interferometer and for the proposed FAST interferometer. To improve measurement quality and reliability, feedback technique have also to be implemented to control interferometer alignment before and during plasma discharge. On this direction we have shown the fast feedback method used to control roof reflector tilt during discharges.

Acknowledgments

This work was supported by the European Communities under the contract of Association between EURATOM/ENEA. The views and opinions expressed herein do not necessarily reflect those of the European Commission.

References

[1] Braithwaite G, Gottardi N, Magyar G, O'Rourke J, Ryan J and Veron D 1989 *Rev. Sci. Instrum.* **60** 2825
[2] Mansfield DK, Park HK, Johnson LC, Anderson HM, Chouinard R, Foote VS, Ma CH and Clifton BJ 1987 *Apl. Opt.* **26** 4469
[3] Carlstrom TN, Ahlgren DR and Crosbie J 1988 *Rev. Sci. Instrum.* **59** 1063
[4] Innocente P and Martini S 1992 *Rev. Sci. Instrum.* **63** 4996
[5] Deng BH, Brower DL,Ding WX, Wyman MD, Chapman BE and Sarff J S 2006 *Rev. Sci. Instrum.* **77** 10F108
[6] Brower DL, Ding W X, Mirnov VV, Van Zeeland MA and Carlstrom TN 2008 *AIP Conf. Proc.* **988** 92
[7] Bagryansky PA, Khilchenko AD, Kvashnin AN, Lizunov AA, Voskoboynikov RV, Solomakhin AL, Koslowski HR and TEXTOR team 2006 *Rev. Sci. Instrum.* **77** 053501
[8] Ding WX, Brower DL, Deng BH, and Yates T 2006 *Rev. Sci. Instrum.* **77** 10F105
[9] Pizzuto A, Annino C, Baldarelli M, Bettinali L, Brolatti G, Crescenzi C, Maddaluno G, Riccardi B, Righetti B G, Roccella M and Semeraro L 2004 *Fusion Sci. and Tech.* **45** 422
[10] Angelini BM, Apicella ML, Buceti G, Centioli C, Crisanti F, Iannone F, Mazza G, Mazzitelli G, Panella M, Vitale V and The FTU Team 2004 *Fusion Sci. and Tech.* **45** 427
[11] Krug PA, Stimson PA and Falconer IS 1987 *J. Phys. E: Sci. Instrum.* **20** 1249
[12] Warr GB, Blackwell BD, Wach J and Howard J 1997 *Fusion Eng. Des.* **34-35** 387
[13] Canton A, Innocente P and Tudisco O 2006 *Appl. Optics* **45** 9105
[14] Canton A, Innocente P, Martini S, TasinatoL and Tudisco O 2001 *Rev. Sci. Instrum.***72** 1085
[15] Mazzotta C, Tudisco O, Canton A, Innocente P, DeBenedetti M, Giovannozzi E, Marocco D, Micozzi P, Monari G and Rocchi G 2006 *Phys. Scr.* **T123** 79
[16] Calabrò G *et al.* 2009 *Nucl. Fusion* **49** 055002
[17] Holtkamp N 2008 Proc. 22nd Int. Conf. on Fusion Energy 2008 (Geneva, Switzerland 2008) (IAEA: Vienna) CD-ROM file OV/2-1 and http://www-naweb.iaea.org/napc/physics/FEC/FEC2008/papers/ov_2-1.pdf
[18] Nishitani T. et al 2008 Proc. 22nd Int. Conf. on Fusion Energy 2008 (Geneva, Switzerland 2008) (IAEA: Vienna) CD-ROM file FT/1-3 and http://www-naweb.iaea.org/napc/physics/FEC/FEC2008/papers/ft_1-3.pdf

14th International Symposium on Laser-Aided Plasma Diagnostics (LAPD14) IOP Publishing
Journal of Physics: Conference Series **227** (2010) 012007 doi:10.1088/1742-6596/227/1/012007

The Motional Stark Effect with Laser-Induced Fluorescence Diagnostic

E. L. Foley and F. M. Levinton

Nova Photonics, Inc. 200 Forrestal Road Princeton NJ 08540

E-mail: `foley@novaphotonics.com`

Abstract. The motional Stark effect (MSE) diagnostic is the worldwide standard technique for internal magnetic field pitch angle measurements in magnetized plasmas. Traditionally, it is based on using polarimetry to measure the polarization direction of light emitted from a hydrogenic species in a neutral beam. As the beam passes through the magnetized plasma at a high velocity, in its rest frame it perceives a Lorentz electric field. This field causes the H-alpha emission to be split and polarized. A new technique under development adds laser-induced fluorescence (LIF) to a diagnostic neutral beam (DNB) for an MSE measurement that will enable radially resolved magnetic field magnitude as well as pitch angle measurements in even low-field (<1 T) experiments. An MSE-LIF system will be installed on the National Spherical Torus Experiment (NSTX) at the Princeton Plasma Physics Laboratory. It will enable reconstructions of the plasma pressure, q-profile and current as well as, in conjunction with the existing MSE system, measurements of radial electric fields.

1. Introduction

The motional Stark effect (MSE) diagnostic was originally developed in 1989 [1]. It has since become the worldwide standard technique for measurement of internal magnetic field pitch angle in tokamaks, and has generated new understanding of plasma stability and transport [2, 3]. The MSE concept relies upon observation of the Balmer-alpha (n=3 to n=2) emission from a neutral hydrogen beam traversing a plasma. Typically, a heating beam is used. As the beam at high velocity, $\vec{v}$, passes through the magnetic field in the plasma, $\vec{B}$, it experiences in its reference frame a Lorentz electric field, $\vec{E} = \vec{v} \times \vec{B}$. This electric field causes the spectral emission to be split and polarized as described by the Stark effect. Traditionally, MSE has used polarimetry to determine the polarization angle of the light, and related that to the magnetic field pitch angle in the plasma. The MSE diagnostic was developed under high magnetic field (>1 T) conditions, and the lower fields of recent high-beta experiments such as the National Spherical Torus Experiment (NSTX) at the Princeton Plasma Physics Laboratory (PPPL) posed challenges to the technique. Nova Photonics has successfully developed narrowband, tunable optical filters to overcome these challenges and installed an MSE system on NSTX that presently provides spatially resolved pitch angle profiles, and has enabled new physics studies [4, 5, 6].

The motional Stark effect with laser induced fluorescence technique under development, MSE-LIF, incorporates a diagnostic neutral beam (DNB) and employs laser-induced fluorescence to achieve measurement of the MSE spectrum with unprecedented precision. The system will have several significant advantages over existing MSE systems. The high level of detail achieved with this diagnostic will allow accurate measurement of line splitting in the MSE spectrum. This measurement will be called 'MSE-LS' for MSE with Line Shift, to be distinguished from the traditional MSE with Line Polarization, 'MSE-LP.' Previous measurements of MSE line splitting have been limited in precision, and not used for equilibrium reconstruction [7, 8, 9]. The MSE-LIF technique discussed in this paper is expected to deliver a unique and extremely valuable measurement that has never been achieved: a radially resolved profile of the magnitude of the Lorentz electric field experienced by neutral beam atoms in the plasma. When accomplished with sufficient precision, this measurement can be used in conjunction with external magnetics to directly reconstruct the plasma pressure profile. Such a result would allow new understanding of magnetic field structure and plasma pressure, as well as enabling study of fast ion populations when used with background thermal pressure measurements. Additionally, the MSE-LIF system will be used to complement the existing MSE system on NSTX to give spatially resolved radial electric field profiles.

The MSE-LIF system can be implemented on experiments that lack heating neutral beams. Magnetic fields as low as a few Gauss can be resolved with the system. Even for experiments which do have a heating beam, MSE measurements cannot typically be made during plasma startup due to wall damage by neutral beam shine-thru and fast ion losses. Operation of the MSE-LIF system with its diagnostic neutral beam will be possible during the plasma startup and current ramp-up to diagnose and study current drive and other startup scenarios such as helicity injection.

Because the wavelength and polarization are set by the injected laser, the MSE-LIF system is insensitive to many issues that complicate calibration of traditional MSE systems. No actual measurement of the polarization or wavelength of collected light is necessary. The traditional MSE-LP measurement can be made with the MSE-LIF system with the incorporation of a polarization rotator on the excitation laser. Polarization effects in coatings on collection optics are not a concern, and the recently reported effect of re-neutralization [10] on the MSE signal is minimal on an MSE-LIF measurement of line splitting, though it can strongly affect polarization measurements.

The MSE-LIF concept has been under development for several years, and has made steady progress towards a final system. The design of the neutral beam injector, done in collaboration with the Plasma and Ion Source Technology Group at Lawrence Berkeley National Laboratory was reported in Review of Scientific Instruments [11]. Extensive LIF measurements in low magnetic fields in neutral gas have been made, and an LIF enhancement phenomenon was observed and reported on [12]. A detailed computer model of the system, unique in its comprehensive calculation of variation of transition probabilities in magnetic fields, including the effects of fine structure, the motional Stark effect and the Zeeman effect as well as collisions with background gas has been developed [13]. Description of the status of development was made in a talk at the 2006 High Temperature Plasma Diagnostics conference, and described in an associated paper [14].

In the following, it will be assumed that the reader has access to these previous references, and more recent work will be emphasized. A significant breakthrough in the MSE-LIF project has been the design of a new laser system optimized for this application. This, combined with detailed computer modeling, has moved the diagnostic development forward to the point of planning for installation on a major experiment, the National Spherical Torus eXperiment (NSTX) at the Princeton Plasma Physics Laboratory.

2. Laser Development

Recent measurements of the Doppler-broadened linewidth of the emission from our diagnostic neutral beam show an energy spread on order 6 GHz. This is equivalent to 100 V of variation out of the 35 kV beam energy. Most of this is thought to be due to straggling in the neutralization process. In order to optimize LIF signal, it is crucial to have a good match of the linewidth of the excitation laser and the beam. In previous work, a narrow (<1 MHz) linewidth dye laser was used for excitation. This laser was only capable of resonating with atoms in a velocity range within the natural linewidth (~100 MHz) of the desired transition. In the experimental configuration of that time, with ~50 V (3 GHz) peak width, this meant only about 3% of the beam atoms could be contributing to the LIF signal at a given time.

In order to boost signal levels, we have undertaken the development of a laser which will allow an excellent match to the beam. The new laser linewidth will be ~6 GHz, and the total laser power will increase significantly, from 300 mW to several Watts.

The new laser system is made possible by a series of recent developments in laser and optics technology. High power laser diodes have become available in the red wavelength range, though commercial applications for lasers at 651 nm are not very common. In order to obtain the power and wavelength we needed for this project, we searched for a development partner who had the proven ability to produce high-quality diodes near this wavelength range, and was interested in working on our particular application. We chose to work with Modulight Inc and are very pleased to report the successful development of a 10 W, 19 emitter diode bar for our project. A photo of a bar and data from Modulight are shown in Figure 1.

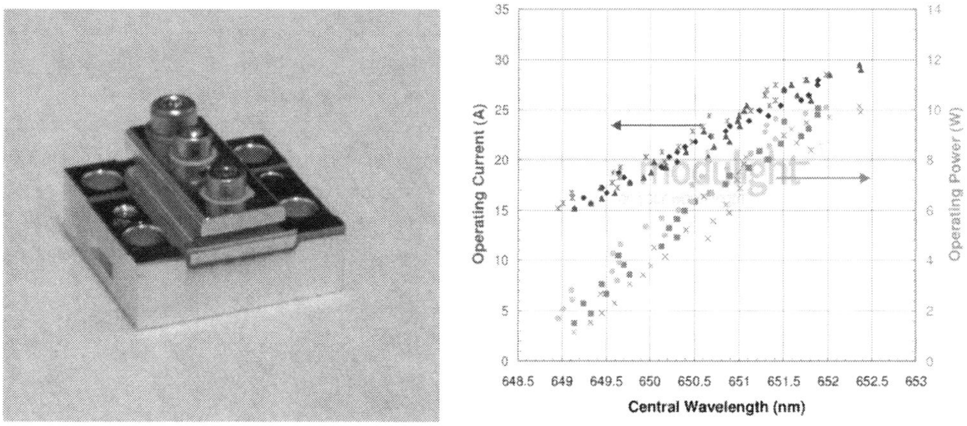

Figure 1. Left: Photo of CS mount 19 emitter diode bar at 651 nm. Right: Plot showing wavelength and power for several Modulight 19 emitter diode bars (courtesy of Modulight, Inc)

The intrinsic linewidth of the diode bars is 2–3 nm, far too wide for our application. In order to narrow the linewidth to match our neutral beam energy spread, we plan to use a volume holographic grating. The VHG acts as a selective mirror, feeding back into the laser only a narrow range of wavelengths, and enabling the laser system to lock and lase mainly in this very narrow wavelength range [15]. High quality VHG's are commercially available from Ondax, Inc.

The 5 to 7 GHz linewidth needed for our application is much more narrow than a typical feedback system. Figure 2 shows a diagram of the optical system and a plot of the narrow acceptance angle for the VHG. Some previous work [16, 17] has demonstrated wavelength narrowing to the $\leq$10 GHz range using a VHG, though the 765–780 nm wavelength range gives a wider VHG acceptance angle, so we expect our work at 651 to be more challenging.

Our laser emitters have a 3.4 degree (FWHM) divergence in the slow axis, and a 35.7 degree (FWHM) divergence in the fast axis. The emitter size in the slow axis dimension is 150 microns, a significant fraction of the 500 micron emitter spacing. It is actually considerably easier to achieve an excellent collimation for the fast axis, because its emitter size is on the few micron scale. An estimate of the total power that can be collimated into the VHG acceptance cone for our laser system gives about 20% in an ideal case. Real systems are never ideal, however,

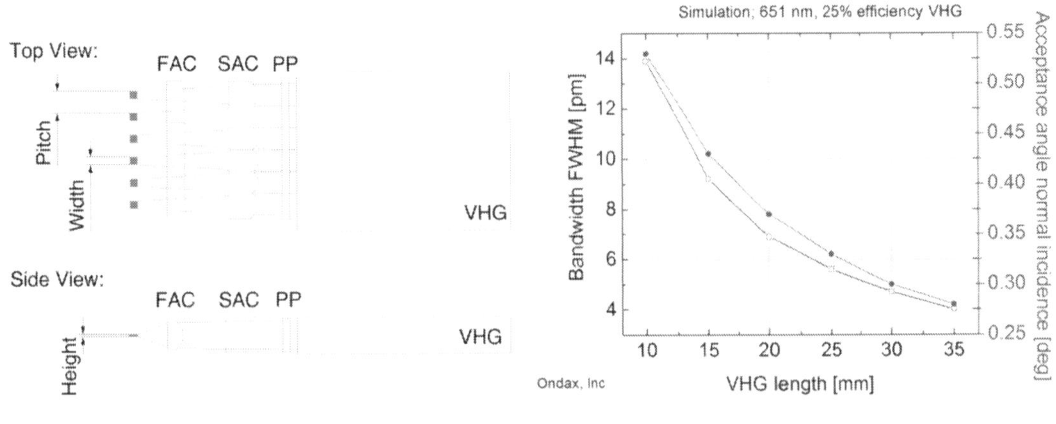

Figure 2. Left: Diagram showing layout of laser diode emitters, fast-axis collimation lens (FAC), slow-axis collimation lens array (SAC), corrective phase plate (PP) and volume holographic grating (VHG). Right: Acceptance angle vs length for volume holographic grating (courtesy of Ondax, Inc)

and slight variations in the diode emission angle and placement, as well as smile effects seen in operation, could significantly reduce that fraction, compromising the system's ability to lock in to the desired wavelength at high power. Because the collimation is so important for this application, we plan to work with compensation plate technology recently developed at Heroit-Watt University [18] and commercialized by PowerPhotonic. Their custom waveplates are individually tailored for each laser diode, and can correct for smile and alignment errors. An example is shown in Figure 3. At present, we are waiting to receive AR coated diode bars for incorporation into our VHG system.

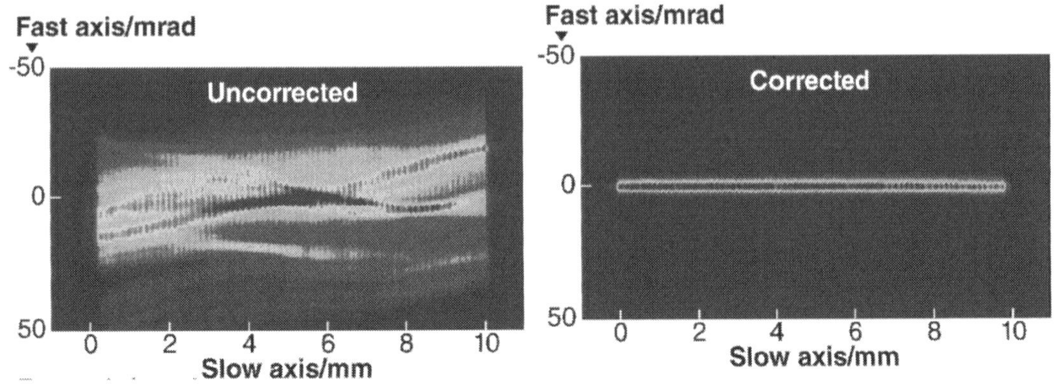

Figure 3. Phase plate correction of diode bar (courtesy of Power Photonic, Inc.)

3. NSTX Installation

Preparations for installation of the MSE-LIF diagnostic on NSTX are in progress. The DNB system will be modified significantly to fit in the available space on the NSTX platform. The hardware for installation on NSTX is shown in Figure 4. Collection optics have been designed, procured and tested, and modifications to the NSTX port cover on which they will be installed are planned for the near future.

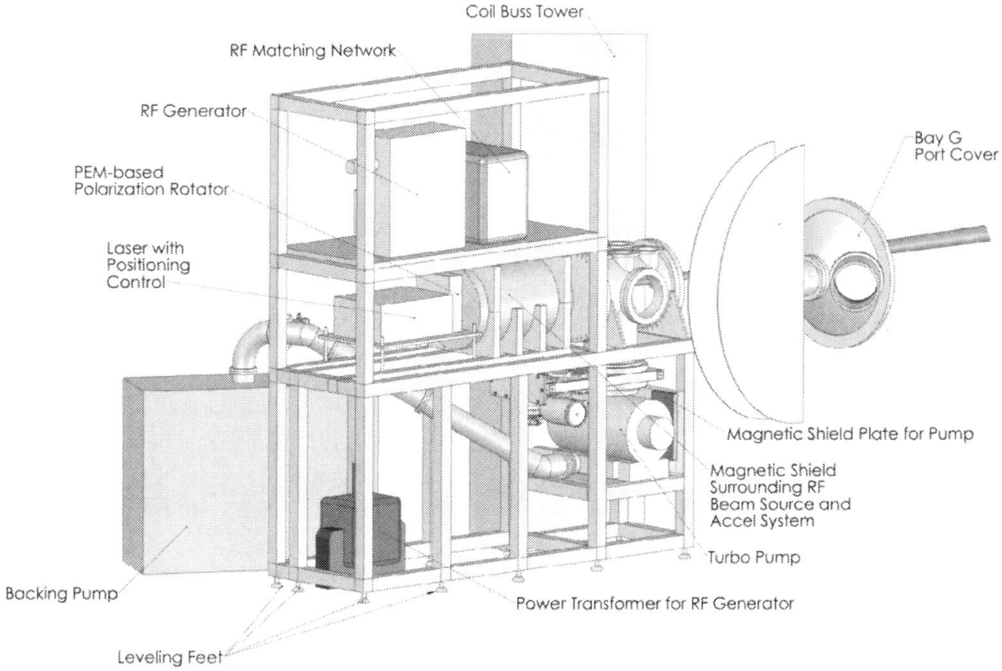

Figure 4. Diagram of MSE-LIF installation on NSTX.

The DNB source will be installed on the NSTX platform where stray fields of hundreds of Gauss develop during plasma operation. A shield to protect both the source and the acceleration and neutralization regions is imperative. A two-layer shield system has been designed, modeled and optimized using 3D simulation software. A diagram of the shield and the simulation results are shown in Figure 5. The design goal of reducing the field by a factor of 1000 in the source, acceleration and neutralization regions has been met.

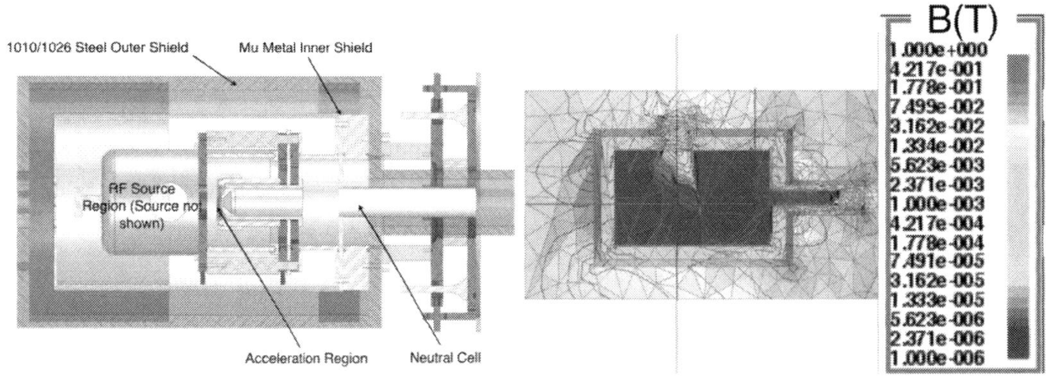

Figure 5. Magnetic shielding design and simulation results showing effectiveness for neutral beam source and neutral cell regions. The cross-section shown in the plot at right is rotated 90 degrees azimuthally from the one shown in the drawing on the left. The simulation plot on the right shows the access for the RF antenna and RF shielding at the top.

The shield involved particular design challenges beyond the reduction of magnetic field, in that the neutral beam source operates at 40 kV, and the region that needed to be shielded is under vacuum. The design presented overcomes these issues by keeping the entire shield at ground potential, and having the shield cross the vacuum boundary. We plan to have the shield nickel-plated to prevent corrosion.

4. Sensitive test of MSE-LS for ITER

The ITER plasma environment poses new challenges for optical diagnostics including the motional Stark effect. A labyrinth of mirrors must be used to transport light from the plasma out for analysis, as a direct line of sight would also transmit dangerous levels of radiation. The plasma will erode and deposit on the first mirror, changing the reflection properties. The traditional MSE measurement of polarization will be particularly susceptible to degradation in this environment. An alternative, of using the MSE-LS, or line shift approach to measure the magnetic field magnitude rather than pitch angle using collisionally-induced fluorescence, has been proposed [19]. Computational studies have demonstrated the ability of the MSE-LS approach to contribute to equilibrium reconstruction as effectively as MSE-LP [20], and the MSE-LIF system on NSTX will provide an excellent test of the concept.

5. Conclusions and Future Work

In the near term, we will be completing development of our 10 W, 651 nm, 6 GHz laser system, and outfitting the diagnostic neutral beam system with a magnetic shield for installation on NSTX. The installation of the MSE-LIF system on NSTX is scheduled for the 2010 opening, and we expect to test the system during the 2011 run year.

Acknowledgments

The authors wish to gratefully acknowledge funding from the U. S. Department of Energy Office of Fusion Energy Sciences Diagnostic Development Grant No. DE-FG02-01ER54616. We thank D. DiCicco and D. Cylinder for excellent technical contributions, and our PPPL collaborators, particularly D. Johnson, B. Stratton and S. Raftopoulos for their ongoing support of this project.

References

[1] Levinton F M *et al.* 1989 *Phys. Rev. Lett.* **63** 2060
[2] Levinton F M, Zakharov L, Batha S, Manickam J and Zarnstorff M C 1994 *Phys. Rev. Lett.* **72** 2895
[3] Levinton F M, Bell R E, Batha S, Synakowski E J and Zarnstorff M C 1998 *Phys. Rev. Lett.* **80** 4887
[4] Levinton F M *et al.* 2007 *Phys. Plasmas* **14** 056119
[5] Levinton F M and Yuh H 2008 *Rev. Sci. Instrum.* **79** 10F522
[6] Yuh H *et al.* 2009 *Phys. Plasmas* **16** 056120
[7] Wolf R C, Eriksson L G, von Hellerman M, König R, Mandl W and Porcelli F 1993 *Nucl. Fusion* **33** 1835–1847
[8] Den Hartog D, Craig D, Ennis D, Fiksel G, Gangadhara S, Holly D, Reardon J, Davydenko V, Ivanov A, Lizunov A *et al.* 2006 *Review of Scientific Instruments* **77** 10F122
[9] Pablant N A, Burrell K H, Groebner R J, Kaplan D H and Holcomb C T 2008 *Rev. Sci. Instrum.* **79** 10F517
[10] Yuh H Y, Levinton F M, Scott S D and Ko J 2008 *Rev. Sci. Instrum.* **79** 10F523
[11] Hahto S K, Hahto S T, Ji Q, Leung K N, Wilde S, Foley E L, Grisham L R and Levinton F M 2004 *Rev. Sci. Instrum.* **75** 355
[12] Foley E L and Levinton F M 2005 *J. Appl. Phys.* **98** 093101
[13] Foley E L and Levinton F M 2006 *J. Phys. B.* **39** 443
[14] Foley E L and Levinton F M 2006 *Review of Scientific Instruments* **77** 10F311
[15] Steckman G J, Liu W, Platz R, Schroeder D, Moser C and Havermeyer F 2007 *IEEE Journal of Selected Topics in Quantum Electronics* **13** 672–678
[16] Meng L, Nizamov B, Madasamy P, Brasseur J, Henshaw T and Neumann D 2006 *Optics Express* **14** 10469–10474
[17] Gourevitch A, Venus G, Smirnov V, Hostutler D A and Glebov L 2008 *Optics Letters* **33** 702–704
[18] Monjardin J F, Nowak K M, Baker H J and Hall D R 2006 *Optics Express* **14** 8178–8183
[19] Foley E L, Levinton F M, Yuh H Y, and Zakharov L E 2008 *Rev. Sci. Instrum.* **79** 10F521
[20] Foley E L, Levinton F M, Yuh H Y, and Zakharov L E 2008 *Nucl. Fusion* **48** 085004

Journal of Physics: Conference Series **227** (2010) 012008 doi:10.1088/1742-6596/227/1/012008

Measurement of neutral flow velocity in an ECR plasma using tunable diode laser LIF spectroscopy combined with saturated absorption spectroscopy

M. Aramaki[1], K. Ogiwara[2], S. Etoh[3], S. Yoshimura[4], and M. Y. Tanaka[2]

[1]Nagoya University, Furo-cho, Chikusa, Nagoya 464-8603, Japan
[2]Kyushu University, Kasuga-koen, Kasuga, Fukuoka 816-8580, Japan
[3]Central Research Institute of Electric Power Industy, Nagasaka, Yokosuka, Kanagawa 240-0196, Japan
[4]National Institute for Fusion Science, 322-6 Oroshi-cho, Toki, Gifu 509-5292, Japan

E-mail: aramaki@nuee.nagoya-u.ac.jp

Abstract. The radial flow of neutral particles in an electron-cyclotron-resonance (ECR) argon plasma has been measured by using a newly developed high resolution laser induced fluorescence (LIF) measurement system. The flow velocity is determined by the Doppler shift of the LIF spectrum. A very high accuracy calibration of an excitation laser has been achieved by installing a saturated absorption spectroscopy unit into the LIF system. We utilized the Lamb dip, which is obtained by the saturated absorption spectroscopy, as a frequency standard for the determination of the flow velocity. The use of Lamb dip is particularly appropriate for the flow velocity measurement, since the position of Lamb dip in the frequency scale is Doppler-shift free and is not disturbed by the motion of reference medium. By utilizing the Lamb dip as the frequency standard, the reliability and stability of the laser frequency calibration is increased. From the radial measurements of LIF spectra of metastable argon atoms at the microwave power of 250 W and 5 kW, it is found that there exists an inward flow of neutral particles in the both plasmas. Both of the radial flow velocity profiles peak around 4 cm from the center, which is comparable with the radius of the boundary of the $E\times B$ rotation and anti-$E\times B$ rotation in the 5 kW discharge. The maximum flow velocity increases with the microwave power.

1. Introduction

The kinetic interaction between plasma and neutral particles has been attracting in the recent years an increasing attention. In particular to understand the static structure of neutral particle distribution in highly-ionized low-pressure plasma, the electron static pressure has to be taken into account [1]. A class of vortices, in which ions rotate in the opposite direction to the $E\times B$ drift, has been observed in our laboratory [2]. This result suggests the existence of a strong force which dominates the electric field. The force acting on ions due to charge exchange collision between the ions and neutrals is considered to play an essential role in generating anti-$E\times B$ rotation. The momentum transfer between ions and neutral particles may change the dynamic behavior of ions. In addition to the static pressure, the dynamic pressure which is caused by the neutral particle flow should be taken into account to clarify the formation mechanism of the anti- $E\times B$ vortices.

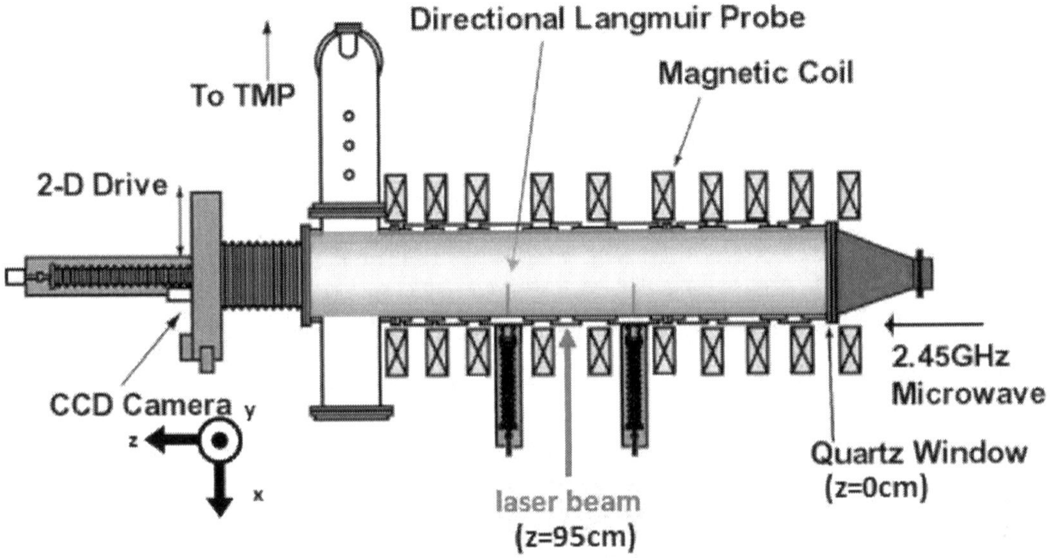

Figure 1. Schematic diagram of HYPER-I device.

To study the role of ion-neutral interaction, it is needed to measure the flow velocity field of the neutral particles. LIF Doppler spectroscopy is the most promising method for the direct measurement of the local neutral flow velocity; however, the lowest neutral-flow velocity that is detectable by the LIF system is limited by the accuracy of frequency calibration and the stability of the excitation laser. Up to date, the flow velocity measurement of the neutral particles has been carried out by Engeln *et al.* for a fast flowing case [3]. However, the LIF Doppler velocimetry for slow neutral particles has not been done so far. The expected neutral flow velocity in our experiment is of the order of 10 m/s or less, and the corresponding Doppler shift is of the order of 10 MHz in visible light range. The newly-developed LIF system utilizes saturated absorption spectroscopy, in which the laser frequency is calibrated by the frequency of a Lamb dip. A set of counter propagating laser beams, which are tuned at same frequency, is used in saturated absorption spectroscopy. The one beam is called a pump beam, and the other beam is called a probe beam. The pump beam is used to pump out the lower state atoms of the absorption transition, and the probe beam is used for absorption spectroscopy. A Doppler absorption spectrum is observed by sweeping the laser frequency. When the pump beam excites the atoms which are travelling at velocity v_0 along the beam line, the probe beam is absorbed by the atoms which are travelling at $-v_0$. At the resonance frequency, the both laser beams are absorbed by the same atoms with $v_0 = 0$. Since the lower state atoms are pumped out by the pump beam, the amount of absorption of the probe beam decreases at the resonance frequency. The decrease of absorption is called a Lamb dip. The line width of the Lamb dip is the order of natural broadening, and is significantly narrower than the Doppler spectrum. The utilization of the saturation spectroscopy for separating overlapped isotopic transitions has been reported [3]. In the present experiment, we utilized the Lamb dip as a frequency standard for the neutral flow measurements [4]. The use of Lamb dip is particularly appropriate for the flow velocity measurement, since the position of Lamb dip in the frequency scale is Doppler-shift free and is not disturbed by the motion of reference medium. By utilizing the Lamb dip as the frequency standard, the reliability and stability of the laser frequency calibration is increased. In addition, Lamb dip appears at the frequency corresponding to the zero velocity in the laboratory frame, which provides the origin of the velocity space. In this study, we adopt the Lamb dip of the $4s[3/2]^o_2(M_J=0)$ - $4p[1/2]_1(M_J=0)$ transition. The $M_J = 0$ to 0 transition is not perturbed by the variation of the magnetic field intensity in the target plasma. This advantage assures long-time stability of the frequency standard without any special requirement for the control of

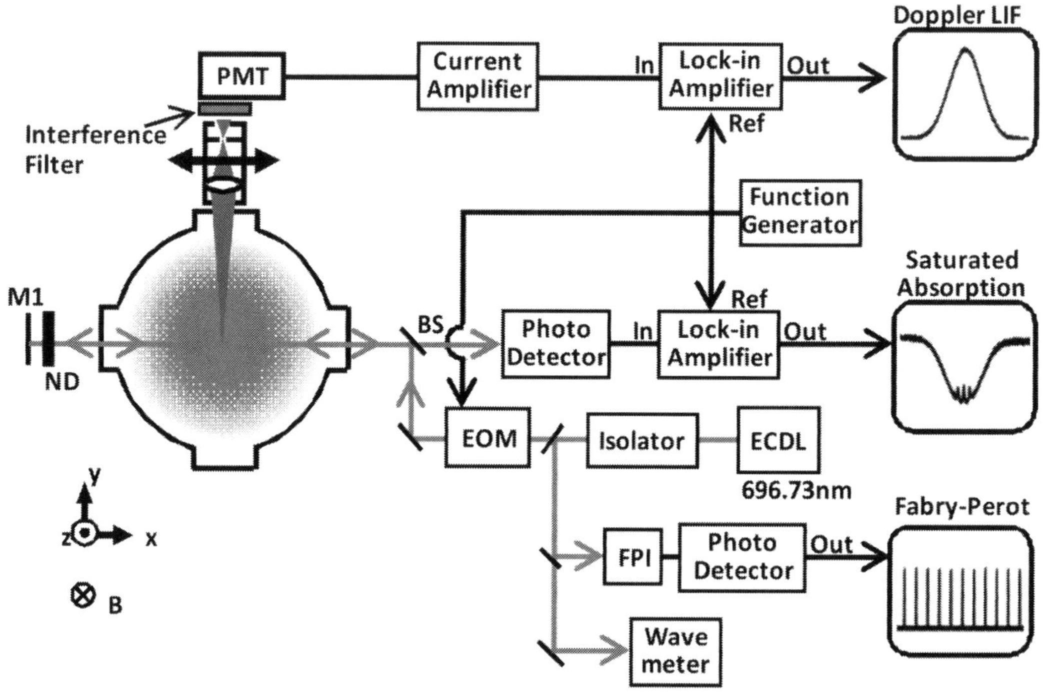

Figure 2. Schematic diagram of LIF Doppler spectroscopy system.

experimental environment. Consequently, the accuracy for frequency calibration is much more improved.

In the following sections, the experimental setup and the new LIF system are explained (Sec.2). The results of system performance and flow velocity measurement in an ECR plasma are presented in Sec. 3, followed by conclusions.

2. Experiments

The experiments have been carried out using the high density plasma experiment (HYPER-I) device at the National Institute for Fusion Science. The HYPER-I device is a linear plasma device equipped with ten magnetic coils (Figure 1). The sizes of cylindrical vacuum vessel are 30 cm in diameter and 200 cm in axial length. Argon plasma is generated by electron cyclotron resonance (ECR) heating with microwave at 2.45 GHz. The input microwave power and the operation pressure were fixed at 250 W or 5 kW and 0.8×10^{-2} Torr, respectively. Optical measurements were performed at the position z = 95 cm from the microwave input window. The detailed descriptions of the HYPER-I device have been reported in detail elsewhere [6, 7].

Figure 2 shows the high resolution LIF Doppler spectroscopy system which has been developed for the measurement of neutral flow velocity. The metastable argon atoms populated in the $4s[3/2]^{o}_{2}$ state are excited to the $4p[1/2]_{1}$ state by an external cavity diode laser (ECDL, Toptica Photonics DL100) tuned at 696.73 nm. Here, we use the Racah (j-l coupling) notation. The output power of the ECDL is 16 mW. The coarse tuning of the frequency of the ECDL is performed using a wavemeter. A Fabry-Perot interferometer (FPI) is used to check the single mode operation of the ECDL, and its fringes are used as frequency markers when scanning the laser frequency. The free spectral range (FSR) of the FPI is 294 MHz. The laser power is modulated by using an electro-optical modulator (EOM, Linos LM 0202). The modulated laser beam is introduced into the plasma along the center chord of the

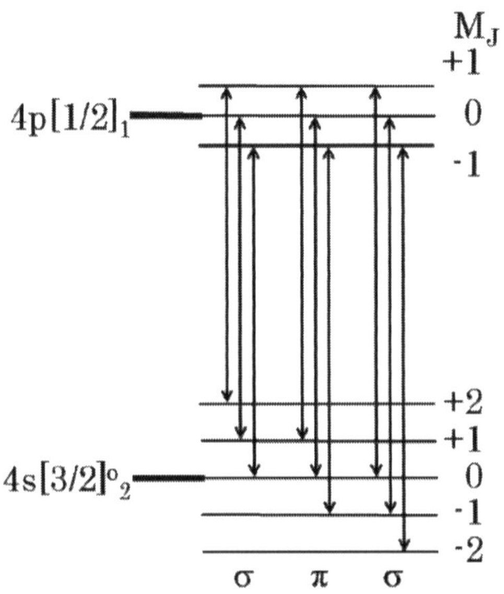

Figure 3. Zeeman sublevels of the $4s[3/2]^o_2$ – $4p[1/2]_1$ transition.

vacuum chamber. In the following, we adopt a coordinate system located at the center of the chamber, and then the laser beam is injected along the negative x-axis. The polarization of the excitation laser is aligned parallel to the magnetic field. A photomultiplier tube (PMT, Hamamatsu R3896) with collection optics is installed on a horizontally-movable stage placed on the top viewing port of the vacuum vessel. The $4s[3/2]^o_2$ - $4p[1/2]_1$ transition is split into the π and two σ components by Zeeman effect (Figure 3). The π components of the $4s[3/2]^o_2$ - $4p[1/2]_1$ transition are excited in the present experiment. The Doppler LIF spectrum of metastable argon atoms has been obtained by sweeping the laser frequency around the resonance center of the π component. The LIF photons of 826.68 nm (vacuum) emitted by the $4p[1/2]_1$ - $4s[1/2]^o_1$ transition have been observed. An interference bandpass filter (825 ± 5 nm) is set between the PMT and the slit to eliminate the plasma-induced emission. Since the spectral bandwidth of laser is much narrower than the Doppler width, the LIF spectrum is proportional to the velocity distribution function. The neutral flow velocity was determined by the Doppler shift of the LIF spectrum. The radial profile of neutral flow velocity is obtained by changing the position of the collection optics on the movable stage. The signal from the PMT is amplified using a current amplifier and then detected by a lock-in amplifier.

The saturated absorption spectroscopy of argon metastable state is carried out simultaneously with the Doppler LIF spectroscopy, where the incident beam of the LIF measurement is utilized as a pump laser. The transmitted beam is attenuated to 0.01% by a neutral density (ND) filter and reflected to re-enter into the plasma along the same path of the pump beam. This backward beam is used as a probe beam of the saturated absorption spectroscopy. The power of the probe beam is weak enough to avoid the saturation of the absorption and the disturbance to the LIF measurement. The absorption signal of the probe beam is detected by a photo diode behind the beam splitter (BS), and finally detected by a lock-in amplifier. The Lamb dip obtained by the saturated absorption spectroscopy is used as a frequency standard.

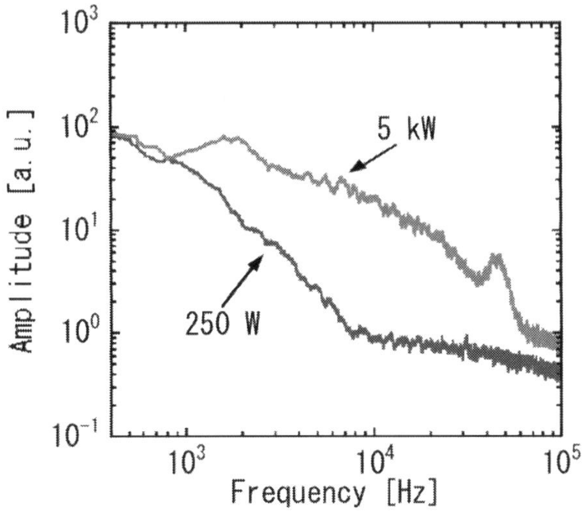

Figure 4. Power spectra of optical emission.

3. Results and Discussions

Optical emission of plasma is the strong noise source in the LIF measurement. Figure 4 shows the power spectra of optical emission of plasma generated at the microwave power of 250 W and 5 kW. Since the power spectrum of the 5 kW discharge emission has strong high frequency component, it was necessary to increase the operating frequency of the EOM to 100 kHz to obtain the sufficiently

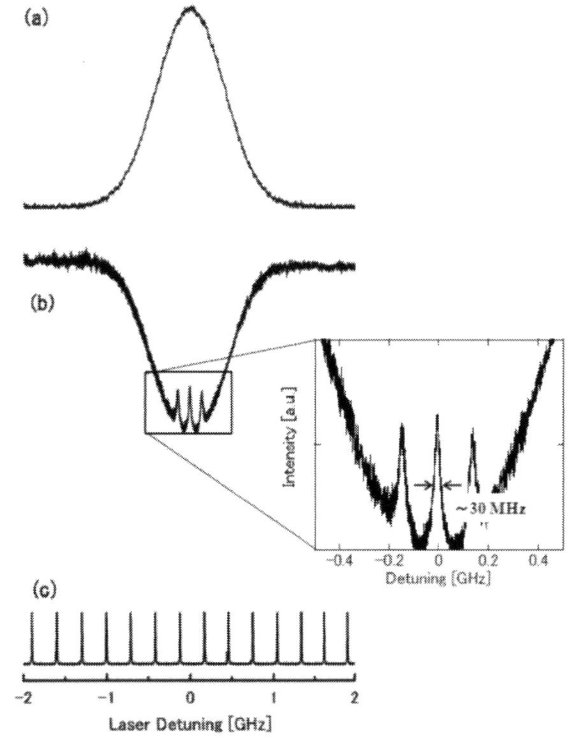

Figure 5.Set of spectra utilized for high accuracy flow detection. (a) LIF spectrum, (b) saturated absorption spectrum, and (c) fringes of FPI.

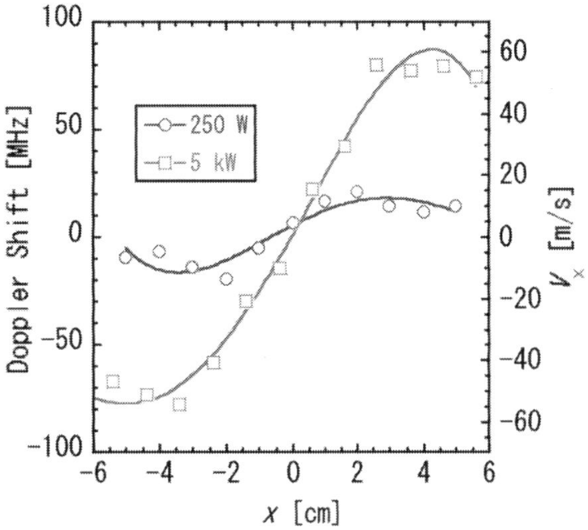

Figure 6. Radial flow velocity profile of neutrals at the microwave power of 250 W (open circles) and 5 kW (open squares).

good signal-noise ratio of the LIF signal. The maximum operating frequency of the EOM is limited by the Einstein's A coefficient of the LIF scheme. Figure 5 shows the set of spectra utilized for high-accuracy neutral-flow detection. The spectra were simultaneously recorded at the microwave power of 250 W. The frequency of the laser beam has been swept around the resonance frequency. Figure 5(a) shows the Doppler LIF spectrum of metastable argon atoms taken at the center of the chamber (x, y = 0 cm, z = 95 cm). The profile of the LIF spectrum is quite well fitted by a Gaussian distribution function. The full width at half maximum (FWHM) of the LIF spectrum is 950 MHz, and the corresponding temperature is 0.033 eV. Figure 5(b) is the saturated absorption spectrum composed of the Doppler-broadened absorption spectrum and the three Lamb dips. It is noted that the line width of the Lamb dip is significantly narrower (~30 MHz) than that of Doppler broadening (~1 GHz), which assures the improvement of accuracy in determining the standard frequency. The splitting of Lamb dips are attributed to the Zeeman splitting of magnetic sublevels of $4s[3/2]^o_2$ and $4p[1/2]_1$ states. The center dip, which is M_J = 0 to 0 transition, is used for the calibration of the absolute value of the laser frequency. Figure 5(c) shows the fringes of the FPI. The cavity length of FPI is fixed during the observation. The frequency interval of each fringe is 294 MHz, and is utilized as the frequency scale. By combining the position of Lamb dip and the fringes of interferometer, we have determined the center frequency of the LIF spectrum. The velocity of the neutral flow has been evaluated from the Doppler shift of the distribution function.

The radial flow velocity profile of the metastable argon atoms has been obtained by measuring the set of spectra at different radial positions. Figure 6 shows the radial flow velocity profile of neutrals along the horizontal axis at the microwave power of 250 W and 5 kW. The sign of Doppler shit is positive in positive x-region and negative in negative x-region. This result means that the neutral particles flow from right to left in the positive x-region and from left to right in the negative x-region, showing the existence of radial inward flow of neutrals in the plasma. To keep the continuity of particle flax, the neutral particles are ionized and pumped out along the axial magnetic field in the plasma column. The plasma rotates in the $E \times B$ direction at 250 W discharge power. On the other hand, in the 5 kW case, the core plasma rotate in the anti-$E \times B$ direction while the peripheral plasma rotate in the $E \times B$ direction. Both of the radial flow velocity profiles peaks around 4 cm from the center, which is comparable with the radius of the boundary of the $E \times B$ rotation and anti-$E \times B$ rotation. The maximum flow velocity increases with the microwave power. The relation between the increase of the flow velocity and the generation of the anti- $E \times B$ will be reported elsewhere.

4. Conclusion

We have developed a high-resolution LIF system with a narrow line-width tunable diode laser. Utilizing a saturated absorption spectrum to calibrate the laser frequency, we have achieved a high accuracy frequency determination. The radial flow velocity of the metastable argon atoms has been measured in an ECR plasma generated at microwave powers of 250 W and 5 kW. It is found that there exists a radial inward flow in the both plasma. Both of the radial flow velocity profiles peak around the boundary of the $E \times B$ rotation and anti-$E \times B$ rotation in the 5 kW discharge. The maximum flow velocity increases with the microwave power. The detailed results will be reported elsewhere.

Saturated absorption spectroscopy is one of the basic methods in precision laser spectrometry. We have introduced this technique into the LIF Doppler spectroscopy system for plasma physics research, and demonstrated that the minimum detectable velocity is significantly improved. LIF Doppler velocimetry with saturated absorption spectroscopy will become a powerful tool for studying dynamical behavior of plasma interacting with neutral flow, which has not been fully explored yet because of the lack of diagnostic tools.

This work was partially supported by the Grant-in-Aid for Scientific Research (A) 19204057 (2007-2009), and also supported by the LHD Project Research Collaboration program at the National Institute for Science NIFS06KOAP016 (2006-2008).

References

[1] D. O'Connell, T. Gans, D. L. Crintea, U. Czarnetzki, and N. Sadeghi, J. Phys. D: Appl. Phys. **41**, 035208 (2008).
[2] A. Okamoto, K. Hara, K. Nagaoka, S. Yoshimura, J. Vranješ, M. Kono, and M. Y. Tanaka, Phys. Plasmas **10**, 2211 (2003).
[3] R. Engeln, S. Mazouffre, P. Vankan, D. C. Schram, and N. Sadeghi, Plasma Sources Sci. Technol. **10**, 595 (2001).
[4] T. Nakano, N. Sadeghi, D. J. Trevor, R. A. Gottscho, and R. W. Boswell, J. Appl. Phys. **72**, 3384 (1992).
[5] M. Aramaki, K. Ogiwara, S. Etoh, S. Yoshimura, and M. Y. Tanaka, Rev. Sci. Instrum. **80**, 053505 (2009).
[6] M. Tanaka, R. Nishimoto, S. Higashi, N. Harada, T. Ohi, A. Komori, and Y. Kawai, J. Phys. Soc. Jpn **60**, 1600 (1991).
[7] M. Y. Tanaka, M. Bacal, M. Sasao, and T. Kuroda, Rev. Sci. Instrum. **69**, 980 (1998).

14th International Symposium on Laser-Aided Plasma Diagnostics (LAPD14) IOP Publishing
Journal of Physics: Conference Series **227** (2010) 012009 doi:10.1088/1742-6596/227/1/012009

Sum frequency generation for studying plasma-wall interactions

Sylvie Roke
Max-Planck Institute for Metals Research, 70569 Stuttgart, Germany, Fax: +49-711-6893612
E-mail: roke@mf.mpg.de

Abstract. Interaction of a plasma with a surface results in chemical and physical restructuring of the surface as well as the plasma in the vicinity of the surface. Studying such a reorganization of the atoms and molecules in the surface layer requires optical tools that can penetrate the plasma environment. At the same time, surface specificity is required. Sum Frequency Generation (SFG) is an optical method that fulfills these requirements. SFG has been developed into a surface specific probe during the eighties and nineties. Nowadays SFG is routinely applied in the research of complex interfaces. In such experiments, liquid/gas, solid/gas, solid/liquid, or liquid/liquid interfaces are probed, and the chemical surface composition, orientational distribution, order and chirality can be retrieved. An application to investigate plasma-wall interactions is feasible too.

1. Introduction

The operation of ITER strongly depends on the (absence of) interaction of the hydrogen plasma with the walls of the reactor [1, 2]. In order to design appropriate reactor walls composed of the most suitable material it is necessary to find out how a plasma interacts with the surface of the walls.

Plasma-surface interaction takes place on a highly confined location in a hostile environment. It typically involves processes such as atomic and molecular collision, surface charging, diffusion, neutralization, photon emission and electron emission. Due to these processes the surface will be restructured, many new chemical species can be formed as well as small dust particles [3].

In order to investigate such processes a multitude of plasma specific and surface specific techniques need to be combined. The Magnum-PSI project at the Institute for Plasma Physics Rijnhuizen aims to study plasma-wall interactions with surface specific methods. Most of the surface specific techniques (such as electron energy loss spectroscopy, X-ray photoelectron emission, Auger spectroscopy, or thermal desorption spectroscopy [4]) are operating either ex-situ (requiring transport through air), or in vacuo when the plasma source is turned off. In order to measure in-situ processes during plasma-wall interaction second-order nonlinear optical methods such as second harmonic and sum frequency generation are suitable. Since these are optical methods that can be applied in-situ during plasma-wall interaction they could be very valuable for controlling the plasma-wall interactions and the accompanied dust formation.

In this Article a description is given of the state of the art of sum frequency generation as well as an outline of how it might be used to measure plasma-wall interactions in-situ. The possible technological complications that might become important in such an implementation are discussed as well.

2. Sum Frequency Generation

Vibrationally resonant sum frequency generation (SFG) from a surface was first demonstrated by Hunt et al. [5] and Harris et al. [6] in 1987 who used SFG to study the structure of pentadecanoic acid on water surfaces and n-octadecylthiol on gold, cadmium stearate on silver and stearic acid on germanium. In a sum frequency generation experiment, two electromagnetic fields, $\mathbf{E}_1(\mathbf{r},t) = \mathcal{E}_1\mathbf{u}_1e^{i(\mathbf{k}_1\cdot\mathbf{r}-\omega_1 t)}$ and $\mathbf{E}_2(\mathbf{r},t) = \mathcal{E}_2\mathbf{u}_2e^{i(\mathbf{k}_2\cdot\mathbf{r}-\omega_2 t)}$, with amplitudes $\mathcal{E}_1$ and $\mathcal{E}_2$, polarization states $\mathbf{u}_1$ and $\mathbf{u}_2$, and frequencies ω_1 and ω_2, are reflected from an interface. The frequency of the field $\mathbf{E}_2$ is tuned around the resonance of the vibrational modes of the interfacial molecules, so that the frequency will be in the infrared (IR) range of 3700 - 1000 cm^{-1} (which corresponds to wavelengths of 2700 nm - 10000 nm or energies of 0.45 - 0.12 eV). The other electromagnetic field is usually kept at a fixed visible (VIS) wavelength (usually 532 or 800 nm, but can in most setups be tuned in the visible wavelength region. Such pulses are readily available from ns, ps, or fs pulsed laser sources. At the interface the optical fields induce a second-order polarization that oscillates at the sum frequency (SF) of the incoming waves:

$$\mathbf{P}_i^{(2)}(\omega_0 = \omega_1 + \omega_2) = \chi_{ijk}^{(2)}(\omega_0 = \omega_1 + \omega_2)\mathbf{E}_{1,j}(\omega_1)\mathbf{E}_{2,k}(\omega_2). \quad (1)$$

This polarization is the source of a sum frequency field that can be emitted from the interface. By solving Maxwell's equations it can be found that the detected intensity is of the following form:

$$I \propto |(\mathbf{u}_0 \cdot \mathbf{L}(\omega_0)) \cdot \boldsymbol{\chi}^{(2)} : (\mathbf{u}_1 \cdot \mathbf{L}(\omega_1))(\mathbf{u}_2 \cdot \mathbf{L}(\omega_2))|^2 I_1(\omega_1) I_2(\omega_2). \quad (2)$$

The magnitude and spectral shape of the intensity is determined by the Fresnel factors $\mathbf{L}(\omega_i)$ that dictate the efficiency of the reflection of the optical beams [7, 8] and the second-order susceptibility tensor $\boldsymbol{\chi}^{(2)}$. $\boldsymbol{\chi}^{(2)}$ is a physical property of the material which must reflect the

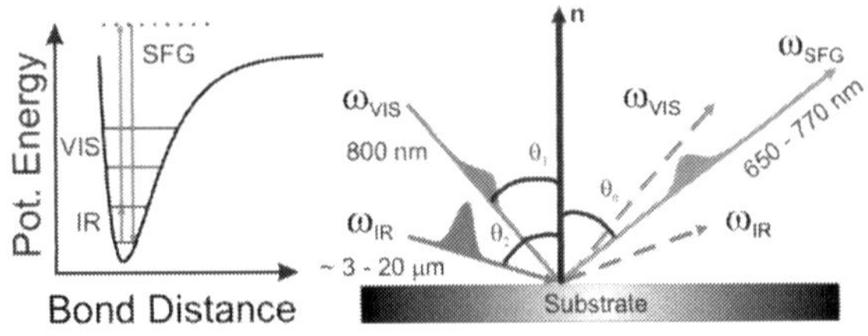

Figure 1. Illustration of infrared visible surface sum frequency generation (SFG). Left: Potential energy diagram of a vibrational transition. The infrared frequency is resonant with the vibrational mode, which renders the technique extremely sensitive to molecular structure. Right: illustration of the technique. An infrared and visible laser pulse incident under angles θ_2 and θ_1 reflect from a substrate. The magnitude of the generated sum frequency field is determined by the second-order nonlinear susceptibility ($\boldsymbol{\chi}^{(2)}$). The angle θ_0 of the outgoing SFG photon is determined by the geometry of the incoming beams [9].

spatial symmetry properties of that material (Neumann's principle). As a consequence (in the electric-dipole approximation) $\chi_{ijk}^{(2)}$ vanishes in bulk media [8]. This means that SFG can often be applied as a surface specific technique, with which only the first few molecular layers situated at the interface are probed. The surface response, identified as $\boldsymbol{\chi}_s^{(2)}$, is usually split up in a frequency dependent resonant (res) and a frequency independent non-resonant (nr) response:

$$\chi_{s,ijk}^{(2)} = \chi_{s,nr,ijk}^{(2)} + \chi_{s,res,ijk}^{(2)} \propto A_{nr}e^{i\Delta\phi} + \sum_n \frac{N_s\langle T_{ia}T_{jb}T_{kc}\rangle R_{n,ab}\mu_{n,c}}{\hbar(\omega_2 - \omega_{0n} + i\Upsilon_n)}, \quad (3)$$

where n refers to a specific vibrational mode, with resonance frequency ω_{0n}, and damping constant Υ_n. This equation relates the resonant surface susceptibility $\chi^{(2)}_{s,res}$ to the hyperpolarizability tensor $\boldsymbol{\beta}^{(2)}_n$ which is the product of the vibrational dipole moment $\boldsymbol{\mu}_n$ with the Raman polarizability $\mathbf{R}_n$ through a transformation matrix $\mathbf{T}$ that transforms from the molecular coordinate frame (a, b, c) to the surface coordinate frame (i, j, k). N_s is the density of surface molecules. A_{nr} is the amplitude of the frequency independent non-resonant response that can be phase lagged with a factor $\Delta\phi$ with respect to the resonant response. The values of ω_{0n}, Υ_n, $\boldsymbol{\mu}_n$, and $\mathbf{R}_n$ depend on the type of molecules, their orientation, chirality and order (see e.g. [10, 11, 12]). This type of structural information is often exclusively interface specific and can be obtained non-invasively, label-free, and in-situ. Table 1 summarizes review papers that have been written about various types of interfaces. Fig. 2 shows example spectra of a polymer/air interface and a liquid/metal interface.

Table 1. Literature reviews about sum frequency generation

General/historic/introductory	[13, 14, 15, 12]
Chiral media	[16, 17, 18]
Kinetics and dynamics	[19, 20]
Solid/solid interfaces	[10, 21]
Solid/liquid interfaces	[10, 21]
Solid/gas interfaces	[22, 23, 24]
Liquid/liquid interfaces	[25, 26]
Liquid/gas interfaces	[27, 28, 29]
Polymers	[22]
Biologically relevant interfaces	[30, 31]
Particle interfaces	[32, 33]

2.1. Typical setup and experiment

In order to generate sufficient electric field amplitudes to generate detectable sum frequency photons, pulsed laser sources are needed. Nowadays, pulsed femtosecond and picosecond lasers can be fabricated routinely and a wide variety of options is available. Since femtosecond sources deliver pulses with a broad frequency bandwith they allow for a simultaneous recording of multiple chemical species and are therefore the preferable choice for studying plasma-wall interactions during plasma operation. Typically a femtoseond system consists of an oscillator that delivers high repetition rate 800 nm pulses that are used as a seed in a chirped pulse amplifier. The amplified pulses can be converted into infrared pulses via optical parametric generation and amplification in combination with difference frequency generation.

The geometry for a surface SFG-experiment is given in Fig. 1. The chosen angles θ_i are determined by the optical properties of the substrate and have to be chosen such that there is an optimal geometry for SFG. This is dictated by the Fresnel factors of the substrate material (Eq. (2)) and the incident polarization of the light (i.e. parallel or perpendicular to the plane of incidence) [10, 7, 11, 12]. As an example, the optical properties of graphite dictate that the incoming visible beam should be incident under more than 75°[1] with respect to the surface normal [37]. For s-polarized light there is no Brewster angle. These considerations would result in $\theta_1 = 80°$ and $\theta_2 = 60°$. For a metal such as platina both angles can be in the range $\theta_{1,2} = 80° - 90°$ for optimal sum frequency generation. In Fig. 3 an illustration of how an SFG setup might be combined with a vacuum machine placed inside a large magnet for generating the plasma is given.

[1] Brewster angle for p-polarized 632 nm light

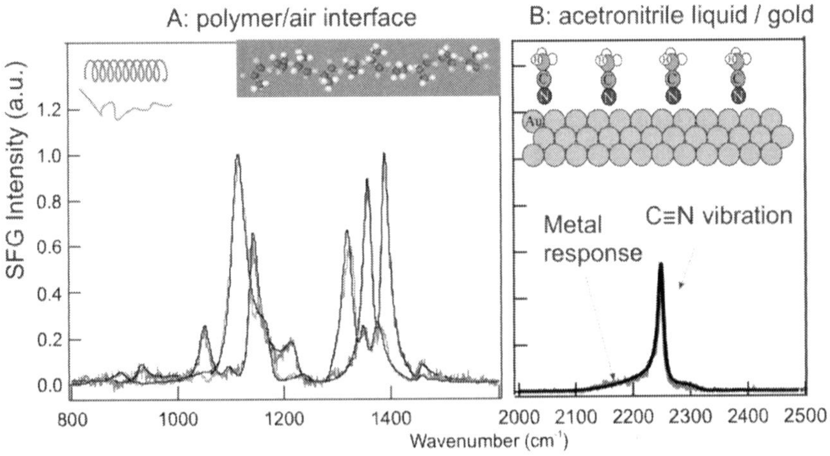

Figure 2. Sum frequency generation from two different interfaces, in different spectral regions. A: The polymer/air interface of the same polymer in a crystalline (red) and amorphous (blue) state. Although the chemical composition is identical, the crystalline polymer film is chiral while the amorphous one is racemic. The SFG spectra show that there is a big difference in the three dimensional surface structure of the two films [34]. B: An SFG spectrum of a liquid (acetonitrile) in contact with a metal (gold). It can be seen that there is interference between a metal originating non-resonant signal and a molecular symmetric resonance of the CN vibrational mode. The mode is red shifted with respect to the liquid value indication a strong interaction with the metal [35, 36]. The typical integration time to record these spectra is 100 s.

3. Probing in-situ surface-plasma interaction

3.1. System and substrates

In a state-of-the-art setup (e.g. [38]), tunable infrared pulses with a duration of $\sim$ 70 fs and an energy of >50 μJ/pulse can be generated in the 2.6-10 μm wavelength range. Using infrared transparent CaF_2 or MgF_2 optics for vacuum ports the accessible frequency range of interest will be 1200-3500 cm^{-1}, which is the region of interest for most chemical compounds. Table 2 summarizes vibrational frequencies of H and D on relevant substrates (S). It can be seen that H atoms on atop positions can be detected readily with SFG. H atoms in more complex bond conformations (like threefold and fourfold sites) exhibit lower resonance frequencies.

Table 2. Relevant frequencies of typical plasma-wall compounds

species	calculated (cm^{-1})	measured (cm^{-1})
Fusion materials		
C-H (atop)	2733 [39]	2838 (H-C(111)^2 [40])
C-D (atop)	-	2115 (D-C(111)^3 [40])
W-H (atop)	-	1738 rare species [41]
W(100)-H (twofold)	-	1070-1300 [42]
B-H	2282	-
Li-H	1359	-
Reference materials		
Pt-H (atop)	2250	2090 polycrystalline Pt electrode [42]
Pt(111)-H (threefold)	-	1230 (asym. stretch)/ 550 (sym. stretch) [42]

3.2. Interference with the light emitted from the plasma

Sum frequency generation relies on the detection of photons and their frequency distribution. A light emitting plasma can therefore interfere with the exclusive detection of the generated sum frequency photons. To circumvent this issue a gated intensified CCD camera needs to be used in combination with fiber coupled optics that is positioned to collect the directionally emitted sum frequency photons. With the gated electronics it is possible to turn on the detection system for only 10 ns. If this electronic gate coincides with the time at which the laser pulses hit the substrate, the continuous incoherent background radiation will not interfere with the coherently generated signal. Furthermore, by using a visible laser source that is tunable from (e.g.) 600 to 800 nm it is possible to vary the frequency of the emitted SF photons while probing the same vibrational resonance. In this way gas emission lines should not interfere with the detection of SF photons.

3.3. Optical probes for dust analysis

The shape and size distribution of the particles in the plasma dust can be imaged and diagnosed with turbidity measurements and speckle pattern analysis. Also, if there are particles in the vicinity of the surface or if the surface is rough, it is possible to generate SF photons not only from the surface but also from the particles (see the right panel of Fig. 3 for an illustration). These photons will not be emitted in the same direction as the reflected phase matched SF beam but rather leave the sample under an angle that depends on the size and shape of the particles [33]. By spatially imaging the fiber output on the camera it is possible to distinguish between particle surfaces and the surface of the substrate. Particles larger than $\sim 1\ \mu$m will create backscattered SF photons [43, 44] than can be detected as well. The ratio between diffuse forward scattered light and backward scattered light can be used to make an estimate of the particle size distribution in the plasma.

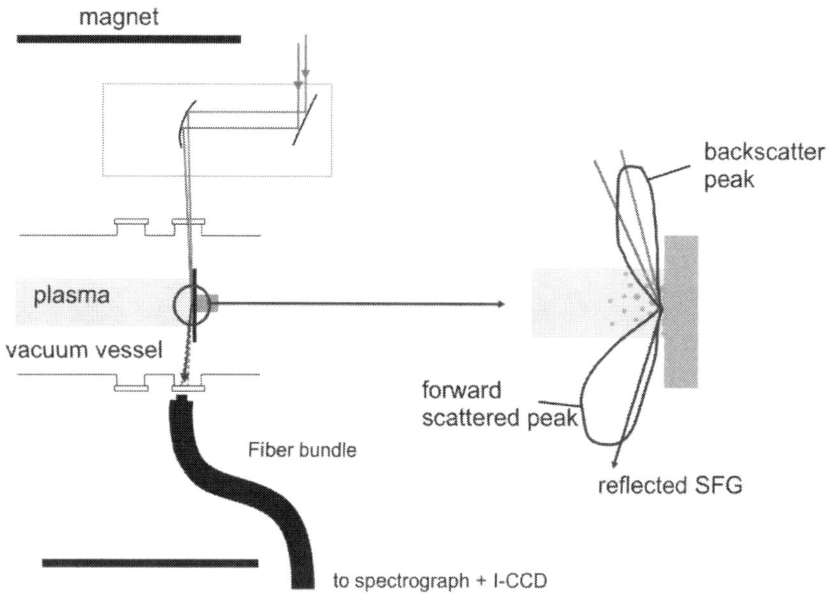

Figure 3. Illustration of a possible sum frequency generation setup inside a vacuum machine surrounded by the coils of a large magnet. The sum frequency photons can be detected using a fiber bundle and a gated intensified CCD camera. The right panel illustrates the expected directions into which scattered SF photons can be emitted from the surface region.

4. Conclusions

We have described the possibility of using the second-order nonlinear optical method of sum frequency generation to study plasma-wall interactions in a setup such as the Magnum-PSI project. Sum frequency generation is an optical method that can be used during plasma operation. Thanks to the use of short pulses with a broad frequency content it should be possible to take snap-shot spectra of the surface during plasma-surface interaction. Further, by detecting as well diffuse (back)scattered light it should be possible to measure the surface structure of the dust particles in the vicinity of the surface as well as estimate their size range.

5. Acknowledgements

This work is part of the research program of the Max-Planck Society.

6. References

[1] Janeschitz, G., Antipenkov, A., Federici, G., Ibbott, C., Kukushkin, A., Ladd, P., Martin, E., and R., T. *Nucl. Fusion 42 (2002) 1420* **42**, 14–20 (2002).
[2] Kleyn, A. W., Lopes Cardozoa, N. J., and Samm, U. *Phys. Chem. Chem. Phys.*, **8**, 1761 – 1774 (2006).
[3] Winter, J. *Phys. Plasmas* **7**, 3862–3866 (2000).
[4] Yates, J. T, J. *Experimental Innovations in Surface Science: A Guide to Practical Laboratory Methods and Instruments.* AIP Press, (1997).
[5] Hunt, J. H., Guyot-Sionnest, P., and Shen, Y. R. *Chem. Phys. Lett.* **133**, 189–192 (1987).
[6] Harris, A. L., Chidsey, C., Levinos, N., and Loiacona, D. *Chem. Phys. Lett.* **141**, 350–356 (1987).
[7] Zhuang, X., Miranda, P. B., Kim, D., and Shen, Y. R. *Phys. Rev. B.* **59**, 12632 (1999).
[8] Butcher, P. N. and Cottor, D. *The Elements of Nonlinear Optics.* Cambridge University Press, (1990).
[9] Shen, Y. R. *The principles of nonlinear optics.* Wiley, New York, (1984).
[10] Bain, C. D. *J. Chem. Soc., Faraday Trans.* **91**, 1281 – 1296 (1995).
[11] Wang, H. F., Gan, W., Lu, R., Rao, Y., and Wu, B. H. *Int. Rev. Phys. Chem.* **24**, 191–256 (2005).
[12] Lambert, A. G., Davies, P. B., and Neivandt, D. J. *Appl. Spectr. Rev.* **40**, 103–144 (2005).
[13] Heinz, T. F. *Nonlinear surface electromagnetic phenomena*, chapter 5. Elsevier, New York (1991).
[14] Shen, Y. R. *Surf. Sci.* **299/300**, 551–562 (1994).
[15] Bloembergen, N. *Appl. Phys. B* **68**, 289–293 (1999).
[16] Sioncke, S., Verbiest, T., and Persoons, A. *Mater. Sci. Eng.* **42**, 115–155 (2003).
[17] Simpson, G. J. *Chem. Phys. Chem.* **5**(9), 1301–1310 (2004).
[18] Belkin, M. A. and Shen, Y. R. *Int. Rev. Phys. Chem.* **24**, 257–299 (2005).
[19] Bonn, M., Hess, C., Funk, S., Miners, J. H., Persson, B. N. J., Wolf, M., and Ertl, G. *Phys. Rev. Lett.* **84**, 4653–4656 (2000).
[20] Kubota, J. and Domen, K. *Anal Bioanal Chem.* **388**, 17–27 (2007).
[21] Vidal, F. and Tadjeddine, A. *Rep. Prog. Phys.* **68**(5), 1095–1127 (2005).
[22] Chen, Z., Shen, Y. R., and Somorjai, G. A. *Annu. Rev. Phys. Chem.* **53**, 437–465 (2002).
[23] Raschke, M. B. and Shen, Y. R. *Curr. Opin. Solid State Mater Sci.* **8**, 343–352 (2004).
[24] Baldelli, S. *ChemPhysChem* **9**, 2291–2298 (2008).
[25] Eisenthal, K. B. *Chem. Rev.* **96**(4), 1343–1360 (1996).
[26] Richmond, G. L. *Chem. Rev.* **102**, 2693–2724 (2002).
[27] Shultz, M. J., Schnitzer, C., Simonelli, D., and Baldelli, S. *Int. Rev. Phys. Chem.* **19**(1), 123–153 (2000).
[28] Richmond, G. L. *Annu. Rev. Phys. Chem.* **52**, 357–389 (2001).
[29] Shultz, M., Baldelli, S., Schnitzer, C., and Simonelli, D. *J. Phys. Chem. B* **106**, 5313–5324 (2002).
[30] Chen, X. Y., Clarke, M. L., Wang, J., and Chen, Z. *Int. J. Mod. Phys.* **19**(4), 691–713 (2005).
[31] Gualtieri, E. J., Haupert, L. M., and Simpson, G. J. *Chem. Phys. Lett.* **465**, 167–174 (2008).
[32] Eisenthal, K. B. *Chem. Rev.* **106**(4), 1462–1477 (2006).
[33] Roke, S. *Chem. Phys. Chem.* **10**, 1380–1388 (2009).
[34] Sugiharto, A. B., Johnson, C. M., Dunlop, I. E., and Roke, S. *J. Phys. Chem. C* **112**, 7531–7534 (2008).
[35] Roke, S., Kleyn, A. W., and Bonn, M. *Chem. Phys. Lett.* **370**(1-2), 227–232 (2003).
[36] Roke, S., Kleyn, A. W., and Bonn, M. *Surf. Sci.* **593**(1-3), 79–88 (2005).
[37] Greenaway, D. L., Harbeke, G., Bassani, F., and Tossati, E. *Phys. Rev.* **178**, 1340–1348 (1969).
[38] Sugiharto, A. B., Johnson, C. M., de Aguiar, H. B., Aloatti, L., and Roke, S. *Appl. Phys. B.* **91**, 315–318 (2008).
[39] Ibach, H. *Electron energy loss spectroscopy and surface vibrations.* Academic Press, New York, (1982).
[40] Chin, R. P., Huang, J. Y., Shen, Y. R., Chuang, T. J., and Reki, H. *Phys. Rev. B.* **52**, 5985–5995 (1995).

Journal of Physics: Conference Series **227** (2010) 012009 doi:10.1088/1742-6596/227/1/012009

[41] Riffe, D. M. and Sievers, A. J. *Phys. Rev. B.* **41**, 3406 (1990).
[42] Christmann, K. *Surf. Sci. Rep.* **9**, 1–163 (1988).
[43] Kerker, M. *The scattering of light and other electromagnetic radiation.* John Wiley and sons, New York, (1969).
[44] de Beer, A. G. F. and Roke, S. *Phys. Rev. B* **79**, 155420–1–9 (2009).

First results and analysis of collective Thomson scattering (CTS) fast ion distribution measurements on ASDEX Upgrade

F Meo[1], M Stejner[1], M Salewski[1], H Bindslev[1], T Eich[2], V Furtula[1], S B Korsholm[1], F Leuterer[2], F Leipold[1], P K. Michelsen[1], D Moseev[1], S K Nielsen[1], B Reiter[2], J Stober[2], D Wagner[2], P Woskov[3], and the ASDEX Upgrade team

[1] Association EURATOM - Risø-DTU, P.O. Box 49, DK-4000 Roskilde, Denmark
[2] Association EURATOM - Max-Planck-Institut für Plasmaphysik, D-85748 Garching, Germany
[3] MIT Plasma Science and Fusion Center, Cambridge, MA 02139, USA

E-mail: femo@risoe.dtu.dk

Abstract. Experimental knowledge of the fast ion physics in magnetically confined plasmas is essential. The collective Thomson scattering (CTS) diagnostic is capable of measuring localized 1D ion velocity distributions and anisotropies dependent on the angle to the magnetic field. The CTS installed at ASDEX-Upgrade (AUG) uses mm-waves generated by the 1 MW dual frequency gyrotron. The successful commissioning the CTS at AUG enabled first scattering experiments and the consequent milestone of first fast ion distribution measurements on AUG presented in this paper. The first fast ion distribution results have already uncovered some physics of confined fast ions at the plasma centre with off-axis neutral beam heating. However, CTS experiments on AUG H-mode plasmas have also uncovered some unexpected signals not related to scattering that required additional analysis and treatment of the data. These secondary emission signals are generated from the plasma-gyrotron interaction therefore contain additional physics. Despite their existence that complicate the fast ion analysis, they do not prevent the diagnostic's capability to infer the fast ion distribution function on AUG.

1. Introduction

Collective Thompson scattering (CTS) offers the capability to measure the fast ion distribution in magnetically confined plasmas which play an important role in magnetic fusion research. A CTS diagnostic detects plasma fluctuations along a wave vector $\mathbf{k}^{\delta} = \mathbf{k}^{s} - \mathbf{k}^{i}$ where $\mathbf{k}^{s}$ and $\mathbf{k}^{i}$ are the wave vectors of the received scattered radiation and the incident probing beam, respectively. The extraction of the component of the ion fluctuations along $\mathbf{k}^{\delta}$ from the scattered radiation ($\omega^{\delta} \approx \mathbf{v}_{ion} \cdot \mathbf{k}^{\delta}$) requires that the radiation is scattered dominantly off fluctuations larger than the Debye length [1] ($(\lambda_D\, \mathbf{k}^{\delta})^{-1} > 1$) which is well satisfied for millimetre-wave probes in tokamaks. The microwave based backscattering CTS system on ASDEX Upgrade (AUG) [2] uses a high power gyrotron as the probe at 105 GHz and is capable of measuring localized 1D ion velocity distributions and anisotropies dependent on the angle of $\mathbf{k}^{\delta}$ to the magnetic field ($\angle(\mathbf{k}^{\delta}, \mathbf{B})$). The schematic in Figure 1 shows an example of a scattering geometry that resolves the fast ion component along $\mathbf{k}^{\delta}$ for $\angle(\mathbf{k}^{\delta}, \mathbf{B}) \approx 120°$ where the

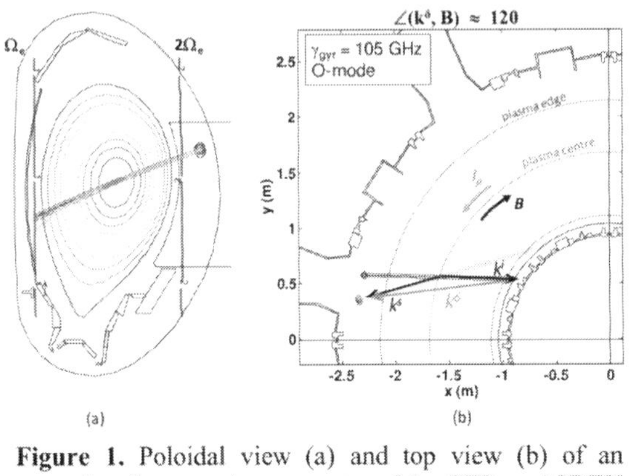

Figure 1. Poloidal view (a) and top view (b) of an example of a scattering geometry of the CTS on ASDEX Upgrade for $\angle(\mathbf{k}^{\delta}, \mathbf{B}) \approx 120°$. The blue and green traces illustrate the probe and receiver beam respectively. The scattering volume, where the beams overlap, is shown in magenta at the plasma centre.

plasma current (I_p) is anti-parallel to the magnetic field. It is important to note that CTS-AUG operates with the electron cyclotron resonance layers (Ω_e and $2\Omega_e$) positioned outside the plasma to avoid gyrotron radiation absorption and minimize the ECE background. The weak scattered signal i.e. the spectral power density (SPD) of ~ 1–30 eV is extracted from the ECE background of typically 10–300 eV, with the aid of modulating the gyrotron with typically 2 ms on-times. A sensitive 50 channel CTS heterodyne receiver system is used, centred near 105 GHz with 10 GHz bandwidth and high frequency resolution of about ~100 MHz for each *channel* (except for the outermost channels). More details on the hardware and the CTS-AUG capabilities can be found in [2]. CTS experiments on AUG have been carried out successfully whereby fast ion distribution results have been attained and presented in section 2. The experiments however, have also uncovered challenges that required additional analysis and treatment of the data. Section 3 gives a brief overview of *secondary emissions* (SE) observed in the measured radiation (raw data) which, to a degree, complicate the analysis, but do not prevent the diagnostic from attaining the fast ion distribution. Section 4 which will describe the principles behind the analysis involved in extracting the spectra from the data which include secondary emission removal from the raw data and the principles involved in the ECE background removal in an H-mode plasma.

2. Results of the confined fast ion distribution for an NBI heated plasma on AUG

Figure 2(a) and (b) show results of the spectral power density and confined fast ion distribution function respectively for an NBI heated H-mode plasma on AUG. The graphs compare two NBI heating configurations using on-axis and off-axis ion beam sources (P_{source} = 2.4 MW). The error-bars shown in Figure 2(a) are one standard deviation of a number of spectral power densities during each heating phase. In this scattering geometry, the k^δ is oriented in the same toroidal direction as the plasma current (I_p) and Figure 2(a) show the frequency up-shift due to the fast ion flow direction as expected from $\omega^\delta \approx v_{ion} \cdot k^\delta$. To attain the fast ion distribution function in Figure 2(b), the measured CTS spectra in Figure 2(a) are fitted using a least squares fitting which takes prior information about parameters, including those from other diagnostics, and implements a Bayesian method of inference using a forward model. More details of the procedure can be found in [3] with its application to AUG in [4].

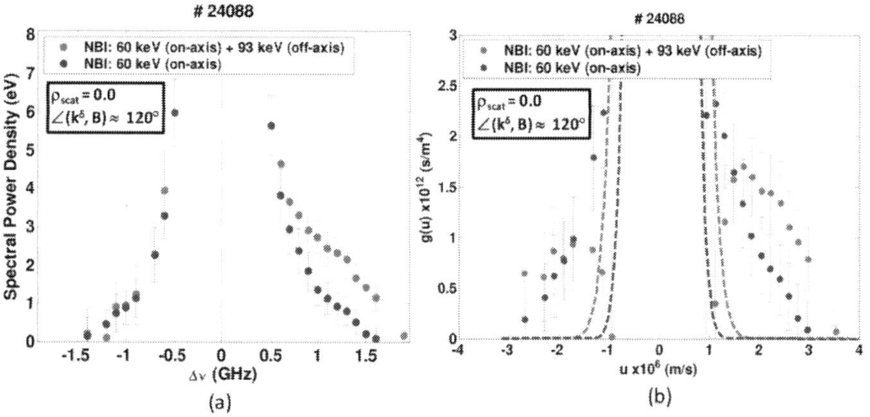

Figure 2. The CTS results; the red (blue) data points on both graphs represents NBI heating configuration with two (one) ion sources. (a) Spectral power density and (b) fast ion distribution function, for a low triangularity standard H-mode plasma with $n_e(0) = 6 \times 10^{19}$ m^{-3}, Bt = 2.6 T. The abscissa in (a) is the frequency and in (b) is the velocity component which is along $\mathbf{k}^\delta$. The g(u) in Figure 2(b) is the projection of the velocity distribution function $f(v_{//}, v_\perp)$ along the direction of $\mathbf{k}^\delta$. More details are described in Reference [4]. The scattering volume is located at the centre of the plasma with $\angle(\mathbf{k}^\delta, \mathbf{B}) \approx 120°$ where Ip and B are anti-parallel. The dashed lines are the bulk ion distribution.

The error bars in Figure 2(b) represent the uncertainty of one standard deviation and includes the uncertainties of parameters from diagnostics other than CTS such as electron/ion temperature, and electron density. Additionally, the uncertainty is such that it could be represented by a scaling factor in the velocity distribution [4]. It can be seen in Figure 2(b) the effect an off-axis beam has on the co-Ip fast ion distribution at the centre of the plasma on AUG.

3. Secondary emissions in the raw data

This section describes the secondary emissions observed in the measured radiation (raw data) during AUG H-mode plasmas that required additional analysis and treatment of the data in order to attain the fast ion distribution function. Reference [2] reported the existence of secondary emissions from the gyrotron itself under certain operating regimes. Notwithstanding operating the gyrotron under the required parameters for a clean spectrum, secondary emissions of a different type appear in the raw data that originate from the antenna front-end and are correlated to the plasma. These signals exist regardless of the antennae orientation or whether the overlap between the probe and receiver is present or not. The intensity values can reach levels of above 3-5 keV (receiver saturation). Nonetheless, the CTS-AUG receiver is sufficiently robust against gain compression and there exists enough isolation between channels in the presence of these secondary emissions. Three types of secondary emissions have been identified each appearing under different experimental conditions and hence believed to be created by different physical mechanisms. Due to limited space, the section will only briefly summarize the characteristics of each and hypothesize on the physics mechanism involved in their creation.

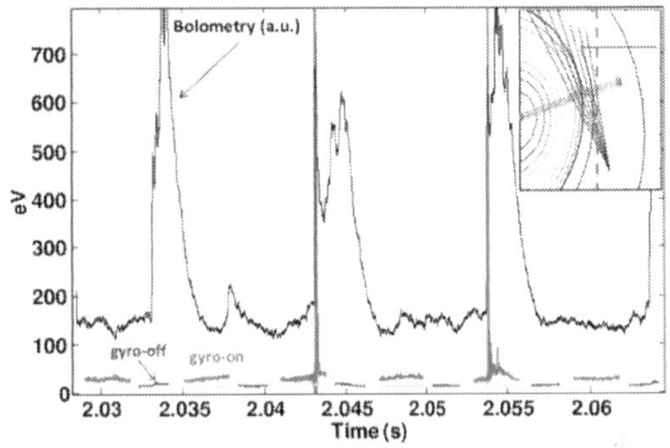

Figure 3. Time trace of CTS raw data sampled in a channel about 400 MHz away from the gyrotron line in an H-mode AUG discharge #23939. Red curves represent samples taken during gyro periods (gyrotron on, CTS + ECE), while blue are samples taken during ECE periods (gyrotron off, ECE only). The green lines are the reconstructed ECE level during the gyro-on periods. The black curve is the raw signal from the outermost viewing channel of the bolometer diode camera located at the same sector as the CTS probe and receiver. The viewing geometry of the bolometer is shown as the dotted lines in the schematic in the insert.

3.1. Secondary emissions from ELMs (SE-ELM)

The most prevalent secondary emission in CTS experiments in AUG H-mode discharges is correlated to ELMs. Shown in Figure 3, the SE signals appear as spikes during the gyro-on periods (in red) and show a clear correlation with the rise of the bolometer signal (shown in black) during an ELM event. These SE-ELM mainly affect the central channels (3 - 5 channels → ± 500 MHz on either side of the gyrotron line). The occurrence and intensity of SE-ELM decrease for channels further away from the gyrotron line and can be asymmetric in frequency. One hypothesis is local heating of ELM filament structures that pass through the $2\Omega_e$ layer in front of the ECRH antenna as they propagate outward. The ELM filaments can carry a significant portion of the pedestal density [3] and can absorb a portion of the gyrotron power when crossing the resonance layer. The high energy electrons created re-emit radiation detected by the neighbouring receiver antenna at μs time scales due to the low confinement time because of to the high parallel conductivity. Regardless of the physics mechanism behind their creation, SE-ELM signals can be avoided in principle by triggering the gyrotron shortly after an ELM is detected by the diode bolometer camera located in the same sector as the ECRH antennae.

3.2. Secondary emissions from NBI ion sources near perpendicular to the magnetic field (SE-NBI$_\perp$)

Figure 4(a) shows two types of secondary emissions generated by a different type of mechanism. The SE-ELM signal, described in the previous section, appears only during the onset of an ELM event as can be seen during the first, fourth, and second-to-last gyro-pulse in Figure 4(a). However, the figure also shows secondary emissions at distinct channels (22-23 & 35-36) on either side of the gyrotron line (ch#29). These signals (SE-NBI$_\perp$) are independent of any ELM event and appear only when the two most perpendicular NBI ion sources (ion sources #1 and #4 in the AUG nomenclature) are on. The spectral power density plotted in Figure 4(b) clearly shows the two peaks at ≈ 670 MHz from either side of the gyrotron line. The two signals are strongly correlated to each other and weakly correlated to the SE-ELM signals present in the more central channels.

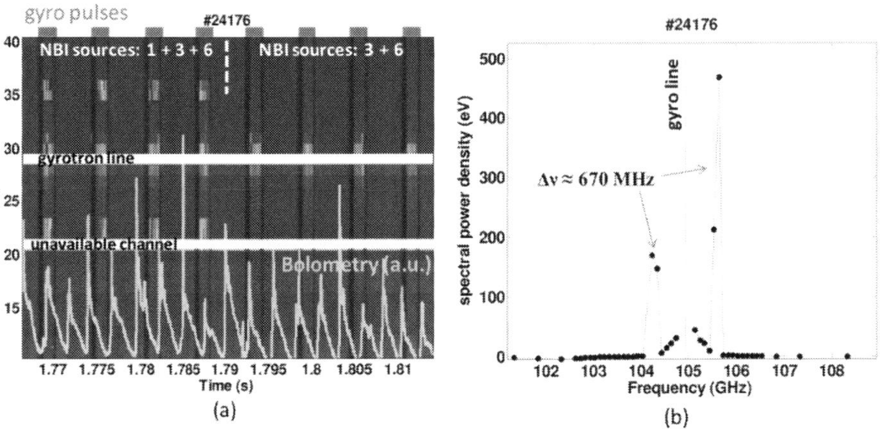

Figure 4. (a) Contour of the raw data from the CTS receiver during the H-mode discharge #24176 heated by a perpendicular ion source (ion source #1). The ordinate is the channel number and abscissa the time in seconds. Channel # 29 corresponds to the gyrotron radiation. The red rectangles at the top indicate the gyro-on time periods. The time trace in green trace superimposed onto the contour is the bolometer diode signal. (b) The spectral power density from discharge 24176 during a near perpendicular NBI ion source.

As with all secondary emissions described in section 2, they are present regardless of the antennae position and overlap. The strong correlation between the two peaks and the symmetry in frequency suggests a wave mixing process between the gyrotron radiation and a wave (or waves) in the plasma with a narrow frequency bandwidth of 600-700 MHz created by a resonant process which occurs when perpendicular NBI ion sources are on. A similar observation was made by a microwave based CTS system on W7-AS where the signal was explained by lower hybrid (LH) instability driven turbulence created by perpendicular fast ions [6] [7]. This well known coupling between LH waves and the perpendicular component of the ions is stochastic in nature and can occur when the wave amplitude is large and when the perpendicular phase velocity ($\omega/k_\perp$) approaches the ion perpendicular thermal velocity [8]. This energy coupling from LH to ions has been measured with charge exchange spectroscopy [9][10].

3.3. *Secondary emissions from ion cyclotron resonance heating (SE-ICRH).*

CTS experiments on AUG discharges with ICRH hydrogen minority heating at f_{ICRH} = 36.5 MHz have revealed a third type of secondary emission shown in Figure 5(a), present only with ICRH heating. Independent from the ELM signal (bolometry), SE-ICRH signals are present at every gyro-pulse and are concentrated in the frequency range close to the centre channels i.e. ±300 MHz corresponding to three channels on either side of the gyro-line. The contour in Figure 5(a) clearly shows the presence of only SE-ELM when the ICRH is switched off (seen at T=4.215 sec). The amplitude of the SE-ICRH in general is much higher than the SE-ELM signal. Experiments were carried out where a portion of the heterodyned IF signal (via a 3dB coupler) was fed to a Tektronix Digital Phosphor Oscilloscope (model DPO 7104) with a fast 120 Mbyte memory frame grabbing capability and 1 GHz bandwidth. Figure 5(b) shows the spectrogram of one gyro-pulse during an ICRH heated plasma discharge. The figure shows the first 500 μs being attenuated by the Voltage Controlled Variable Attenuator (VCVA) and the frequency chirping of the gyrotron. The spectrogram clearly shows harmonic structures that follow the gyrotron frequency chirp and the drift due to thermal expansion of the gyrotron cavity. The frequency between each harmonic is about 36.5 ± 0.5 MHz which is equal to the ICRH generator frequency. In addition, bursts at ≈ 300 μs intervals shown Figure 5(b) are also observed by the CTS receiver. These harmonic signals around the gyrotron separated by f_{ICRH} were also observed on the CTS at TEXTOR [11]. This suggests a non-linear wave mixing between the gyrotron and the ICRH waves. This interaction is likely to be located in the edge region where parasitic absorption of ICRH waves exists thus creating rectified sheaths on opened field lines [12]. It is not clear yet if the amplitude of SE-ICRH is dependent on the ICRH coupling or other plasma parameters. The nonlinear wave mixing between two very different frequencies is believed to be very weak. However, it is important to note that the CTS detector measures powers that are ~ 10^{10} to 10^{12} times weaker than the power of the gyrotron radiation.

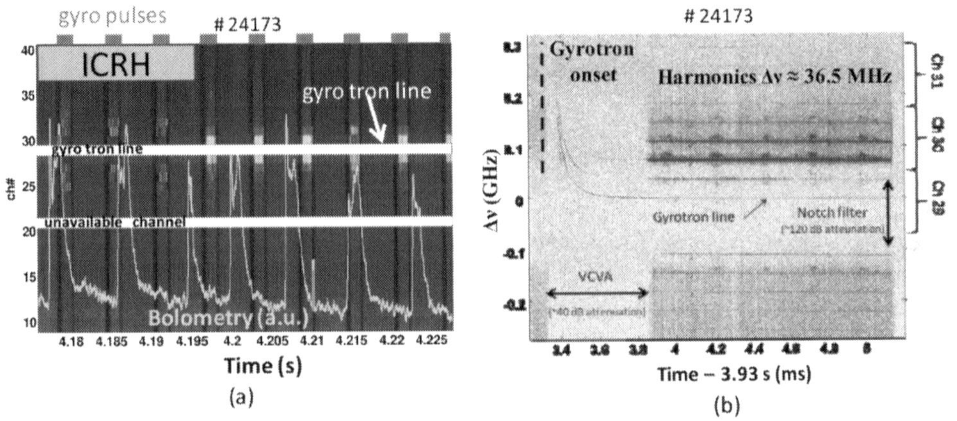

Figure 5. (a) Contour of raw CTS data for H-mode AUG discharge. The ICRH minority hydrogen heated discharge (f_{ICRH} = 36.5 MHz) ends at T=4.195 sec. The ordinate is channel number and abscissa the time. The red rectangles at the top indicate the gyro-on time periods. The green time trace superimposed onto the contour is the bolometer signal. (b) spectrogram of one gyrotron pulse in same discharge during ICRH heating. The spectrogram intensity has not been calibrated using the system's instrument function. The braces on the right axis represent the corresponding channel of the CTS receiver.

4. Signal analysis

4.1. Secondary emission rejection

Each of the SE signals described in section 2 are well constrained in a narrow frequency range, hence it is possible to process the data to attain the fast ion distribution function. The first two types of secondary emissions in section 2 usually affect only a part of a gyro pulse. Hence several methods have been used to remove portions of gyro-pulses that contain secondary emissions. In the most successful method, each gyrotron pulse is divided into smaller parts where the normalized signal fluctuation level, defined as the standard deviation of the signal normalized by the mean for the pulse part, is used to identify and reject those pulse parts which cause outliers in the upper tail of the distribution of normalized fluctuation levels. Evidently tolerances on identifying outliers are not generic and require manual testing and verification. Improvements include using the ECE background of neighbouring data during the gyrotron off period as a normalizing factor. A wavelet approach is being considered.

4.2. Updated ECE background subtraction procedure

In order to remove the ECE background for extraction of the scattered radiation, the ECE background measured during the passive ECE-only periods (gyro-off) is used to predict and remove the background during the gyro periods (gyro-on). The background of a CTS channel is well correlated with the outermost channels of the CTS receiver ($\Delta\nu \approx \pm 5$ GHz) where no scattered signal is measured nor expected. The procedure involved estimating the background of the gyro-on pulse for each channel, by the use of the adjacent gyro-off pulse(s) of the same channel and the use of the signal in the outermost channels during the gyro-on periods. However, the ECE background emanating from an H-mode AUG plasma has a high variability in time due to ELMs and other MHD events making this procedure problematic on AUG - contrary to the relatively quiescent TEXTOR plasma where this approach was very robust [13]. A new approach on AUG was adopted where for each channel, correlation coefficients are created from a variety of data sets during the gyro-off periods and used to predict the channel background during the gyro-on periods. The data used in the correlation coefficients are the outmost channels of the CTS and from other diagnostics such as plasma position, D_α radiation, and plasma density in the pedestal and scrape-off layer. This method was tested on a time window of the CTS receiver that passively viewed the plasma (with no gyrotron pulses), hence only ECE radiation. By assigning chosen time intervals considered as *pseudo-gyro-on* periods, the background of the *pseudo-gyro-off* periods are reconstructed and compared to the actual ECE data. Such tests have shown that the standard deviation of the difference between the two is less than 1 eV and is lower than the standard deviation of a series of spectral power densities from actual data measured during a steady state plasma phase.

5. Conclusion

Fast ion distribution results have been attained using the CTS-AUG. Challenges inherent to the CTS-AUG such as secondary emissions in the raw data necessitated additional study and analysis in order to attain the fast ion information. This was possible due to the non-random nature of the SE signals in frequency (unlike the secondary emissions from the gyrotron itself [2]). In some instances, the exclusion of entire channels was required in the inference procedure. Preliminary studies haves shown that the exclusion of certain channels do not significantly affect the determination of the fast ion distribution function. There is a possibility that a microwave based fast ion CTS diagnostic on ITER [15][16][17] may encounter new SE signals from alpha driven instabilities, but many of the possibilities such as Alfvén eigenmodes would be too low in frequency to effect the fast ion CTS spectra. In the event that new high frequency SE is found, it is likely as shown in Figure 5 that measurements with a high frequency resolution (in conjunction with the CTS receiver) can identify the SE and thus facilitate removal from the SPD data The CTS diagnostic on AUG will become a valuable experimental tool in understanding fast ion physics in confined plasmas. Knowledge gained in achieving the physics and technical objectives of such a system is an important step toward a fast ion CTS diagnostic on ITER.

References

[1] Salpeter E E, 1960 *Phys. Rev.* **1528,** 120
[2] Meo F et al 2008 *Rev. Sci. Instrum.* **79**, 10E501
[3] Bindslev H 1999 Rev. Sci. Instrum. **70** 1093–1099
[4] Salewski M et al, 2010 *Nucl. Fusion* **50** 035012
[5] Herrmann A, Schmid A, Müller H W, Maraschek M, Neuhauser J and the ASDEX Upgrade Team, 2008 *Proceedings of the 22nd IAEA Fusion Energy Conference, Geneva, Switzerland*
[6] Suvorov E V et al, 1998 *Nuclear Fusion*, **38**, No. 5 661-671
[7] Shalashov A G,Suvorov E V, Lubyako L V, Maassberg H and the W7-AS team, 2003 *Plasma Phys. Control. Fusion* **45** 395–412
[8] Karney C F F, September 1978 *Phys. Fluids* **21** (9)
[9] Ryter F et al, 1986 *Proceedings of the 15th European Conference on Controlled Fusion and Plasma Heating, Schliersee* edited by G. Briffod and M. Kaufman (European Physical Society, Petit-Lancy, **10c**, Pt. I, p. 101
[10] Nemoto M et al, 1991 *Phys Rev Lett*, **67**, 70
[11] Woskov P P *et al* 2006 *Rev. Sci. Instrum.* **77** 10E524
[12] Brambilla M at IPP Garching, private communicaiton
[13] Bindslev H et al, 2006 *Phys Rev Lett* **97,** 205005
[14] Nielsen S K et al 2008 *Physical Review E* 77 016407
[15] Meo F et al *Rev.* 2004 *Sci. Instrum.* **75** 3585–3588
[16] Leipold F et al, 2009 *Rev. Sci. Instrum.* **80**, 093501
[17] Salewski M et al, 2009 *Nucl. Fusion* **49** 025006

Laser Thomson Scattering, Raman Scattering and Laser-Absorption Diagnostics of High Pressure Microdischarges

Vincent M Donnelly[1], Sergey G Belostotskiy[1], Demetre J Economou[1] and Nader Sadeghi[2]

[1] Department of Chemical and Biomolecular Engineering, University of Houston, Houston, TX 77204, USA
[2] Laboratoire de Spectrométrie Physique (UMR C5588), Université J. Fourier de Grenoble, B P 87, F-38402 Saint-Martin d'Hères Cedex, France
E-mail: vmdonnelly@uh.edu

Abstract. Laser scattering experiments were performed in high pressure (100s of Torr) parallel-plate, slot-type DC microdischarges operating in argon or nitrogen. Laser Thomson Scattering (LTS) and Rotational Raman Scattering were employed in a novel, backscattering, confocal configuration. LTS allows direct and simultaneous measurement of both electron density (n_e) and electron temperature (T_e). For 50 mA current and over the pressure range of 300 – 700 Torr, LTS yielded $T_e = 0.9 \pm 0.3$ eV and $n_e = (6 \pm 3)\cdot 10^{13}$ cm^{-3}, in reasonable agreement with the predictions of a mathematical model. Rotational Raman spectroscopy (RRS) was employed for absolute calibration of the LTS signal. RRS was also applied to measure the 3D gas temperature (T_g) in nitrogen DC microdischarges. In addition, diode laser absorption spectroscopy was employed to measure the density of argon metastables (1s5 in Paschen notations) in argon microdischarges. The gas temperature, extracted from the width of the absorption profile, was compared with T_g values obtained by optical emission spectroscopy.

1. Introduction

Microdischarge plasmas or microplasmas are non-equilibrium gas discharges operating at pressures of hundreds of Torr and having dimensions of tens to hundreds of micrometers. Their discharge scaling parameter $p{\cdot}L$ (p is gas pressure and L is characteristic dimension) is similar to that of larger plasmas. Microdischarges have many unique applications, such as UV light sources, plasma display panel cells, ozone sources, and biomedical applications.

Microdischarge diagnostics are important for understanding basic microdischarge operation, however, conventional methods (e.g., probes, microwave interferometry) are difficult or impossible due to the small size and high pressure. In contrast, optical diagnostics are convenient for characterization of microdischarges. In this paper, measurement of basic microplasma parameters (i.e., electron density (n_e), electron temperature (T_e), neutral gas temperature (T_g) and density of excited metastable species) using laser spectroscopy is reported. The techniques include Laser Thomson Scattering (LTS), Rotational Raman Scattering (RRS) and Diode Laser Absorption Spectroscopy (DLAS).

2. Experiment

A direct current microplasma source in symmetric slot-type parallel-plate geometry was chosen for investigation [1]. The interelectrode gap could be varied from d = 300 to 600 μm and the electrode surface area was 5 x 0.5 mm^2. In the rotational Raman scattering experiments the discharge was operated in pure N_2 (400–700 Torr) at typical voltages of 360 – 400 V between the electrodes at currents of 5 – 30 mA. In laser Thomson scattering experiments the discharge was in Ar. The discharge current was set at I = 50 mA, resulting in a voltage between the electrodes in the range 310-350 V (as pressured was varied). Higher currents resulted in glow-to-arc transitions. A schematic diagram of the experimental apparatus is presented in Figure 1. The microdischarge was mounted on a holder that was connected to an x-y-z linear stage. This arrangement allowed the discharge to be precisely moved with respect to the stationary laser beam and light collection optics, so that 3D spatially resolved measurements could be carried out.

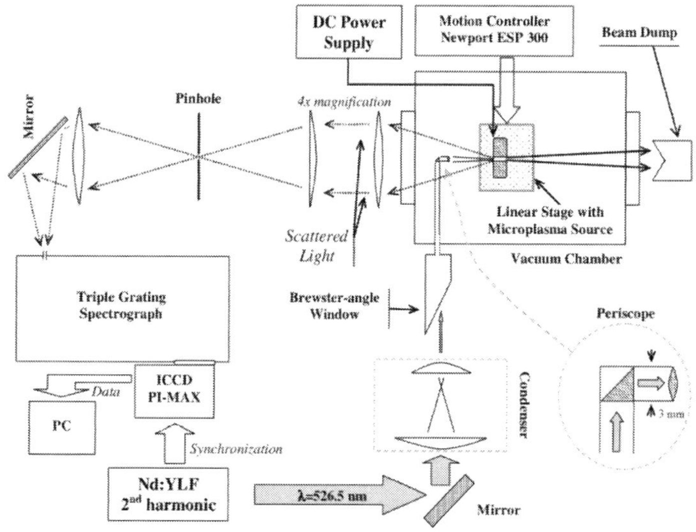

Figure 1. Schematic diagram of the experimental apparatus for laser scattering spectroscopy.

The pulsed, frequency doubled Nd:YLF laser had a wavelength of 526.5 nm, a spectral linewidth of 0.05 nm (FWHM), a pulse frequency of 3 kHz, a pulse duration of 100 ns, and an average power of 6 W. The laser beam was guided into a periscope that focused the beam into the microdischarge with a ~40 μm FWHM diam.

In most laser scattering experiments light is detected at an angle of θ =90° with respect to the laser beam axis [2-5]. In a slot-type microdischarge with a long, deep, and narrow gap, the z-axis [1] is the most appropriate direction of propagation of a tightly focused laser beam, as well as for detection of the scattered light. Therefore, a backscattering geometry (i.e., θ = 180°) was used. Since the laser beam did not pass through the objective lens and window used to view Raman scattering, the background radiation from scattering of the beam was minimized. Scattered light was collected by two lenses and focused on a 100 μm diameter pinhole spatial filter with 4X magnification. The transmitted light was then focused by a third lens (with 2X demagnification) onto the 100 μm entrance slit of the Triple Grating Spectrometer (TGS), similar to one used by van de Sande [2, 3].The TGS provides very efficient spectral filtering of the scattered light at the laser (central) wavelength, while transmitting the light shifted with respect to the laser wavelength. Different combinations of spatial filters and entrance slits were evaluated for rejecting scattered laser light, while maintaining an adequate portion of the Raman spectrum with sufficient resolution. For rotational Raman spectroscopy, a wider spatial filter (750 μm) could be used since the Raman peaks were sufficiently shifted from the laser wavelength. However, the spectral width of the Thomson signal ($\propto \sqrt{T_e}$) is relatively small due to the low electron temperatures (~3.5 nm for T_e = 1 eV), hence, a 500 μm-wide filter was the best compromise. With the laser stray light reduced by a factor ~10^4, the third grating dispersed the light over 10.9 pixels/nm on an intensified charge-coupled device (ICCD). The ICCD was gated ON for 300 ns, synchronized with the laser pulse, to reject most of the background emission from the plasma. Spectra accumulated over typically 100 s ($3 \cdot 10^5$ laser pulses) were recorded and processed by a program that implemented a nonlinear least squares fit to the experimental data. Data points in the spectral range blocked by the spatial filter were not included in the fitting.

The synthetic Raman spectra were a convolution of the theoretical Raman spectrum (100 rotational levels of N_2 were accounted for) with the measured apparatus function [1]. The apparatus function was

Journal of Physics: Conference Series **227** (2010) 012011 doi:10.1088/1742-6596/227/1/012011

represented by a 0.35 nm FWHM Gaussian. A second, broad Gaussian (2.7 nm FWHM) accounted for a relatively small background, perhaps from scattering off the optics or the face of the ICCD.

The synthetic Thomson spectrum was modelled by a Gaussian function, assuming Maxwellian electron energy distribution function (EEDF) [6]. Absolute calibration (required for extracting the absolute electron density n_e) was done using Raman scattering from nitrogen. Before and after a set of experiments, the chamber was evacuated and then filled with pure N_2. The calibration constant was found by fitting the Raman signal to a synthetic spectrum at room temperature.

The apparatus used for tunable diode laser absorption spectroscopy is shown schematically in Figure 2. An External Cavity Diode Laser (ECDL) in the Littman configuration working around λ = 801.5 nm was used as the light source. The width of the laser line (<10 MHz) was much smaller than the typical bandwidth of the absorption line (>1 GHz). Therefore, the absorption profile was recorded by scanning the laser frequency across the absorption line.

The laser beam from the ECDL passed through a beam splitter. Part of the beam was guided to a Fabry-Perot interferometer (1 GHz free spectral range) to perform *in-situ* calibration of the laser frequency. The rest of the beam was attenuated, to <100 μW, by neutral density filters, directed through the interelectrode space of the microdischarge, and then through an aperture. The aperture suppressed emission from the plasma by reducing the collection solid angle. A set of mirrors was used to magnify (11X) and image the laser light on a linear CCD array, which allowed spatially resolved (across the electrodes) measurement in a single acquisition. The experiment was controlled by LabVIEW software.

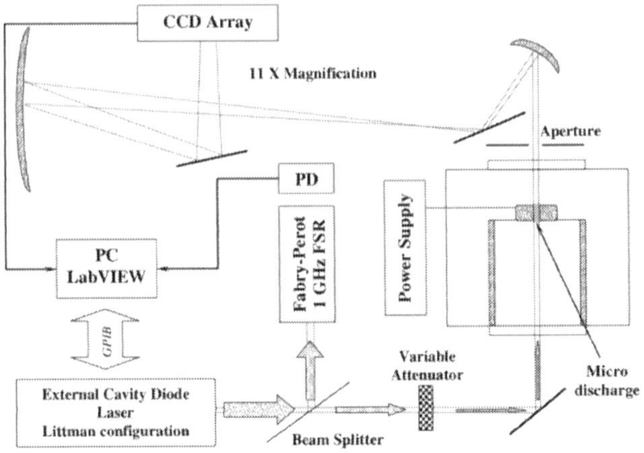

Figure 2. Experimental setup for tunable diode laser absorption spectroscopy.

To avoid possible long-term drift of the laser power or frequency, reference (I_0) and absorption (I) intensities were collected by pulsing the plasma (~1 Hz), and recording I_0 with plasma OFF, and I with plasma ON, for each laser frequency. Such a procedure is important for spatially resolved measurements, since the spatial distribution of the laser intensity across the beam may vary with laser frequency. The transition from the lowest metastable state $Ar(2p_8) \leftarrow Ar(1s_5, {}^3P_2)$ (Einstein A coefficient = $9.28 \cdot 10^6$ s^{-1} [7, 8]) at λ = 801.48 nm was used for absorption measurements.

3. Results and Discussion

Rotational Raman scattering measurements were carried out in a pure N_2 microdischarge with the interelectrode gap of d = 300 μm.

Figure 3 presents typical net spectra (with the background subtracted) recorded with the plasma ON (top) and OFF (bottom). Note that the background dominates only in the region close to the central (laser) wavelength (~3 nm wide), but it is small compared to the signal in the wings of the spectra. The background-dominated region was excluded from the fit. The gas temperature was determined by fitting the experimental spectrum with a synthetic (model) spectrum, with T_g as the adjustable parameter. The net spectrum with the plasma OFF was fit first to obtain the proportionality constant relating the measured signal intensity to the relative computed intensity. The analysis also confirmed that the model correctly predicted the gas temperature which, for plasma OFF, was the ambient temperature. This same proportionality constant was then used to fit spectra recorded with the plasma ON. The good fit to the observed plasma-ON spectra showed that N_2 was the dominant species with a number density that varied as $1/T_g$ at constant pressure, as expected.

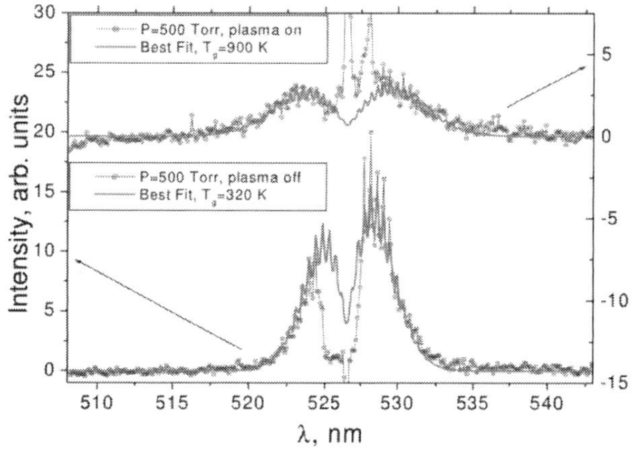

Figure 3. Net (background substracted) Raman spectra and best fits for P = 500 Torr: (Top) – plasma ON (I = 20 mA), best fit T_g = 900 K; (Bottom) – plasma OFF, best fit T_g =320 K (ambient temperature).

Using this approach, spatially resolved T_g profiles in N_2 DC microdischarges were obtained. The results of these measurements can be found in [1].

The LTS measurements were carried out in a pure Ar microdischarge with the interelectrode gap at d = 600 μm. An example of the net (background subtracted) Thomson spectrum is presented in Figure 4. Despite the presence of noise, the Gaussian fit was reasonably good (note that the central part of the spectrum (near the laser line) was suppressed by the mask in the TGS and therefore it was excluded from the fit).

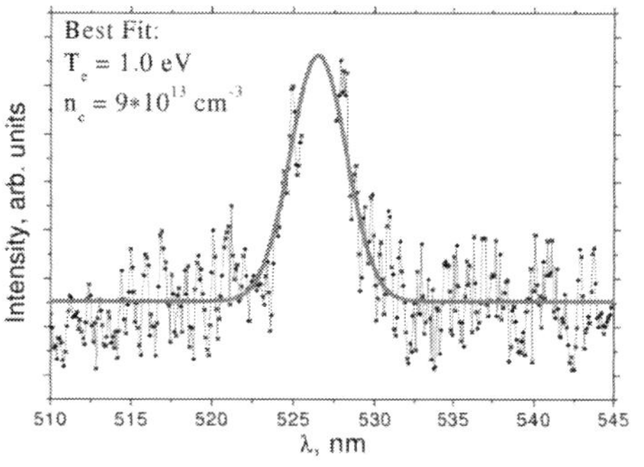

Figure 4. An example of net (background subtracted) Thomson spectrum. The red line is the best fit resulting in T_e=1.0 eV and n_e=9·10^{13} cm^{-3}.

Due to the rather long duration of the laser pulse (~100 ns), gating the ICCD did not provide sufficient suppression of the background plasma emission. The problem of background plasma emission also precluded the use of photon counting to improve the signal-to-noise ratio (the ICCD noise was relatively small compared to the plasma emission). These problems were compensated by signal averaging over a very large number (0.3 million) of laser pulses.

The electron temperature was extracted from the width of the Gaussian fit [6].

The values of n_e and T_e shown in Figure 4 are quite typical for this kind of microdischarge [9, 10] and are consistent with the predictions of a 1-D mathematical model, described in [10].

The n_e data obtained by LTS were compared with those obtained by Optical Emission Spectroscopy (OES) [11] using Stark broadening of Balmer-β emission resulting from H atoms due to the water impurities. Electron densities obtained by both techniques are compatible (see Figure 5).

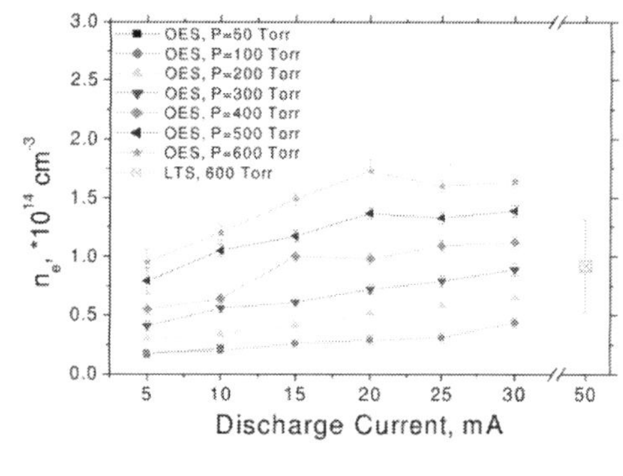

Figure 5. Comparison of n_e measurements obtained by LTS and OES.

Diode Laser Absorption measurements were carried out in a pure Ar microdischarge with an interelectrode gap of d = 300 μm. Before every experiment, the dark current of the CCD array as well as the plasma emission (although it was highly suppressed by the aperture) were recorded. The measured intensity was automatically corrected by subtracting these background signals.

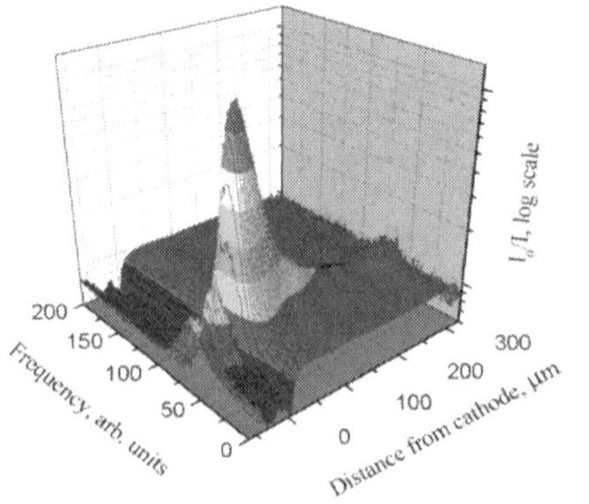

Figure 6. Profile of the ratio I_0/I as a function of interelectrode distance and laser frequency (logarithmic scale).

Figure 6 shows the ratio of transmitted laser intensity with plasma OFF (I_0) and with plasma on (I) – I_0/I (on a logarithmic scale) – as a function of interelectrode distance and laser frequency. One can obtain spatially resolved profiles of the argon metastable density by integrating the profile of Figure 6 with respect to the laser frequency at different spatial locations.

Figure 7 shows the Ar(1s$_5$) density profiles at different currents and pressures. The profiles are consistent with the properties of the cathode sheath and negative glow, where most of metastable production occurs. The peak metastable density increases with increasing pressure, but decreases with increasing discharge current. The latter trend was attributed to the gas heating [12]. As the pressure increases, the Ar(1s$_5$) density gradient becomes stronger and the profile moves closer to the cathode [12].

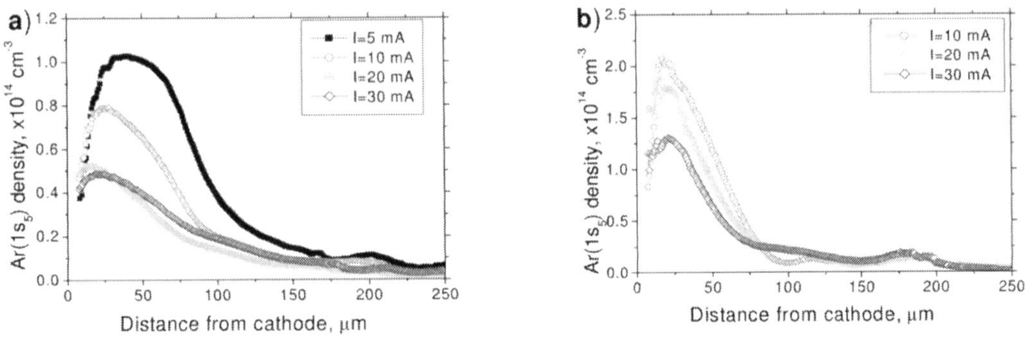

Figure 7. Spatial profiles of argon metastable density (a) at 100 Torr and different currents; (b) at 300 Torr and different currents.

The gas temperature T_g was extracted by analyzing the absorption lineshapes. Experimental profiles were fit to a Voigt function with T_g as the fitting parameter.The results of T_g measurements obtained by absorption spectroscopy are presented in Figure 8. This Figure also shows a comparison of the results obtained by DLAS and those obtained by OES [11]. OES measurements were based on the rovibrational bands of the N_2 first positive system. Here, gas temperatures are compared at the spatial location of the peak argon metastable density (~50 μm from the cathode) and for P = 100 Torr. The same trend with discharge current is observed with both methods. However, the values obtained with OES are slightly higher than those obtained by absorption spectroscopy.

This discrepancy can be explained by the influence of the trace gas admixture on the microdischarge characteristics. Absorption spectroscopy was done in pure argon microdischarges, while OES was done in microdischarges of argon with a trace of N_2. The insert of Figure 8 shows that, even small additions of nitrogen (0.5%), result in higher discharge voltage for the same current (compared to pure argon), and therefore higher power deposition. This may explain the higher T_g values obtained by OES.

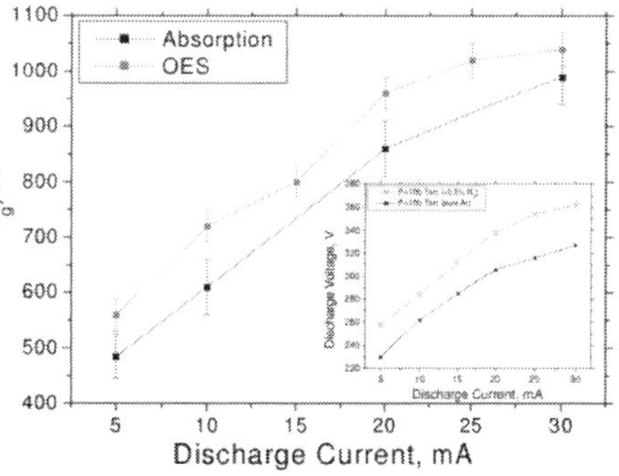

Figure 8. Comparison of T_g measurements obtained by absorption spectroscopy and OES. Insert: I-V characteristics of a pure argon microdischarge (black) and of an argon microdischarge with 0.5% of N_2 (red); P = 100 Torr.

4. Conclusions

Parallel plate slot-type DC microdischarges were characterized using laser spectroscopy. Laser Thomson Scattering (LTS) was employed for simultaneous direct measurements of electron density (n_e) and electron temperature (T_e) in argon microdischarges. n_e values obtained by LTS were consistent with those found by Optical Emission Spectroscopy. Diode Laser Absorption Spectroscopy (DLAS) was used for spatially resolved measurements of argon metastable Ar($1s_5$) density in argon microdischarges. The analysis of the absorption lineshapes also provided the gas temperature. By fitting the absorption lineshapes to Voight functions, the gas temperature was extracted as the fitting parameter. The values of T_g obtained by DLAS were found to be in reasonable agreement with those obtained by OES, showing the same increasing trend of the gas temperature with current.

The authors thank the Department of Energy (DE-FG02-03ER54713), and the University of Houston GEAR program for financial support of this work.

References

[1] Belostotskiy S G, Wang Q, Donnelly V M, et al. 2006 *Applied Physics Letters* **89**
[2] van de Sande M J and van der Mullen J J A M 2002 *Journal of Physics D: Applied Physics* **35** 1381-1391
[3] van de Sande M J 2002 "Laser Scattering on Low Temperature Plasmas" (Eindhoven: Eindhoven University of Technology).
[4] Kono A and Nakatani K 2000 *Review of Scientific Instruments* **71** 2716-2721
[5] Muraoka K, Uchino K and Bowden M D 1998 *Plasma Physics and Controlled Fusion* **40** 1221-1239
[6] Belostotskiy S G, Khandelwal R, Wang Q, et al. 2008 *Applied Physics Letters* **92**
[7] Wiese W L, Brault J W, Danzmann K, et al. 1989 *Physical Review A* **39** 2461-2471
[8] "NIST: Atomic Spectra Database".
[9] Noguchi Y, Matsuoka A, Bowden M D, et al. 2001 *Japanese Journal of Applied Physics Part 1-Regular Papers Short Notes & Review Papers* **40** 326-329
[10] Wang Q 2006 "Plasma Diagnostics and Modelling of Direct Current Microplasma Discharges at Atmospheric Pressure" (Houston: University of Houston, Department of Chemical and Biomolecular Engineering).
[11] Belostotskiy S G, Ouk T, Donnelly V M, et al. *Journal of Applied Physics* 2010 **107** accepted for publication
[12] Belostotskiy S G, Donnelly V M, Economou D J, et al. 2009 *IEEE Transactions on Plasma Science* **37** 852-858

Diode laser cavity-ringdown absorption spectroscopy for measuring $N_2(A^3\Sigma_u^+)$ density in plasmas

Y Horikawa[1], **K Kurihara**[2] **and K Sasaki**[3]

[1]Department of Electrical Engineering and Computer Science, Nagoya University, Nagoya 464-8603, Japan
[2]Research and Development Center, Toshiba Corporation, Yokohama 235-8522, Japan
[3]Plasma Nanotechnology Research Center, Nagoya University, Nagoya 464-8603, Japan

E-mail: sasaki@nuee.nagoya-u.ac.jp

Abstract. We developed a system of cavity-ringdown absorption spectroscopy employing a cw diode laser for measuring the absolute density of $N_2(A^3\Sigma_u^+)$ in plasmas. We achieved a sensitive detection limit of 10^{-7} for the absorbance. The saturation of absorption was avoided simply by switching off the laser beam when the cavity length was detuned slightly from the length corresponding to the perfect resonance. The absolute $N_2(A^3\Sigma_u^+)$ density was deduced from the absorbance of the $B^3\Pi_g(v'=2)-A^3\Sigma_u^+(v"=0)$ band by comparing the absorption spectrum with spectral simulation, where we assumed the same values for the translational and rotational temperatures.

1. Introduction

The metastable $A^3\Sigma_u^+$ state of molecular nitrogen plays important roles in discharge kinetics as well as plasma-surface interaction in material processing. We are interested in plasma-aided nitriding of silicon for synthesizing silicon nitride films with high dielectric constants, which are necessary for gate-insulation films in metal-oxide-semiconductor field-effect transistors [1]. The active species which play the principal role for surface nitriding have not been identified yet, and $N_2(A^3\Sigma_u^+)$ is a candidate for the nitriding precursor. The difficulty in the experimental investigation of the nitriding precursor is due to the difficulty in the measurement of the $N_2(A^3\Sigma_u^+)$ density. Although the $N_2(A^3\Sigma_u^+)$ density is measured by laser-induced fluorescence [2, 3], the calibration of the absolute density is not an easy task. On the other hand, the measurement by simple absorption spectroscopy is not sensitive because of the small transition probability of the first positive system. Although intracavity laser absorption spectroscopy is a sensitive absorption method [4], its applicability is limited by the geometry of the experimental apparatus. In this work, we developed a system of cavity-ringdown absorption spectroscopy (CRDS) [5, 6] for measuring the $N_2(A^3\Sigma_u^+)$ density in an inductively-coupled plasma source.

2. Experimental apparatus

A schematic illustration of the experimental apparatus is shown in figure 1. Inductively-coupled plasmas (ICP) were produced in a cylindrical vacuum chamber with a diameter of 30 cm. The vacuum chamber had a quartz window at the top, and a one-turn antenna and a matching circuit were placed above the quartz window. An rf power supply at 13.56 MHz was connected

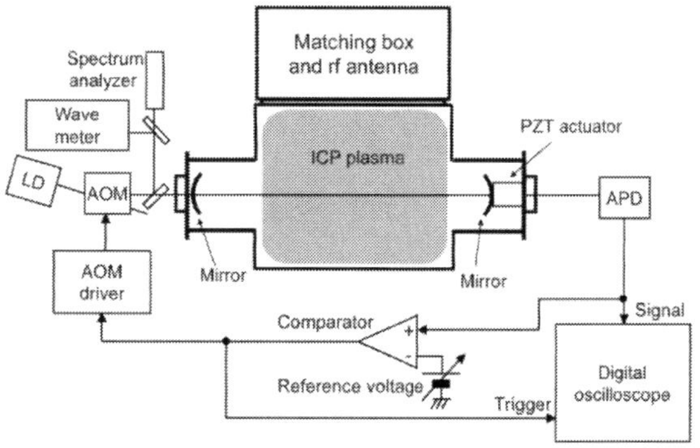

Figure 1. Schematic illustration of the system of cavity-ringdown absorption spectroscopy installed in an inductively-coupled plasma source.

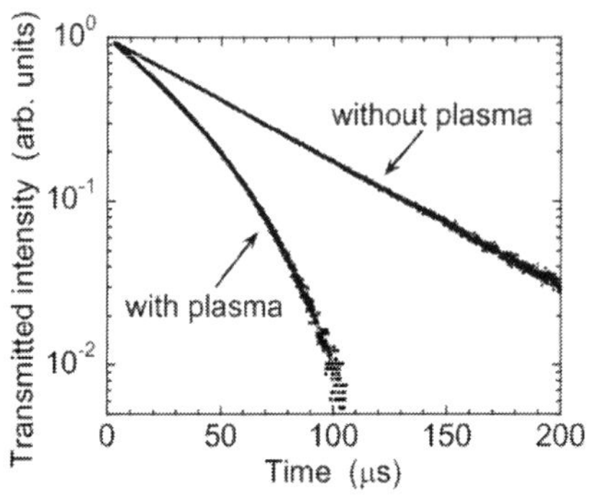

Figure 2. Decay curves of the transmitted laser intensity with and without the plasma. The laser beam was switched off when the cavity length was tuned to the almost perfect resonance.

to the matching circuit. Two long tubes were attached to the vacuum chamber, and the plasma was placed inside a cavity consisting of two concave mirrors with high reflectivities. The cavity length was 1 m, and the radii of the cavity mirrors were 3 mm. The distance between the quartz window and the optical axis of the cavity was 15 cm. A piezo-electric transducer supported one of the cavity mirrors. All the optics including the mirror holders were installed inside the chamber. The mirror holders had motor-drive systems, and the angle of the mirrors were adjusted from the outside of the chamber.

The light source was a cw diode laser. The diode laser beam was diffracted using an acousto-optic modulator (AOM). The wavelength of the diffracted laser beam was tuned around 771.1 nm, which corresponded to the $B^3\Pi_g(v'=2) - A^3\Sigma_u^+(v"=0)$ absorption band of N_2 (the first positive system). The wavelength was monitored using a wave meter and a spectrum analyzer. The diffracted laser beam was injected into the cavity, and the intensity of the laser beam transmitted through the cavity was measured using an avalanche photo diode (APD). When the cavity length was modulated using the piezo-electric transducer, the signal from APD gave us the transmission curve of the Fabry-Perot interferometer. The resonance between the cavity length and the wavelength of the laser beam was detected using a comparator with a reference voltage. The output of the comparator was used for stopping the laser beam by switching off the AOM device. The temporal decay of the transmitted laser intensity after the truncation was recorded in a digital oscilloscope. The absorption spectrum was obtained by repeating the measurement with sweeping the laser wavelength.

3. Results and discussion

The absorbance in CRDS is given by

$$\alpha l = \left(\frac{1}{\tau} - \frac{1}{\tau_0}\right)\frac{L}{c}, \tag{1}$$

where $1/\tau$ and $1/\tau_0$ are the decay frequencies of the transmitted laser intensities with and without the plasma, respectively, α is the absorption coefficient, l and L are the lengths of the plasma and the cavity, respectively, and c is the speed of light. Figure 2 shows an example of the temporal decay of the transmitted laser intensity observed in the empty cavity. This decay

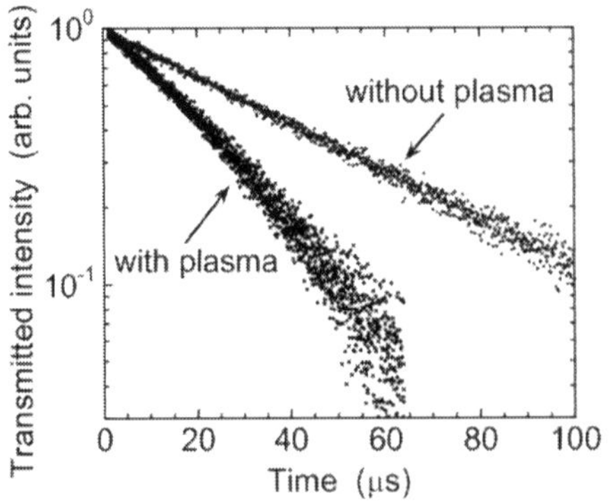

Figure 3. Decay curves of the transmitted laser intensity with and without the plasma. The laser beam was switched off when the cavity length was slightly detuned from the perfect resonance.

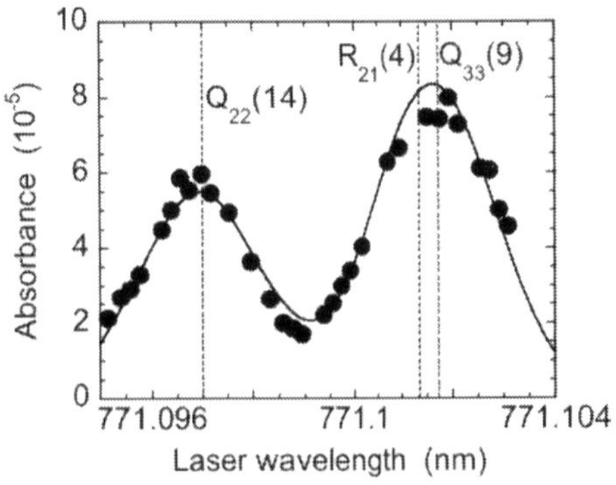

Figure 4. Absorption spectrum observed in a nitrogen plasma at an rf power of 300 W and a pressure of 60 mTorr.

curve was obtained by switching off the diode laser beam when the cavity length was tuned to the almost perfect resonance (*i.e.*, the transmitted laser intensity was almost the maximum). The laser wavelength was 771.1016 nm, which corresponded to the overlap of the $R_{21}(4)$ and $Q_{33}(9)$ absorption lines. The decay curve observed in the empty cavity was fitted well with an exponential function, and the decay frequency was evaluated to be $1/\tau_0 = 1.756 \times 10^4$ s^{-1}, corresponding to 99.997% for the reflectivities of the mirrors. We repeated the measurement of the decay frequency in the empty cavity for 10 times. As a result, the scattering of the decay frequency, which was evaluated by the standard deviation of the 10 measurements, was 30 s^{-1}. Therefore, the CRDS system developed in this work was compatible with a small absorbance of 10^{-7}.

We must be careful about saturation of absorption in cw laser CRDS because of the strong laser field inside the cavity. A decay curve in the presence of a nitrogen plasma is also plotted in figure 2. The rf power and the gas pressure were 300 W and 20 mTorr, respectively. As shown in figure 2, the decay curve in the presence of the plasma was not fitted with an exponential function. The decay frequency was accelerated with time. The small decay frequency in the initial part is attributed to the saturation of absorption due to the strong laser field inside the cavity. Since the laser intensity decreased with time, the saturation of absorption was moderated in the latter half of the decay curve, resulting in the acceleration of the decay frequency.

We used a simple way for avoiding the saturation of absorption in this work. Figure 3 shows decay curves with and without the plasma, when the reference voltage to the comparator was 30% of that in figure 2. In this case, the cavity length at the truncation was detuned slightly from the perfect resonance. In the case of the slightly detuned truncation, as shown in figure 3, the decay curve in the presence of the plasma was roughly fitted with an exponential function because of the weak laser intensity inside the cavity. We operated the CRDS system by adjusting the reference voltage to the comparator in order to avoid the saturation of absorption.

Figure 4 shows an absorption spectrum observed in a nitrogen plasma at an rf power of 300 W and a pressure of 60 mTorr. The absorbance observed experimentally was below 10^{-4} as shown by the solid symbols in the figure. There were three absorption lines of $Q_{22}(14)$, $R_{21}(4)$, and $Q_{33}(9)$ in the wavelength range shown in figure 4. The locations of the absorption

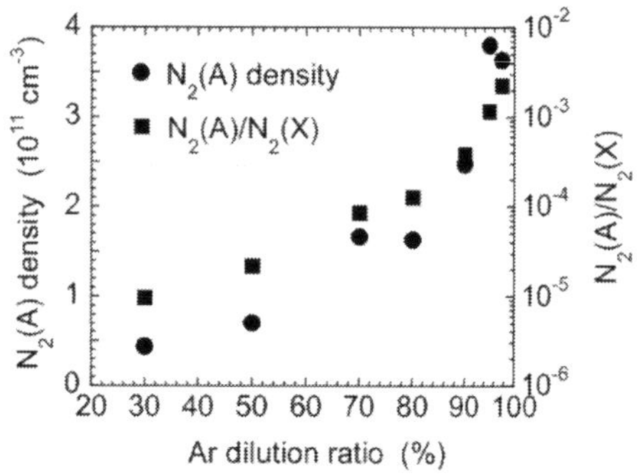

Figure 5. Absolute density of $N_2(A^3\Sigma_u^+)$ in N_2/Ar mixture plasmas as a function of the percentage of Ar. The rf power and the total pressure were 300 W and 100 mTorr, respectively. The ratio of the $N_2(A^3\Sigma_u^+)$ density to the density of ground-state molecular nitrogen is also plotted.

lines are indicated by the broken lines in the figure. The solid curve shows the result of a spectral simulation, where we assumed the same values for the translational and rotational temperatures of $N_2(A^3\Sigma_u^+)$. As shown in figure 4, the experimental spectrum was fitted well by the spectral simulation. From the comparison between the experimental and simulated spectra, we evaluated the absolute density and the temperature of $N_2(A^3\Sigma_u^+)$ to be 1.1×10^{11} cm^{-3} and 740 K, respectively.

We have reported the absolute $N_2(A^3\Sigma_u^+)$ densities in ICP nitrogen plasmas in a separate paper [7]. In this paper, we report the absolute $N_2(A^3\Sigma_u^+)$ density in nitrogen-argon gas mixture plasmas. Figure 5 shows the absolute $N_2(A^3\Sigma_u^+)$ density as a function of the percentage of Ar. The rf power and the total pressure were 300 W and 100 mTorr, respectively. It was interesting that the $N_2(A^3\Sigma_u^+)$ density increased with the percentage of Ar, in spite of the decrease in the partial pressure of N_2. Accordingly, the ratio of the $N_2(A^3\Sigma_u^+)$ density to the density of ground-state molecular nitrogen increased significantly with the percentage of Ar. The significant increase is possibly attributed to the energy transfer from Ar metastable atoms to N_2. Further investigations are necessary to understand the kinetics of $N_2(A^3\Sigma_u^+)$ in nitrogen-argon mixture plasmas.

4. Conclusions

We have developed a CRDS system employing a cw diode laser for measuring the absolute $N_2(A^3\Sigma_u^+)$ density in plasmas. The system has a high sensitivity corresponding to an absorbance of 10^{-7}. We have succeeded in evaluating the absolute $N_2(A^3\Sigma_u^+)$ density from the absorption spectrum of the $B^3\Pi_g(v' = 2) - A^3\Sigma_u^+(v" = 0)$ absorption band. We measured the absolute $N_2(A^3\Sigma_u^+)$ densities in nitrogen-argon gas mixture plasmas and found that the $N_2(A^3\Sigma_u^+)$ density increased significantly with the percentage of Ar. The CRDS system and the measurement results will be utilized for investigating the principal precursor for nitriding of silicon surfaces.

Acknowledgments

The authors are grateful to Nader Sadeghi for his valuable advices about the CRDS system.

References

[1] Sekine M, Saito Y, Hirayama M and Ohmi T 1999 *J. Vac. Sci. Technol. A* **17** 3129
[2] De Benedictis S, Dilecce G and Simek M 1993 *Chem. Phys.* **178** 547
[3] Kramesa B, Glenewinkel-Meyer Th and Meichsner J 2001 *J. Appl. Phys.* **89** 3115.
[4] Foissac C, Campargue A, Kachanov A, Supiot P, Weirauch G and Sadeghi N 2000 *J. Phys. D: Appl. Phys.* **33** 2434
[5] Busch K W and Busch M A ed. 1999 *Cavity Ringdown Spectroscopy* (Washington, DC: American Chemical Society)
[6] Hancock G, Peverall R, Richie G A D and Thornton L J 2006 *J. Phys. D: Appl. Phys.* **39** 1846
[7] Horikawa Y, Kurihara K and Sasaki K submitted to Jpn. J. Appl. Phys.

Development of terahertz gyrotrons and their application to CTS on LHD

T Saito[1], T Notake[2], Y Tatematsu[1], A Fujii[1], S Ogasawara[1], La Agusu[1], T Idehara[1], S Kubo[3], T Shimozuma[3], K Tanaka[3], M Nishiura[3], K Kawahata[3], V N Manuilov[4]

[1] FIR FU, University of Fukui, 3-9-1 Bunkyo, Fukui-shi, 910-8507, Japan
[2] Riken, Sendai, Japan
[3] National Institute for Fusion Science, 322-6 Oroshi-cho, Toki-shi, 509-5292, Japan
[4] Nizhny Novgorod State University, Nizhny Novgorod, 603600, Russia

saitot@fir.u-fukui.ac.jp

Abstract. Development of a high-power, sub terahertz pulse gyrotron has started in FIR FU for application to CTS on LHD. A new electron gun was designed to produce a laminar electron beam with a good quality. Second harmonic $TE_{6,5}$ and $TE_{8,5}$ modes were selected as oscillation modes well isolated from other competing modes. Single mode oscillation has been confirmed for both $TE_{6,5}$ and $TE_{8,5}$ modes. The maximum power is larger than 50 kW for $TE_{6,5}$ mode (0.349 THz) and about 40 kW for $TE_{8,5}$ mode (0.390 THz). These powers are new records of second harmonic gyrotron in this frequency range. Feasibility study of CTS with a 0.4 THz pulse gyrotron from a high density plasma in LHD was carried out. The CTS condition is satisfied for a wide operation regime and scattering angles large enough for good spatial resolution. Ray tracing calculation shows that the launched scattered beams are propagated almost straight. CTS spectra calculated with a newly developed code indicates that a large signal to noise ratio can be obtained against ECE for use of a 100 kW gyrotron.

1. Introduction

Recently, application of high frequency gyrotrons as high power sources in the sub terahertz region has been spreading to various fields. Among those, development of a sub terahertz high power pulse gyrotron as a power source of collective Thomson scattering (CTS) diagnostics on high density plasmas is a challenging task. We have started this study for CTS on the large helical device (LHD).

For a power source of CTS, fusion grade gyrotrons with frequencies from tens of GHz to 140 GHz are considered and experiments with these gyrotrons have started [1]. However, electromagnetic waves with these frequencies suffer from a strong plasma dispersion effect when they are applied to a high density plasma. Moreover, high level background electron cyclotron emission (ECE) is a large noise source. Use of a sub-terahertz gyrotron will resolve these problems. In advance of gyrotron development, the condition of CTS was examined. Figure 1 shows the region in which the CTS condition $\alpha = 1/k\lambda_D > 1$ holds in space of scattering angle and wavelength of the probe beam [2]. When we set the gyrotron frequency at 0.4 THz, the CTS condition is satisfied for a sufficiently large scattering angle over a wide operation range of LHD; N_e up to several times 10^{20} m^{-3} and T_e in the range of several keV. This frequency is larger than the fifth harmonic electron cyclotron frequency at

2.75 T and background ECE as well as plasma dispersion effects can be substantially reduced. On the other hand, difficulty of gyrotron development increases with frequency and CTS power decreases with frequency. Therefore, 0.4 THz is an appropriate choice. The target power is set to be 100 kW at the moment.

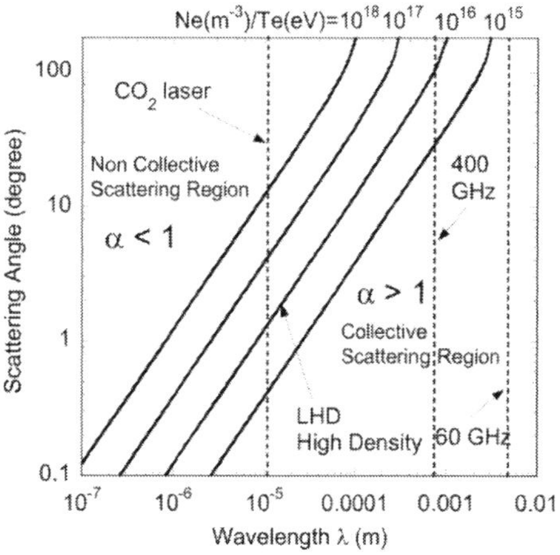

Figure 1. Collective scattering region is shown for the wavelength of the power source. Parameter is the ratio of the electron density in m^{-3} to the electron temperature in eV.

Research center for development of far-infrared region, University of Fukui (FIR FU) is developing high frequency gyrotrons up to 1 THz [3]. We have started the challenging task of development of a high-power, sub terahertz pulse gyrotron [2, 4]. In this paper we report the performance of the developed pulse gyrotron. Preliminary feasibility study of CTS on LHD with a 0.4 THz pulse gyrotron is also presented.

2. Development of sub terahertz high power pulse gyrotron

The design of the developed gyrotron consists of three aspects. Use of a newly designed electron gun is the first aspect [5]. This electron gun generates a laminar electron beam with a good quality for the beam voltage V_k of 60 kV, the beam current I_b of 7 A and those of 70 kV, 10 A. The second aspect is selection of the oscillation modes that are well isolated from other competing modes. In particular, isolation from fundamental modes is very important for a second harmonic mode to oscillate as a single mode. We have searched many candidate modes and finally selected two modes $TE_{6,5}$ and $TE_{8,5}$. The electron beam optics is the third aspect and it is closely connected with the mode selection because the electron beam radius R_b at the cavity should coincide with the peak radius of the coupling coefficient between the electron beam and the wave field. In this design the second peak of the coupling coefficient is used owing to limitation of the electron beam optics. The cavity dimensions are 2.99 mm in radius and 12 mm in length of the straight section [6]. The resonance frequency is 350 GHz for the $TE_{6,5}$ mode and 392 GHz for the$TE_{8,5}$ mode, respectively. A theoretical model [7] predicts more than 30% perpendicular efficiency even for the second peak coupling and the oscillation power of about 80 kW for the pitch factor of the electron beam of 1.3.

Figure 2 (a) shows oscillation intensities as functions of the magnetic field at the cavity with V_k as a parameter. The beam current is fixed at 5 A. Figure 2 (b) indicates oscillation intensities measured through a high pass filter with the cutoff frequency of 303 GHz. Then, peaks in figure 2 (b) indicate second harmonic oscillation. The modes were identified as the $TE_{6,5}$ and $TE_{8,5}$ modes respectively from frequency measurement with a Fabry-Perot interferometer. In the lower field side to the $TE_{8,5}$ mode peak, the $TE_{17,2}$ mode may oscillate. Single mode oscillation was checked for each mode by using the Fabry-Perot interferometer without the high pass filter. Single mode oscillation of the two modes was confirmed for operation up to V_k = 60 kV and I_b = 11 A.

Oscillation power was measured with a water load installed just outside of the vacuum window. The present gyrotron has no internal mode convertor and the oscillation power is transmitted to the vacuum window through a straight section with an up taper. The normal pulse width for power measurement is 1 μs and the repetition frequency is 10 Hz. Beam current dependence of the output power of the $TE_{6,5}$ and $TE_{8,5}$ modes were measured. The beam voltage was set at around 57 kV. Since the oscillation power of each mode depends on R_b, the gun coil current was adjusted at each value of I_b. Single mode oscillation was confirmed for each data point. The output power increases with I_b although it rather saturates for I_b larger than 8 A. The maximum power so far obtained is larger than 50 kW for $TE_{6,5}$ mode (0.349 THz) and about 40 kW for $TE_{8,5}$ mode (0.390 THz) [8]. The measured values were corrected considering the transmission coefficients of the vacuum window and the water loads. According to the reported power of second harmonic gyrotrons [9], these values are new records in this frequency range.

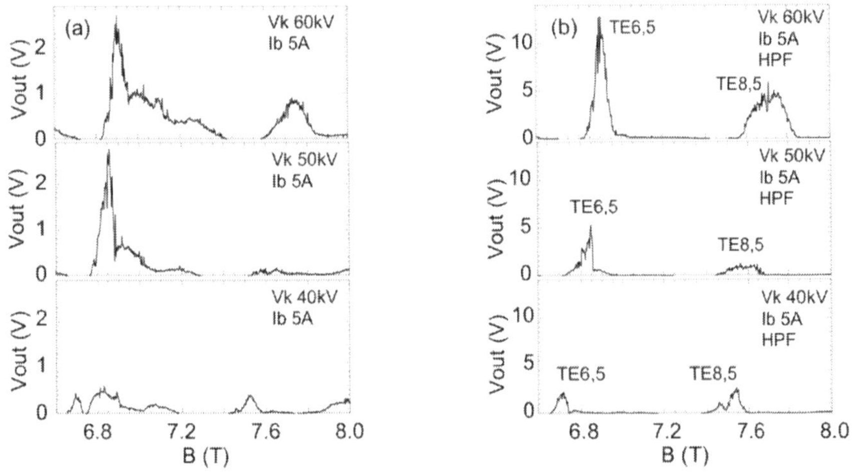

Figure 2. Oscillation intensities are plotted as function of the magnetic field at the cavity with the beam voltage as a parameter. (a) Oscillation intensities were measured at an open end of a waveguide connected to the output window of the gyrotron. (b) A high pass filter with the cutoff frequency of 303 GHz was inserted in between the waveguide end and the detector.

The experimentally obtained maximum total efficiency is 14 % for the $TE_{6,5}$ mode at I_b of 5 A. This value is lower than that expected from calculation with the theoretical model. Moreover, the efficiency further decreases for larger value of I_b. One possibility of the low efficiency is degradation of the quality of the electron beam. Indeed, V_k of 60 kV is rather low for large beam current. Improvement of the beam quality needs higher value of V_k. However, the available voltage is limited at about 60 kV at the moment. We have designed and fabricated a new gyrotron aiming at 100 kW around 0.4 THz with second harmonic oscillation by further improvement of the operating mode and the cavity design. Experiment of this gyrotron has just started. Efforts for higher cathode voltage operation will be done.

3. Feasibility study of collective Thomson scattering on LHD

Two available ports on the LHD vacuum vessel are tentatively chosen for injection and receiving antennas. For this choice, scattering angle θ_s is 104 degrees, which is large enough to obtain high spatial resolution. Beam radius is calculated assuming a Gaussian beam. When the beam radius is focused to $w_0 = 1$ cm at the scattering point, the radius w_i at the injection port is 5 cm for 0.4 THz. This value is small enough that the antenna can be installed without difficulty.

Next, ray tracing calculations are carried out. A probe beam with frequency of 0.4 THz propagates through a plasma of the peak electron density of 10^{20} m^{-3} and reaches the center of plasma without refraction. Waves scattered from different points along the incident beam also reach the receiving antenna port straight.

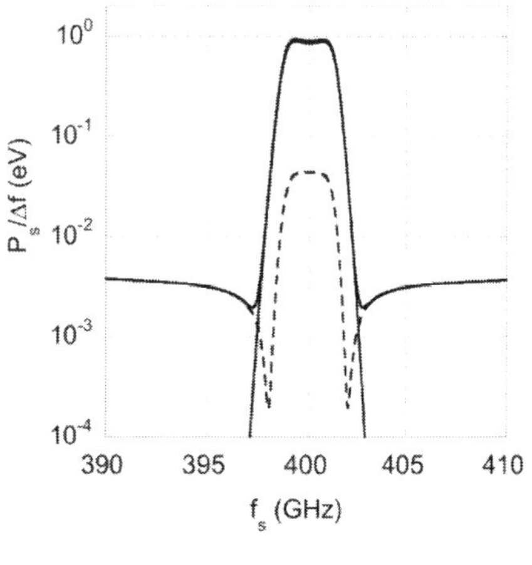

Figure 3. Scattering power spectrum in unit band frequency for T_e = Ti = 2 keV, N_e = 10^{20} m^{-3}. Solid and dashed lines indicate power spectra due to motions of electrons collective and non collective with ions, respectively.

Scattering power spectrum observed at the receiving antenna is evaluated as [10]

$$P_s = P_{in} N_e r_e^2 \Gamma S \Delta f \frac{\lambda_i^2}{\sqrt{\pi} w_0 \sin\theta_s}$$

where P_{in} is the gyrotron power, N_e is the electron density, r_e is the classical electron radius, λ_i is the wavelength of the incident wave, Δf is the detection band width, Γ is the geometrical factor [11] and S is the scattering form factor. An example of power spectrum is show in figure 3, where P_{in} = 100 kW, N_e = 10^{20} m^{-3} and w_0 = 1 cm are assumed. Solid and dashed lines plot the power spectra stemming from electron motion collective with ions and thermal electron motion, respectively. Collective scattering is dominant and the spectrum of P_s gives us ion information. The calculated power per unit frequency band is about 1.0 eV. The background ECE level calculated with a radiation transfer code is less than 0.1 eV for the peak electron temperature of 2 keV [4]. With the pulse width of 100 μs and 100 GHz detection band width, a signal to noise ratio of the order of 100 is expected for the 2.75 T standard operation of LHD.

Thus, 0.4 THz, 100 kW gyrotron is one of promising candidates of power source for CTS from high density plasma in LHD. However, the necessary power should be determined from the total sensitivity of the CTS diagnostic system.

References

[1] Bindslev H et al., 2006 *Phys. Rev. Lett.* **97** 205005
[2] Notake T et al., 2008 *Rev. Sci. Instrrum.* **79** 10E732
[3] Bratman V et al., 2009 *IEEE Trans. Plasma Sci.* **37** 36
[4] Saito T et al., 2008 *Conference Digest of IRMMW and THz*, W5D28
[5] Manuilov V N et al., 2008 *Int. J. Infrared and Milli. Waves* **27** 1103
[6] Notake T et al., 2009 *Plasma and Fusion Res.* **4** 011
[7] Danly B G and Temkin R J 1985 *Phys. of Fluids* **29** 561
[8] Notake T et al., submitted to *Phys. Rev. Lett.*
[9] Thumm M 2008 *Wissenschaftliche Berichte FZKA7392, Research report of Research Center, Karlsruhe Universit*
[10] Sheffield J 1975 *Plasma scattering of electromagnetic radiation* (Academic Press)
[11] Bretz N 1987 *J. Plasma Phys.* **38** 79

Journal of Physics: Conference Series **227** (2010) 012014 doi:10.1088/1742-6596/227/1/012014

Initial result of collective Thomson scattering using 77 GHz gyrotron for bulk and tail ion diagnostics in the Large Helical Device

M Nishiura[1] , S Kubo[1], K Tanaka[1], N Tamura[2], T Shimozuma[1], T Mutoh[1], K Kawahata[1], T Watari[1], T Saito[3], Y Tatematsu[3], T Notake[4] and LHD experiment group[1]

[1]National Institute for Fusion Science, 322-6 Oroshi, Toki, Gifu 509-5292 Japan
[2]Nagoya University, Furo-cho, Chikusa-ku, Nagoya, 464-8601 Japan
[3]FIR FU, University of Fukui, 3-9-1 Bunkyo, Fukui, 910-8507 Japan
[4]RIKEN, 519-1399 Aoba, Aramaki, Aoba-ku, Sendai, 980-0845 Japan

E-mail: nishiura@nifs.ac.jp

Abstract. The collective Thomson scattering (CTS) technique has been utilized with the backscattering configuration in the collective scattering regime to diagnose the velocity distribution functions in the Large Helical Device (LHD). The receiver was equipped with 16 channels and the first test has been carried out using the eight channels for scattered radiation and these channels cover a few GHz frequency shift from the 76.95 GHz probe beam. During the discharge, the electron density and temperature at the central region of the LHD are $1\times10^{19}m^{-3}$, and 1.0 keV, respectively. The probing beam with rectangular wave modulation is injected by 50 Hz in order to be distinct from the background electron cyclotron emission (ECE). The scattered radiation is resolved successfully at each channel of CTS receiver system. The detected signals of bulk ion and electron components are by a factor of 10 ~ 10^2 larger than the background ECE signal. We found that the measured spectra are in reasonably agreement with the theoretical spectra calculated by using the reliable measured electron temperature and density for input parameters. The CTS receiver system will be improved to obtain more accurate velocity distributions in high temperature plasmas.

1. Introduction

Confined energetic particle diagnostics is a major concern to thermo nuclear fusion for maintaining and controlling the self burning. One of the strong tools considered is a collective Thomson scattering diagnostic. Recent developments of powerful radiation sources, like mega watt class gyrotrons, made it possible to detect the scattered electromagnetic waves by fluctuations in plasmas [1, 2, 3]. For the bulk and tail ion temperature measurements in the LHD, the collective Thomson scattering (CTS) diagnostic has been proposed using the existing electron cyclotron resonance heating (ECRH) by gyrotron with the frequency of 77 GHz. The CTS is also assessed numerically using the injector and receiver geometry of the LHD with accessible plasma parameters[4]. The results show that with the CTS technique is possible to obtain useful information on both the thermal and fast ions from the scattered radiation. However the careful analysis is still required, because it is found that the shape and radiation power of scattered spectra depend strongly on the geometrical and experimental conditions. In addition, electron cyclotron emission (ECE) should be taken into account as a background noise for

CTS diagnostic. The ECE levels in some scattering geometries and magnetic configurations are estimated from the calculation on the line of sight [5]. The study found that the ECE noise becomes the minimum for O-mode radiation at the magnetic field of 2.2 Tesla. Even though the ECE level at 2.75 Tesla is relatively high compared with that of 2.2 Tesla, the preliminary results in a LHD discharge, and found a more effective CTS signal in a LHD discharge.
In this paper we show the progress in calculated spectra based on the density fluctuation and the estimated output signals by using the receiver configuration with 8 channels. The output signals estimated here are compared with those of the measured scattered radiations. These results are discussed, and the receiver system will be improved toward the forthcoming LHD campaign in autumn on 2009.

2. Collective region in LHD

The CTS diagnostic requires the spectrum analysis of the fluctuation wave vector $\boldsymbol{k}^{\delta}$ to obtain the thermal and fast ion distribution functions with the relation $\boldsymbol{k}^{\delta} = \boldsymbol{k}^{s} - \boldsymbol{k}^{i}$. Here $\boldsymbol{k}^{s}$and $\boldsymbol{k}^{i}$ are the incident and the scattered radiations, respectively. Since the existing ECRH system is utilized as an initial trial, the port locations for probing and receiving beams are limited to the backscattering and near perpendicular to the toroidal magnetic field. Figure 1 shows the collective scattering region, where the electron and the ion distribution functions are distinguished from the scattered radiations. The Salpeter parameter $\alpha = 1/|\boldsymbol{k}^{\delta}|\lambda_D = 1/(2\,|\boldsymbol{k}^{i}|\,\lambda_D \sin(\theta/2))$ as a function of scattering angle for 77 GHz probing radiation at different plasma parameters. Here λ_D is the Debye length. The Salpeter parameter should be larger than a unity. In any case, we found that α is more than a unity for the scattering angle from 0 to π radians in LHD discharges with the electron temperature of 1~10 keV and the density of $5\mathrm{x}10^{18}$~10^{20} m^{-3}.

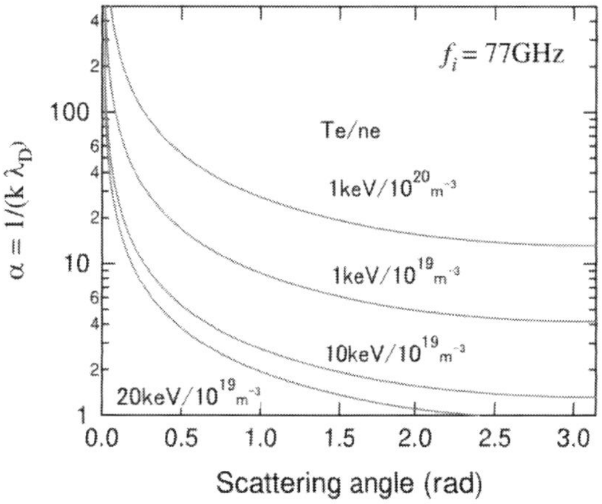

Figure 1. The Salpeter parameter α as a function of scattering angle for CTS diagnostics in LHD. Gyrotron frequency of 77 GHz is used.

3. Scattering geometries in LHD

The CTS diagnostics in LHD has the potential to give us the information on the bulk and the fast ion velocity distribution function. Figure 2 shows the probing and the scattered beams in the same poloidal cross section of LHD. The variation to the toroidal angle is not so large for both beams. Therefore the measured velocity distribution function is in the perpendicular direction and not in parallel due to the backscattering and near perpendicular geometries. This geometry was set to the LHD discharge of #91756, which is shown in the later section. The scattering volume is located at z = 0.5 m from the mid plane of LHD, and is located at the most probable value of the normalized minor radius r/a ~ 0.7 [5]. The probe and receiving beams are Gaussian in shape and have a beam waist of a few cm. The width of beam overlap extends from r/a = 0.5 to 1.0. When the z is increased, the bulk and tail levels decrease, as is shown in figure 2 right. Our present CTS diagnostic neglect the width of beam overlap. The toroidal magnetic field at the magnetic axis on R=3.6 m becomes 2.75 Tesla.

The small amount of the impurity has a possibility to change the CTS spectrum shape. The precision of measured ion and electron temperatures becomes worse due to the shape changes. In the typical case, the impurity carbon C^{6+} is added to 10% of the electron density. In this paper, the impurity effect is neglected, but it should be included for more accurate measurement of T_i.

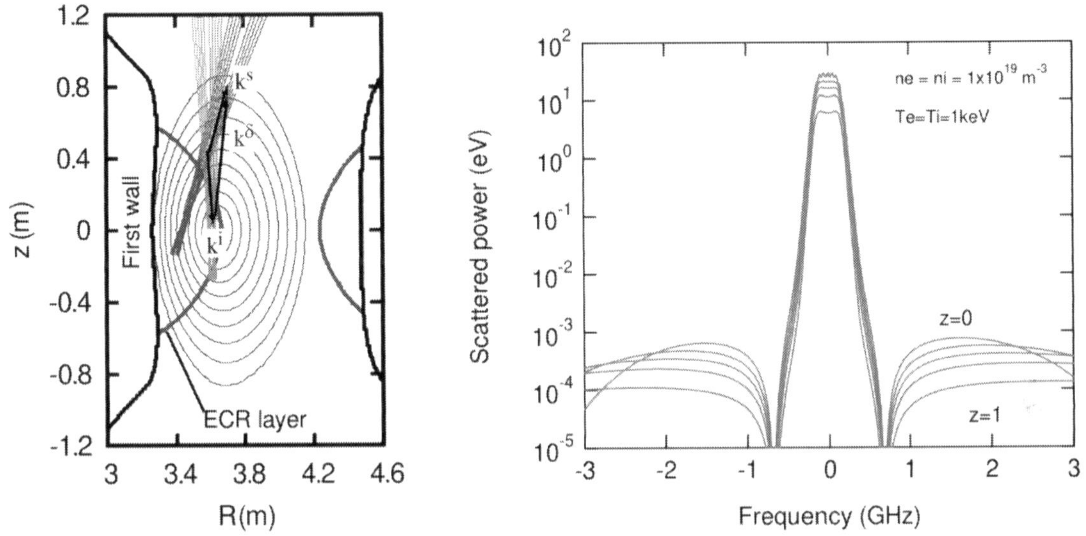

Figure 2. The incident and scattered wave vectors at the magnetic field of 2.75 Tesla at the magnetic axis of 3.6 m in LHD(left figure). Different scattering geometries are shown where the scattering volume is localized at Z = 0.0, 0.3, 0.5, 0.7, and 1.0. ECR layers and magnetic flux surfaces are drawn. The CTS spectra are varied with the same scattering geometries as the left figure. The electron density n_e and the ion density n_i are equal to 1×10^{19} m^{-3}, and their temperatures are set to 1 keV.

4. CTS spectra and their output at detectors of a receiver system

The CTS spectra are calculated based on the collective scattering theory, proposed in the references of [6, 7]. In the previous paper [4], the fluctuation power is provided by a scattering form factor and the geometrical factor $\Gamma \sim 1$. This assumption is valid for the rough estimate. However the geometrical factor can change in the X-/O-mode and their frequencies. For more accurate CTS spectra, the X-/O-mode electromagnetic wave in plasmas of the incident and the scattered beams was calculated, based on the references of [6, 7, 8]. However the theory does not treat the complete electromagnetic fluctuation, and it will be included in the future. The incident and scattered waves of O-mode are chosen in the experiments, because of free from the cut off in this condition, hence less prone to reflection

The receiver system employed is a high sensitive heterodyne radiometer. It consists of mainly the notch filter, a pin switch, the mixer with local oscillator, filter bank with band pass filters, and crystal detectors. The details are noted in the references of [4, 5]. The specification of the notch filter at the centre frequency of 76.95GHz is the attenuation of 120 dB with the bandwidth of 200 MHz, and -3dB with bandwidth of 400 MHz.

The scattered electromagnetic wave propagates through the notch filter, and is converted at the mixer in the range from 300 MHz to 10 GHz. The following filter bank separates the wave into 16 channels, which can detect the scattered radiations with the bandwidth of 100 MHz for bulk component channels and 200 MHz for high energy ones. The local oscillator frequency of 74 GHz was chosen originally, but in the experiment we had to use the alternative oscillator frequency of 77.845 GHz temporally, thus only 8 channels are used.

In LHD discharge #91756, the CTS probing beam is injected at the modulation frequency of 50 Hz to subtract the background ECE radiation from the scattered radiation. The background ECE radiation signal is averaged for a few ms from the trailing edge during the gyrotron off periods. During t = 0.9 ~ 1.8 sec the discharge is sustained by ECRH without energetic neutral beams. The spatial profiles of the electron temperature and density are measured by incoherent Thomson scattering. The reliable parameters become $T_e = 0.5$ keV and $n_e = 1\times10^{19}$ m^{-3} at R = 4.1 m. The measured and expected CTS spectra are plotted for the resolved 8 channels in figure 3. The calculated CTS spectrum is obtained by assuming that the ion and the electron temperatures are equilibrium condition. It is found that the expected CTS spectrum for the temperatures of 0.5 keV is in better agreement with the measured one, while the discrepancy becomes larger for the temperatures of more than 1.0 keV. The spectrum shape is in agreement. However we still need to check the validity of absolute signal levels.

The CTS diagnostics is demonstrated successfully. We expand the number of filter bank channels from 8 to 32 with using the original local frequency of 74 GHz in the forthcoming experimental campaign.

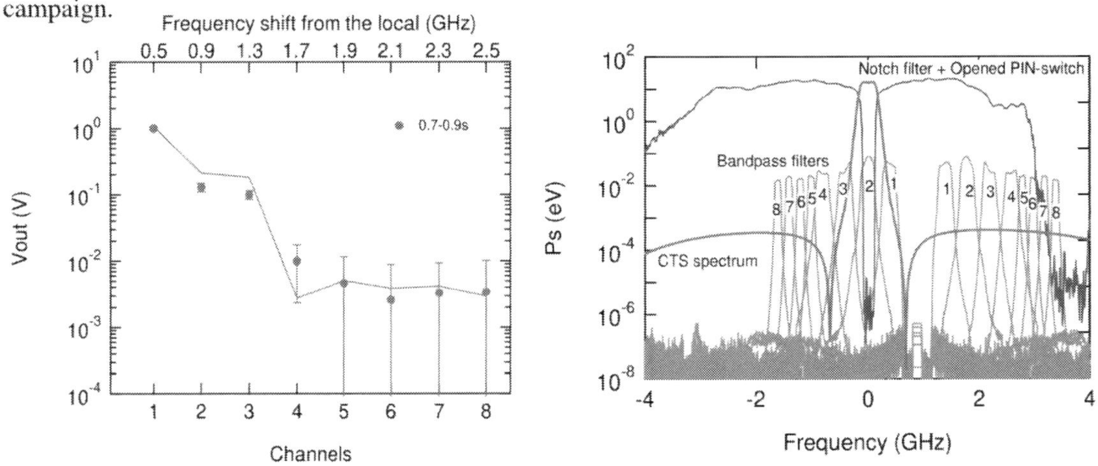

Figure 3. Measured CTS spectrum (closed circles) is resolved into 8 channels during t = 0.7 ~ 0.9 s (left). Each channel is the sum of both side band signals. The calculated CTS spectrum (line) is plotted for $T_e = T_i = 0.5$ keV. Right figure shows that the characteristics on the notch filter including the PIN-switch and the 8 channel band pass filters are overlaid at the local oscillator frequency of 77.845 GHz, which corresponds to 0.895 GHz on the abscissa.

Acknowledgment

We would like to thank Dr. Korsholm and CTS group of RISØ national laboratory for useful discussions.

5. References

[1] Bindslev H, Hoekzema J A, Egedal J, Fessey J A, Hughes T P, Machuzak J S, Phys. Rev. Lett. **83**, 3206.

[2] Korsholm S B, Bindslev H, Meo F, Leipold F, Michelsen P K, Michelsen S, Woskov P, Westerhof E, FOM ECRH team, Oosterbeek J W, Hoekzema J, Leuterer F, Wagner D, ASDEX Upgrade ECRH team, Rev. Sci. Instrum. **77**, 10E514.

[3] Meo F, Bindslev H, Korsholm S B, Tsakadze E L, Walker C I, Woskov P, and Vayakis G, Rev. Sci. Instrum. **75**, 3585.

[4] Nishiura M, Tanaka K, Kubo S, Saito T, Tatematsu Y, Notake T, Kawahata K, Shimozuma T, Mutoh T, Rev. Sci. Instrum. **79**, 10E731.

[5] Kubo S, Nishiura M, Tanaka K, Shimozuma T, Tatematsu Y, Notake T, Saito T, Yoshimura Y, Igami H, Tamura N, accepted for publication, Plasma Fusion Res. in 2009.

[6] Vahala L, Vahala G, and Sigmar D J, Nuclear Fusion **26**, 51.

[7] Hughes T P and Smith S R P, Nuclear Fusion **28**, 1451.

[8] Hughes T P and Smith S R P, J. Plasma Physics **42**, 215.

Pulsed polarimetry techniques for fusion plasmas

R J Smith

University of Washington, Box 352250, Seattle, Washington, 98195, US

E-mail: smith@aa.washington.edu

Abstract. Pulsed polarimetry techniques are described for the determination of the *local* magnetic field distribution, $\boldsymbol{B}(\boldsymbol{r})$, in magnetically confined plasmas. Pulsed polarimetry is a non-perturbative Lidar-*like* technique that exploits both the Thomson scattering and magneto-optic Faraday effects to measure $B_{||}$ at the position and along the sightline of a polarized light pulse propagating in the plasma. The implementation of pulsed polarimetry on high performance magnetized plasmas of relevance to magnetic fusion falls naturally into three categories based on the size and optical activity of the plasma and present-day laser and detector options: 1) large tokamak plasmas typified by ITER and DEMO, 2) *meter*-sized high energy density plasmas typified by the Magnetized Target Fusion (MTF) program and 3) *cm*-sized plasmas typified by wire Z-pinches. Plausible pulsed polarimetry systems are presented for each category along with the current interest in obtaining the local field measurement for the respective programs.

1. Introduction

The confining magnetic field (poloidal field, $\boldsymbol{B}_\theta$, for tokamaks) is a key parameter in magnetically confined plasmas. A determination of $\boldsymbol{B}$ is equivalent to measuring $\boldsymbol{j}$. The magnetic field touches on all facets of plasma performance: stability, dynamics and transport (energy and particle confinement). Despite this significance, a robust non-perturbative $\boldsymbol{B}$ diagnostic has eluded the plasma physics community. Stellar techniques exist for density, n_e, temperatures, T_i and T_e, ion flow velocity, $\boldsymbol{V}_i$, with high spatial resolution, precision and bandwidth. So much so, that for tokamaks, $\boldsymbol{B}_\theta(\boldsymbol{r})$ has been reasonably constrained. Conventional polarimetry[1] is the mainstay $\boldsymbol{B}$ diagnostic. The Motional Stark effect (MSE) diagnostic measures the *local* magnetic pitch angle.[2] Grad-Shafranov equilibrium solvers reconstruct the magnetic equilibrium in great detail throughout the plasma volume. However, polarimetry measures a *non-local* sightline averaged $\langle n_e B_{||} \rangle$, not B directly, MSE requires a dedicated, particle beam and is difficult to implement on next step tokamaks and equilibrium solvers are slow and rely heavily on internal measurements for accuracy. Considering the entire magnetic fusion energy (MFE) program, knowledge of *local* $\boldsymbol{B}$ is far from satisfactory. Outside of the tokamak program, the diagnostic choices for constraining $\boldsymbol{B}$ are disappointingly few. These plasmas can be highly dynamic on *ns*-μs time scales, with difficult noise problems. Diagnostics must have exceptional bandwidth. One cannot simply ignore these confinement concepts which hold the records in fusion reactivity, broadly known as high energy density (HED) plasmas.

2. The pulsed polarimetry method

Pulsed polarimetry[3] is a non-perturbative Lidar-*like* remote sensing technique based on both the Faraday and Thomson scattering (TS) effects. The measured polarization azimuth, α, of the backscatter is directly related to $n_e B_{||}$. With n_e independently measured, $B_{||}$ is determined. Unlike conventional Lidar, α is a *non-local* quantity. Both plasma scattering and plasma dispersion theory at optical frequencies are relevant. For instance, the cold (T_e=0) plasma dispersion theory determines the refractive index for O and X-mode propagation ($\boldsymbol{k}\cdot\boldsymbol{B}$=0) with $\boldsymbol{E}$ || and $\perp$ to $\boldsymbol{B}$, respectively, and the circularly polarized modes ($\boldsymbol{k}\cdot\boldsymbol{B}\neq 0$). In this model, $\alpha(l)$, is given by equation (1) for f ($=c/\lambda$), well above sightline cutoff.

$$\alpha(l) = 2\text{x}2.63^{-13}\,\lambda[m]^2 \int_0^l n_e[m^{-3}]\, B_{\|}[T]\, ds \text{ [rad]} \tag{1}$$

Which can be inverted to give $B_{\|}(l)$, $$B_{\|}(l) = \frac{1.9x10^{12}}{\lambda^2 n_e(l)} \frac{d\alpha}{dl} \tag{2}$$

For incoherent scattering, the scattering level is given by equation (3) with collection solid angle, $\Delta\Omega$, pulse energy, E_{pulse} and spatial resolution, Δs. For the coherent regime, $\alpha_e > 1$, a suitable relation between $n_e(l)$ and $I(l)$ is needed. Finite-T_e effects modify the scattered distribution and introduce depolarizing effects but backscatter faithfully preserves the polarization of the pulse.[4]

$$I(l) = E_{pulse} r_e^2 \Delta\Omega\, n_e(l) \Delta s / \tau_{pulse} \text{ [W]} \tag{3}$$

If $\boldsymbol{k}\cdot\boldsymbol{B}$~0 or $B_\perp$ is strong, the Cotton Mouton (CM) effect can introduce an ellipticity, $\delta(l)$. If this effect is strong, α may no longer be directly related to $n_e B_{\|}$ and both $\alpha(l)$ and $\delta(l)$ are then needed to resolve $\boldsymbol{B_\theta}$. In this paper, only direct detection is considered. Measureable quantities are contained in the backscatter's Stokes vector, S. An important property of S for incoherent scattering is additivity: $S=\Sigma_i S_i$ for any partition of the scattering volume. Individual electron radiators determine S.[4]

3. Pulsed Polarimetry techniques for fusion relevant devices

3.1. ITER tokamak (4 m size)

The implementation of pulsed polarimetry on large, low density plasmas is best served with a photodiode detection system. Tokamaks are the most demanding application given the low n_e (weak scattering levels) and $\alpha(L_p)$ must be large (~1) to compensate for the low SNR. The FIR range of 50-400 μm is suitable and familiar, lying above the strong ECE background. FIR laser and detector technologies plausibly allow 30 *cm* spatial resolution, equivalent to JET's TS Lidar system for a 4 *m* device.[5] Plausible pulsed polarimetry sightlines on ITER are shown in figure 1.

To achieve a 30 *cm* resolution, detector bandwidths of 400 *MHz* and a τ_{pulse} ~300 *ps* are required.[3] The LiHe cooled *Nbn* hot electron bolometers can achieve this. Quantum well (dot) infrared detector (QW(D)IP) technologies are also promising which offer focal plane arrays. Detectors are small, ~λ in size. Radiation is coupled to the detector using antennas or a condensing lens. A laser system based on a CO_2 pumped FIR laser can plausibly reach the required power level (~100 *MW*) and pulse lengths.

The measurement duration, 25 *ns* over 4 *m*, can provide prompt real-time feedback for detection and control of destructive MHD instabilities so important for machine protection. For advanced tokamak scenarios, the detection *and localization* of the instabilities would be of importance.

Pulsed polarimetry is also a complete diagnostic. Finite-T_e effects can be corrected using the measured $T_e(l)$ profile. Also, both $B_{\|}$ and $\boldsymbol{B}_\perp$ can be sensed. Diagnostic integration is therefore high.

As regards polarimetry, pulsed polarimetry uniquely spatially resolves B, defeats refraction, eliminates retro-reflectors, allows arbitrary sightlines, removes ambiguities from the presence of both the Faraday and CM effects and allows a λ closer to sightline cutoff. The important inboard sightlines become, in principle, accessible as shown in figure 1.

As regards interferometry, profile information can only be provided by a multi-chord system or a sophisticated reflectometer. A single mid-plane tangentially viewing pulsed polarimeter measures n_e(R), T_e(R) and potentially the plasma diamagnetism (the distributed Shafranov shift).

3.2. Magnetized target fusion (MTF) plasmas (m-size)

If the required spatial resolution is ~*cm* or less, photodiodes become problematical with very small detector areas for the required bandwidth. Streak cameras are a better choice with 10 *ps*, even 1 *ps* response times. The slit entrance aperture allows 2-d intensity imaging or a spectral (T_e) resolution.

The MTF program seeks to reach fusion reactor conditions by compressing a field reversed configuration (FRC) plasma with a metal liner.[6] The plasmas are dynamic on a ~10 μs time scale with n_e ~$10^{23\text{-}25}$ m^{-3}, B ~5-500 T, T_e ~ 0.1-10 *keV* and L_p ~ 30-6 *cm*. The compression time scale is ~10 μs and energy transport models predict that a fusion Q >1 will be reached under the most pessimistic transport model. However the plasma-Al liner boundary may develop a R-T instability as the liner is expected to form a plasma at B >300 *T* from the induced surface current density. The challenge is to characterize the plasma state, transport and MHD stability over this enormous range of n_e, T_e and B. For τ_{pulse} ~20 *ps* and 5 *ps* detector response time, T_R, Δs ~4 *mm* ($=c/2\surd(\tau_{pulse}^2+T_R^2)$). Figure 2 shows a MTF pulsed polarimeter system. The internal distributions of (n_e,T_e,B) for FRCs are unknown, and transport poorly understood. As n_e increases, the resolution of all parameters improves. The measurement interval is far shorter than τ_E ~1 μs. A multi-pulse system would resolve the dynamics.

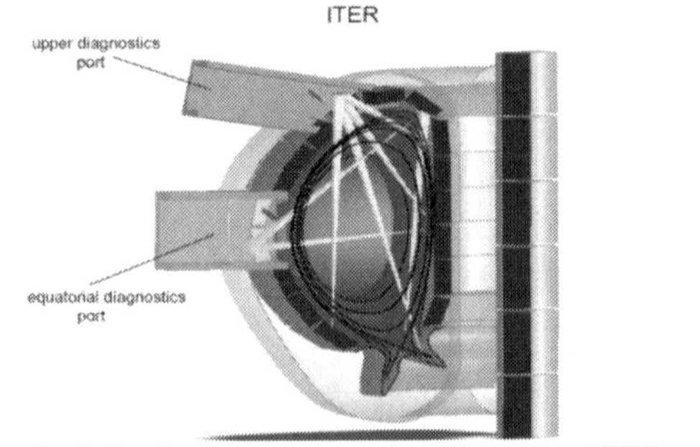

Figure 1. Pulsed polarimetry sightlines on ITER. Divertor access is possible by extending the pulse path. The mid-plane sightline provides tangential viewing sightlines out of the poloidal plane.

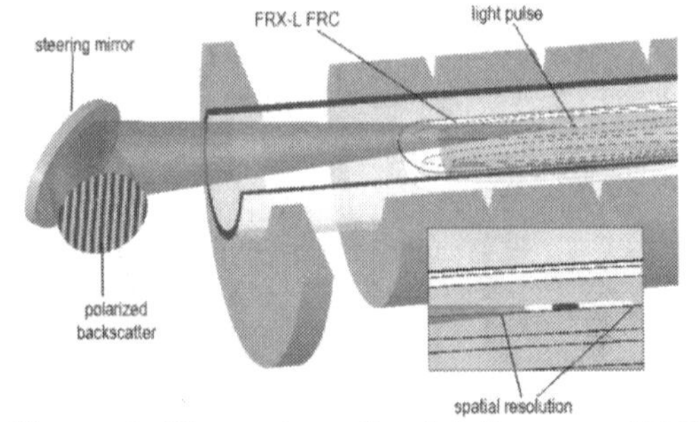

Figure 2. Illustration of a sightline on a FRC. Backscatter is shown with the orientation of the polarization. The spatial resolution is determined by the detector response time.

Measurements of ∇T_e, ∇n_e contribute fundamentally to an understanding of energy and particle transport. Liner-plasma interactions can be measured. The measurements directly aid numerical simulations. Most importantly, the MTF concept would have experimental validation in addition to measured neutron flux. Pulsed polarimetry could elevate FRC research to that of the tokamak.

3.3. Z-pinch and laser produced plasmas (cm size)

At the extreme range of MTF: n_e=3x10^{25} m^{-3}, B=500 *T*, L_p= ~2 *cm*, streak camera detection becomes impractical. A high speed photographic method is advocated here.[7] The backscatter is directed through a scattering cell and an ultra-fast shutter based on the optical Kerr effect (OKE) is used to freeze this intensity distribution to be recorded by a CCD camera.

The *s* and *p* backscatter components are both imaged from a depth-wise narrow pulse as shown in figure 3. The shutter, consisting of a Kerr cell sandwiched between crossed polarizers, is gated ON by an intense pulse also supplied by the probe laser. The shutter time is given by the longer of the pulse duration or the relaxation time of the medium. The CS_2 medium has a 1.8 *ps* relaxation time producing a Δs ~270 μm and ~400 *fs* times are possible.[8] The method can be used to generate 2-d images with the transverse dimension being a spatial or spectral dimension. A relativistic longitudinal smearing is present. This can be dealt with by pre-distorting the pulse prior to scattering or using narrow-depth pulses prepared with anamorphic optics. Using scattering to visualize scattering incurs a *SNR* penalty but the penalty is acceptable and the gains in detection performance warrant it.

A plausible laser system for probing the precursor phase of a wire Z pinch plasma (n_e=1.4x10^{19} cm^{-3}, B=20 *T*, T_e=100 *eV*) with sub-*mm* spatial resolution consists of a 10 J pulse at 1057 *nm* and τ_{pulse}=350 *fs*. A shutter time of 2 *ps* provides a Δs<500 μm yielding a $\alpha(\Delta s)$ of 1.2° at λ=528 *nm*. For a collection half angle, $\theta_{1/2}$, of 3°, the scattering energy level over Δs is 4.7 *nJ* or 1.2x10^{10} *ph*. Considering the scattering cell: over 2Δs, 3.1x10^8 *ph* are scattered with an α_{cell} = 4 cm^{-1}. A collection angle of 30° collects over 7% of the re-scatter (2.1x10^7 *ph*). The SNR is 4.6x10^3 allowing angular resolutions to a few 0.01° or a few percent accuracy in $B_{||}$ from photon statistics. A consideration for this technique is the light leakage when the shutter is OFF. The OFF/ON transmission ratio can be low ~10^{-5}. In principle, two OKE shutters can be used in series. Light leakage is also present in image intensifiers. Inverse bremsstrahlung is a consideration and would be mistaken for a decreasing $n_e(l)$ but pulsed polarimetry can be corrected using measured $n_e(l)$ and $T_e(l)$ if Z_{eff} can be estimated.

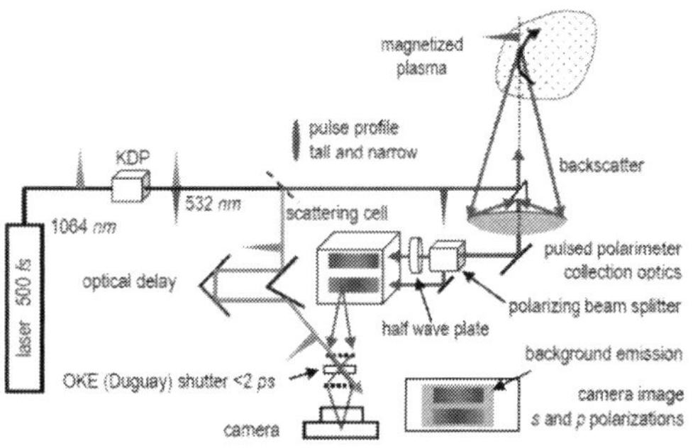

Figure 3. A schematic for a pulsed polarimeter using an imaging detection system based on an OKE ultra-fast shutter. A 2-d image of an extended scattering volume is obtained for each polarization state. A waning of intensity in one channel would correspond to a waxing of intensity in the other. Background emission is obtained over a longer time interval.

Wire array Z pinches have produced record X-ray levels, ~300 TW[9] and may play a role in magnetic fusion as an X-ray source for inertial confinement (ICF).[10] The dynamics of wire Z pinches by which the plasma is assembled on axis is driven by magnetic force and not fully understood. Instabilities developing in the phase of wire ablation form implosion bubbles and ablation jets with <1 *mm* spatial scales. Field measurements with sub-*mm* and sub-*ns* resolutions are needed.

4. Summary

Pulsed polarimetry provides remote sensing of *local* B, n_e, T_e across the full spectrum of magnetically confined plasmas with fusion relevant parameters. Photodiode detection and FIR lasers are amenable to the low density tokamak plasmas which require modest spatial resolution (30 *cm*) for their ~4 *m* size; streak camera detection and lasers in the NIR and visible are appropriate to the high density, high field plasmas of the MTF program providing sub-*cm* resolution for their 30 *cm* size; and high speed imaging using an ultra-fast shutter to resolve sub-*mm* features on wire Z pinches.

Pulsed polarimetry is uniquely suited to the measurement needs of future burning tokamaks (DEMO), by alleviating the need for retro-reflectors and fixed sightlines of a conventional polarimeter or the dedicated beam of an MSE diagnostic which also will have difficulty providing inboard measurements. The attributes of high diagnostic integration in one port, advantageous operation in the FIR range with regards to first mirrors, arbitrary sightlines, and robustness to the hostile environment of a burning plasma implied by an optical remote sensing technique typifies pulsed polarimetry.

The pulsed polarimetry technique has seemingly wide applicability to future confinement scenarios at the highest parameter ranges. An instrumental sensitivity $\sim\lambda^2 n_e B L_p$ (line integrated rotation angle over the plasma) guarantees ever better spatial resolution for the more challenging future plasma scenarios. Higher n_e raises the scattering level allowing Δs to be reduced at a fixed detection SNR, a stronger magnetic field raises the $n_e B$ product providing comparable sensitivity to α at shorter wavelengths for the reduced Δs, and the faster dynamics of the future plasma is always accommodated as τ_E is necessarily longer than a L_p/c transit time.

5. Acknowledgements

This work has been supported by the Office of Fusion Energy Sciences.

References

[1] Hutchinson I H 2002 *Principles of Plasma Diagnostics* (Cambridge University Press, Cambridge) 133-41
[2] Thomas D M *et al.* 2008 *Fusion Sci. Tech.* **53**(2) 487-527
[3] Smith R J 2008 *Rev. Sci. Instrum.*, **79**, 10E703
[4] Segre S E, Zanza V 2000 *Phys. of Plasmas*, **7**, 2677-84
[5] Salzmann H *et al.* 1988 *Rev. Sci. Instrum.*,**59**(8), 1451
[6] Intrator T 2004 *Phys of Plasmas,* **11**(5), 2581
[7] Duguay M A *et al.* 1971 *App. Optics*, **10**(9), 2162
[8] Yasui T *et al.* 1998 *Nonlinear Optics* '98. Materials, Fundamentals and Applications Topical Meeting (Cat. No.98CH36244), 218-20
[9] Deeney C *et al.* 1998 *Phys.Rev. Lett.* **81** 4883
[10] Sanford T W *et al.* 1999 *Phys. Rev. Lett.* **83**, 5511

Journal of Physics: Conference Series **227** (2010) 012016 doi:10.1088/1742-6596/227/1/012016

Electron density measurement for plasmas by Terahertz time-domain spectroscopy

A Ando[1], H Kitahara[1], T Kurose[1], K Kitano[1], K Takano[2], M Tani[3], M Hangyo[2] and S Hamaguchi[1]

[1] Center for Atomic and Molecular Technologies, Osaka University, Osaka, 565-0871, Japan
[2] Institute of Laser Engineering, Osaka University, Osaka, 565-0871, Japan
[3] Research Center for Development of Far-Infrared Region, Fukui University, Fukui, 910-8507, Japan

E-mail : ando@ppl.eng.osaka-u.ac.jp

Abstract. A new plasma diagnostic tool of terahertz time-domain spectroscopy (THz-TDS) has been developed. The THz-TDS system allows one to obtain the electron density and collision frequency of a plasma simultaneously from the phase shift and absorption rate of THz waves that are transmitted through the plasma. The line-averaged electron densities of Ar inductively coupled plasmas (ICPs) and CH_4/Ar capacitively coupled plasmas (CCPs) were evaluated by the system and found to be 10^{17}-10^{18} m^{-3} for Ar ICPs and 10^{15}-10^{16} m^{-3} for CH_4/Ar CCPs, depending on the discharge conditions. These results have demonstrated the feasibility of electron density measurements by THz-TDS for reactive plasmas.

1. Introduction

Terahertz time-domain spectroscopy (THz-TDS) has been applied to the measurement of electron densities in steady-state inductively coupled plasmas (ICPs) and capacitively coupled plasmas (CCPs). The ultimate goal of this study is to develop a new electron density diagnostic system that is appropriate for high-density plasmas, to which commonly used Langmuir probe or microwave interferometry techniques for plasma density measurement cannot be applied. Especially of our interest is to apply THz-TDS to low-gas-temperature atmospheric-pressure plasmas, which have recently employed for a wide range of applications in nonconventional plasma processing (For example, [1].)

Since plasma frequencies of high-density plasmas may lie in a sub THz region, the electron density and the collision frequency in such a plasma can be estimated from the phase shift and the transmissivity of THz waves that is transmitted through the plasma. In THz-TDS, both absorption and phase spectra of THz waves are obtained from Fast Fourier transformation of the time-dependent signals of an electromagnetic (EM) wave that is transmitted through a medium. Jamison *et al.* estimated time dependent electron densities and collision frequencies of transient pulsed helium discharges by THz-TDS [2]. In our study, on the other hand, we have demonstrated for the first time that THz-TDS can be applied also to the electron density measurement of steady-state discharges generated by inductively coupled plasma (ICP) or capacitively coupled plasma (CCP) systems.

The phase of an electromagnetic wave shifts when it passes through a plasma. The phase shift of a THz wave passing through a plasma may be evaluated from the following expression;

$$\Delta\Phi = \frac{\omega L}{c}\left\{1-\left(1-\frac{\omega_p^2}{\omega^2+\nu^2}\right)^{\frac{1}{2}}\right\} \quad \text{with} \quad \omega_p = \left(\frac{n_e e^2}{\varepsilon_0 m_e}\right)^{\frac{1}{2}}, \tag{1}$$

where ω is the frequency of the THz wave, L is the length of the plasma that transmits the THz wave, ω_p is the electron plasma frequency, n_e is the electron number density and m_e is the electron mass. Equation (1) shows that the electron density of a plasma averaged along the light path length can be estimated from the phase shift obtained by the transmission measurement of a THz wave. In THz-TDS, the electron density can be obtained without the information on the electron temperature unlike the probe method.

2. Experimental apparatus

2.1. Terahertz time-domain spectroscopy (THz-TDS)

We have used a THz-TDS system at Institute of Laser Engineering of Osaka University [3]. A schematic layout of the THz-TDS system used in this study is shown in figure 1. A laser beam emitted from the femtosecond laser (Tsunami, New Port/ Spectra Physics Inc.) is polarized by a half-wave plate and split by a beam splitter into the pump beam and the probe beam. Average power of the pump beam at the emitter is reduced to 12mW by a variable neutral density (ND) filter. The probe beam is then delayed by an optical time delay mirror and reaches the detector. The pump beam is focused to the gap of a dipole-type photoconductive antenna, which is installed on the emitter and biased at a 60 V (peak-to-peak) 10 kHz alternating voltage. THz waves radiated from the emitter are collimated to an about 1 inch diameter and directed to the plasma chamber by an off-axis parabolic mirror. THz waves passing through the plasma chamber are then directed to the detector by another off-axis parabolic mirror. When the THz wave and the pump beam arrive at the detector simultaneously, the magnitude of the instantaneous electric field of the THz wave is recorded.

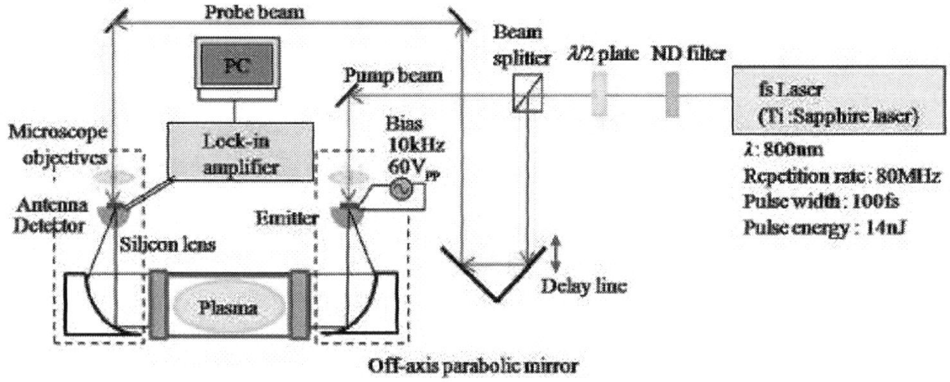

Figure 1. The schematic layout of THz time-domain spectroscopic system.

2.2. Plasma chambers

A schematic diagram of the plasma chamber used for ICP discharges is shown in figure 2. The system consists of a vertical quartz tube (100mm in length and 56mm in outer diameter) and a horizontal quartz chamber (140mm in length and 56mm in outer diameter). A copper antenna to generate ICP plasmas is wound around on one side of the cross chamber. A schematic diagram of the plasma chamber used for CCP discharges is shown in figure 3. The electrodes which surround the quartz tube chamber (250mm in length and 56mm in outer diameter) are cupper half tubes of 180mm in length. A

13.56MHz radio frequency (RF) power is applied to the electrode of each system through an impedance matching network. The plasma chamber is set in an aluminium box that prevents RF noise generated by the plasma formation system from affecting the measurement system. The polyethylene windows or the high resistivity Si windows, through which THz waves can pass with little loss, are attached to the both ends of the plasma chamber.

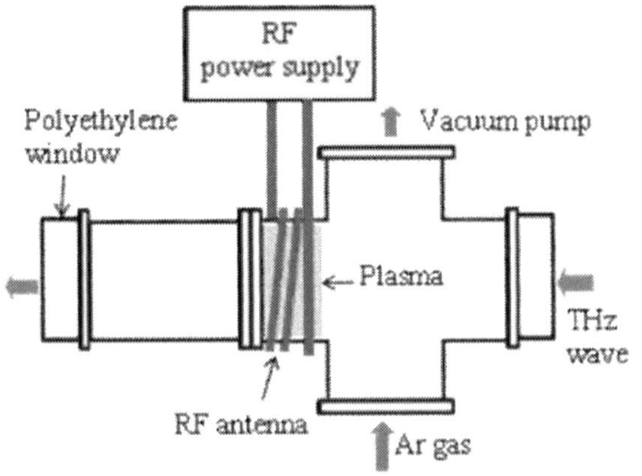

Figure 2. Schematic layout of plasma chamber for ICP.

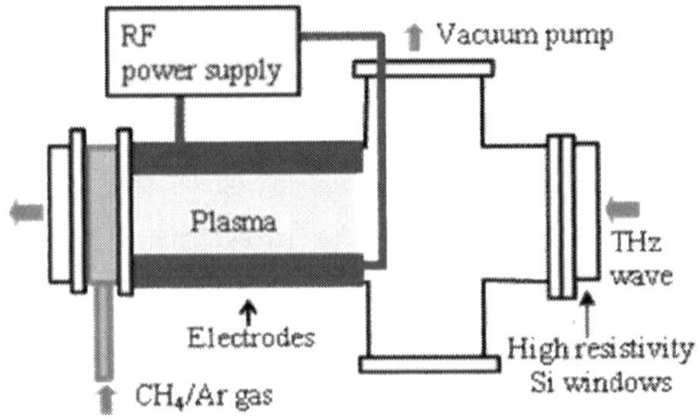

Figure 3. Schematic layout of plasma chamber for CCP.

3. Experimental results

The electron densities of Ar ICP plasmas and CH_4/Ar CCP plasmas have been measured by the THz-TDS system described above. The operating pressures were 25Pa and 100Pa and the operating power was in the range from 20W to 100W for Ar ICP plasma. Figure 4 shows the phase shift of a THz wave pulse transmitted into an Ar ICP plasma at 25Pa and 40W. The phase shift of the THz wave was evaluated from the difference between the phases of THz wave transmitted into the plasma and that into the gas (without a plasma). The electron densities of the Ar ICP plasmas were estimated by fitting the equation (1) to the phase shift, as shown in figure 4. The line averaged electron densities of the Ar ICP plasmas for different RF powers at 25Pa and 100Pa are plotted in figure 5. Since it is expected that the collision frequency ν is lower than the THz wave frequency ω for the Ar ICP plasmas, ν was neglected when equation (1) was used for fitting. The plasma length L in equation (1) was assumed to be 40mm, which was consistent with the independently performed density distribution measurement by a Langmuir probe [4]. Since the measurement error of the phase shift *per se* in a typical THz-TDS system is less than 1%, the main source of electron-density measurement errors is the uncertainty of the plasma length L, as easily seen in equation (1). Although the error bars are not given in figure 5, it should be construed that the (line averaged) electron density given in figure 5 may contain errors of about ±10% arising from the difficulty of evaluating the plasma length L exactly.

The phase shift of THz pulses that are transmitted into a CH_4/Ar CCP plasma is shown in figure 6. CH_4/Ar CCP plasmas were generated with the flow rates of 6 sccm for CH_4 and 12 sccm for Ar at the total pressure of 28Pa. The RF input power was 100W. The plasma length L in equation (1) was estimated to be about 180 mm. The electron density obtained from the phase shift given in figure 6 is 4.4×10^{15} m^{-3} [5].

4. Conclusions

THz-TDS has been applied to the measurements of electron densities of Ar ICP plasmas and CH_4/Ar CCP plasmas. Unlike probe measurement, THz-TDS is a non-intrusive method for electron density measurement. Compared with microwave interferometry, which is another non-intrusive method for electron density measurement in plasmas, THz-TDS uses much higher frequencies and shorter wavelengths. This suggests that THz-TDS is a useful non-intrusive method for electron density measurement for high -density small plasmas.

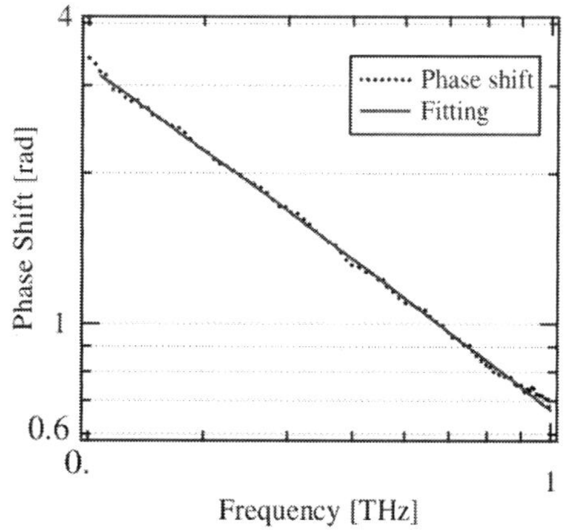

Figure 4. Phase shift of THz pulses through an Ar ICP plasma at 25Pa and 40W.

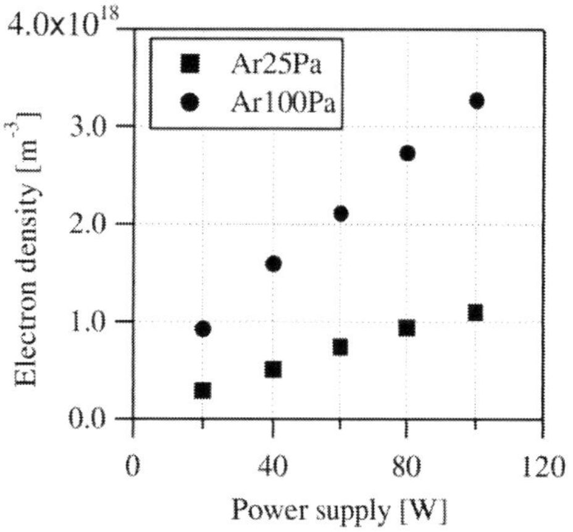

Figure 5. Electron densities of Ar ICP plasmas measured by THz-TDS as function of RF power at 25Pa and 100Pa.

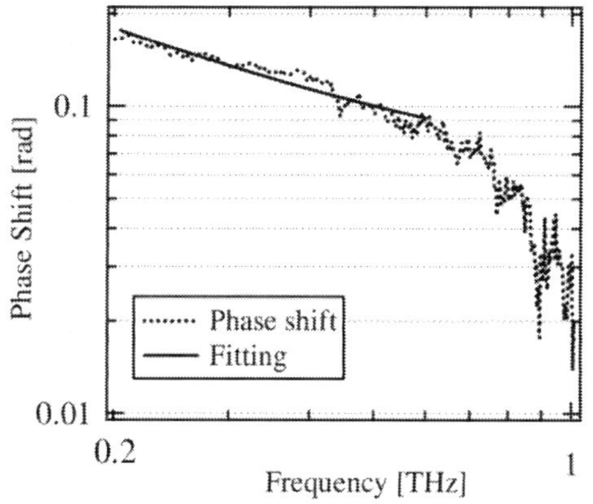

Figure 6. Phase shift of THz pulses through a CH_4/Ar CCP plasma.

Acknowledgments

This research has been funded in part by the Grants-in-Aid for Scientific Research from the Ministry of education, culture, sports, science and technology (MEXT) of Japan.

References

[1] A Fridman "*Plasma Chemistry*" Cambridge University Press (2008).

[2] S P Jamison, J Shen, D R Jones, R C Issac, B Ersfeld, D Clark and D A Jaroszynski , *J. Appl. Phys.* **93**, 4334 (2003).

[3] M Hangyo, T Nagashima and S Nashima, *Meas. Sci. Technol.* **13**, 1727 (2002).

[4] K Kitano, K Kurose, A Ando, V Raymond, H Kitahara, K Takano, M Tani, M Hangyo and S Hamaguchi *2nd International Symposium on Atomic Technology (ISAT-2) Book of Abstracts* P-35 (2007).

[5] A Ando, H Kitahara, T Kurose, K Kitano, K Takano, M Tani, M Hangyo and S Hamaguchi, *3rd International Symposium on Atomic Technology (ISAT-3) Book of Abstracts,* PP-4-97 (2009).

Journal of Physics: Conference Series **227** (2010) 012017 doi:10.1088/1742-6596/227/1/012017

Time resolved LIF with a fast-scanning diode laser

F Skiff and V Patel

Department of Physics and Astronomy
University of Iowa, Iowa City, IA 52242 USA

E-mail: Frederick-skiff@uiowa.edu

Abstract. Ion phase-space resolved fluctuations carry detailed information pertaining to transport as well as to the plasma degrees of freedom. Using a 500 mW diode laser that can scan over Doppler broadened Argon ion lines in less than 50 microseconds we observe fluctutations and correlations with an adjustable bandwidth by means of a comb-filter. The high laser intensity coupled with fast scanning makes optical pumping between the Zeeman sublevels observable. This opens a new strategy for optical tagging as well as observing correlations between fluctuations at different ion velocities. The experiments are performed in a 2 meter length 0.1 meter diameter CW Argon gas discharge created by a 10 MHz inductive plasma source in a uniform 1 kG magnetic field. The plasma density is typically 10^9 cm^{-3}, the electron temperature is 2 eV and the ions have a temperature of 0.1 eV. Under these conditions the ions are weakly collisional.

1. Introduction

Coherent detection is useful for the study of reproducible [1] and especially harmonic fluctuations [2] using laser-induced fluorescence (LIF). Recently particle-velocity-resolved plasma cross-power spectra and higher-order spectra have been measured using the cross-correlation approach [3]. In the earliest measurements, fluctuations in the fluorescence light are cross-correlated between two points along a single laser beam. Extending this technique, correlations between fluctuations at *different* ion velocities requires two separate tunable laser systems. An alternative strategy for measuring fluctuating distribution functions which we explore here is to rapidly scan the wavelength of a single laser so as to obtain a measurement of a distribution function in a time short compared to the periods of the fluctuations of interest. The advantage of the technique is that it can be performed using a single laser and a single detection system. Furthermore, it is not necessary to chop the laser so long as the rapid scan extends past the Doppler broadened line so that the level of collisionally induced fluorescence light can be determined. Of course, as the laser intensity is increased in an effort to increase the photon count rate one invariably has to deal with optical pumping and its effects on the measurement. Here we present first results of using a fast-scanning diode laser system to measure fluctuating distribution functions with a bandwidth that is equal to the scan frequency (each period of the scan frequency measures the distribution function twice), but adjustable through the use of a comb filter. Difficulties encountered in implementing the technique are discussed. The role of optical pumping on the results is observed and compared to a collisional-radiative simulation of the experiment. We note that fast scanning LIF has been done previously with the intent of measuring a distribution function in a single pulsed plasma discharge, but was not used to study fluctuations [4].

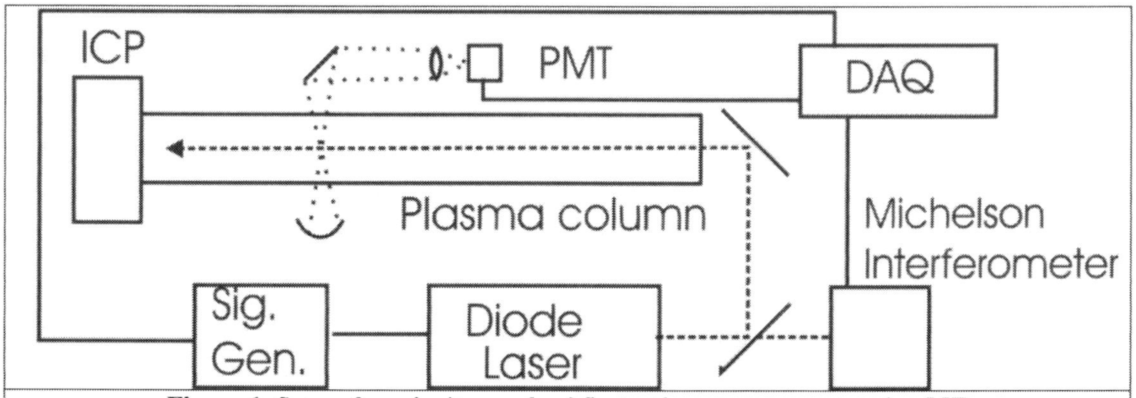

Figure 1. Set-up for velocity-resolved fluctuation measurements using LIF.

2. Experimental set up

The experiments are performed in a weakly-collisional inductively coupled RF gas discharge at 10 MHz. The plasma is a cylinder of 3 meter length and 0.1 meter diameter immersed in a straight magnetic field along the axis that is 1kG. The neutral pressure is regulated to 0.2 mTorr of Argon. The plasma density is typically 10^9 cm^{-3} of singly ionized Argon with an electron temperature of 2 eV and an ion temperature of 0.1 eV. The experimental set-up for cross correlated LIF is shown in figure 1. For these experiments we used a Toptica DL-100 0.5 W tuneable diode laser operating at 668nm which is a metastable line with fluorescence observed at 442nm. This transition is widely used for diode laser LIF [5]. The Toptica diode laser can be modulated rapidly, using the internal ramp, (which drives a piezo-stack on the grating used in the external cavity and can also scan the diode current for enhanced stability during the grating scan). Because the nominally triangular ramp signal does not necessarily correspond to the actual grating position or laser frequency, and there may be other electronic phase shifts, we monitor the instantaneous relative laser wavelength using a home-made Michelson interferometer with an output that is digitized at the same time as the LIF signal. Signals are digitized at 1 MHz sample rate for files of duration 1 second.

3. Initial Results and analysis

The fluctuations in the PMT signal at low frequency are dominantly caused by the background light (collisionally induced fluorescence). By constructing a comb filter in the frequency domain that is a sum of bandpass filters centered on each of the harmonics of the scan frequency it is possible to have a measurement of the distribution function with a specified bandwidth. Fourier reconstruction of the filtered data is shown in Figures 2 and 3 for a 10 Hz and 1 Hz fluctuation bandwidth respectively and a scan period of 176 μs. With such strong filtering it is clear that the distribution functions, that are nominally recorded every 80 microseconds, only manifest variations over longer timescales. By subtracting this average distribution from a less filtered time series it is possible to look at the fluctuation $\delta f = f - \langle f \rangle$. In the plasma cylinder the fluctuations are mostly due to m=1 convectively stabilized dissipative drift instability driven by the radial electron pressure gradient. These fluctuations occur near the electron drift frequency (~1.5 kHz).

Below we will return to the curious feature that the distribution function appears to depend on direction of scan. Also, the laser light was linearly polarized so that both of the (σ polarized) Zeeman components are visible. Nevertheless, one can look at a given point on the distribution function and look at the fluctuations as a function of the scan number. In order to overcome the photon statistics fluctuations (the LIF photon count rate is close to 10^7 per sec) we calculate the cross correlation:

$$C(z_2 - z_1, v_2, v_1, t_2 - t_1) = \langle f(z_1, v_{\|1}, t_1) f(z_2, v_{\|2}, t_2) \rangle$$

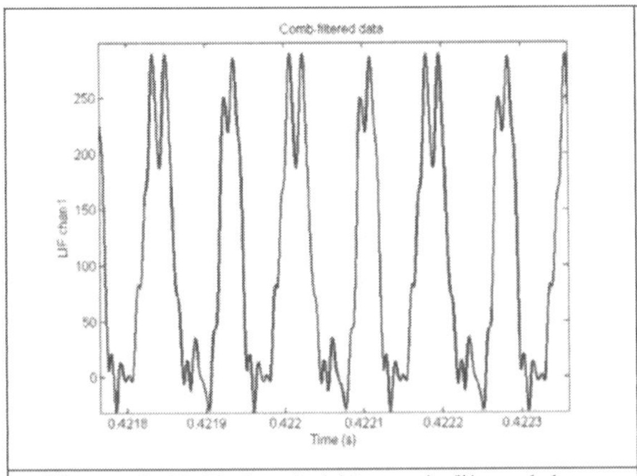

Figure 2. 10 Hz bandwidth comb-filtered data.

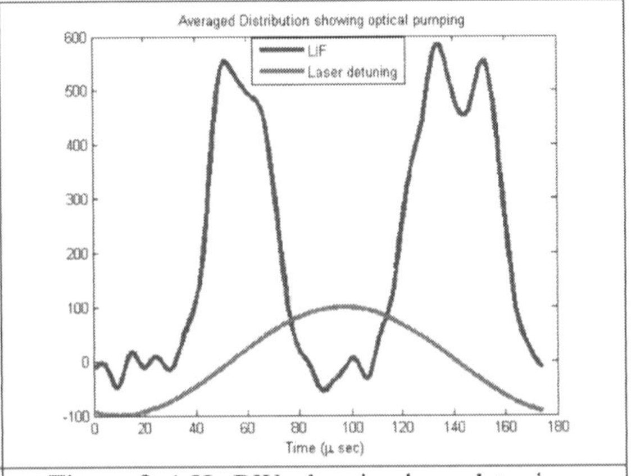

Figure 3. 1 Hz BW, showing laser detuning.

As a point of departure, we can calculate a velocity-integrated cross correlation integrating *C* over the single variable $v_1=v_2=v$. The result of computing this integrated correlation function from the data in a single 1 second record is shown in figure 4. A velocity-resolved power spectrum from fast-scanning is shown in Figure 5.

4. Complications

There are a number of complications that arise in implementing the fast-scanning technique which we need to briefly discuss. Firstly, because diode lasers, including the Toptica laser used here, have not been used extensively in this fast-scanning mode it is not known if one will significantly shorten the life of the piezo stack by scanning the laser at high frequency and with large amplitudes. Running the scan for more than a second appears to produce a thermal drift which could be related to heating of the piezo.

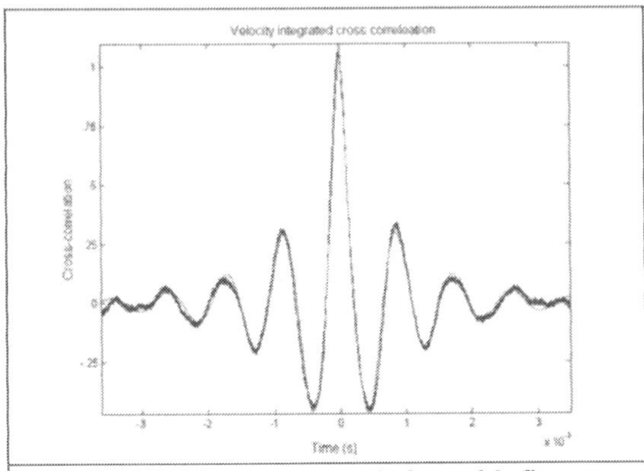

Figure 5. Cross-correlation with fit.

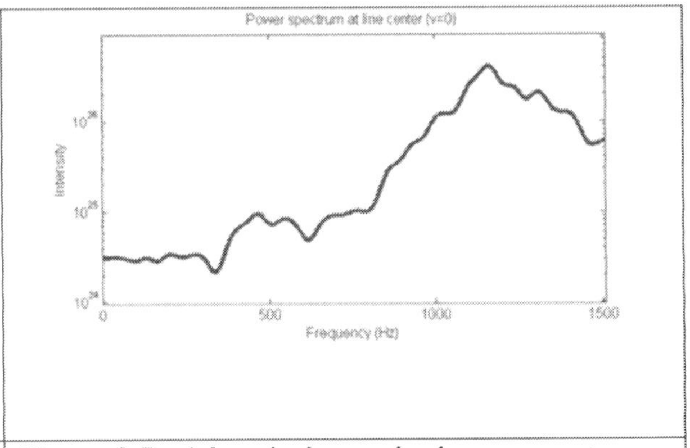

Figure 6. Particle-velocity resolved power spectrum.

Secondly, it is necessary to monitor in detail the actual laser frequency and mode purity because scanning can cause the laser to mode hop. During fast scanning it is not possible to use a conventional spectrum analyzer because they scan too slowly. We built a simple Michelson interferometer to monitor the output wavelength in real-time and observed occasional irregularities. We also monitored the laser output power with good time resolution in an effort to catch mode hops. We used the internally generated triangular ramp for the scan, but it is evident from the data that the mechanical scan becomes sinusoidal as the scan frequency is increased (Figure 3). For increased stability, the diode laser controller uses a "feed-forward" slight modulation of the diode current during a piezo scan to inhibit mode hopping. This feed-forward is also a dynamical system which can sometimes be used in fast scanning but can introduce complications in the wavelength-vs-time dynamics. At some points the Michelson output becomes difficult to interpret (the direction of phase progression is ambiguous). What we did was to work on the laser alignment (and oscillator diode temperature) to the point that the laser was stable and scanning well and all the signals made sense, but there were situations where the laser output was not easy to interpret. For this reason it may be best to avoid using the "feed-forward" signal when possible.

Finally, there is the complication of optical pumping which usually occurs at some level in LIF measurements. Even when the laser is circularly polarized, there are six components to each of the σ polarized groups for the 668nm metastable line in Ar II. Each of these lines corresponds to a coupling beween particular magnetic sublevels of the upper and lower states of the ion. If fluorescence is observed then the ion has decayed to a third level (producing fluorescence at 442 nm in this case), but there is also the possibility of optical pumping between the magnetic sublevels which becomes noticeable as the laser intensity in increased – and produces an apparent shifting of the lines (depending on the direction of scan) evident in figures 3 and 4. This optical pumping effect (related to "spin-tagging" [6]) can be simulated using a simple collisional-radiative model for the three-level system which includes spontaneous decay, electron collisional excitation as well as laser pumping. It may be possible to exploit the sublevel redistribution due to optical pumping as an implementation of a particle labelling technique.

5. Summary

We have experimentally explored the technique of periodic fast-scanning using a tuneable diode laser to measure fluctuating ion velocity distribution functions in a linear magnetized plasma column. Despite the complications it has been possible to obtain cross-correlation functions and power spectra that are particle velocity resolved. Optical pumping between the magnetic sublevels of the metastable state is observed and can be qualitatively reproduced from a collisional-radiative model.

Acknowledgments

Work supported by the NSF-DOE collaboration on basic plasma science. DOE grant number DE-FG02-99ER54543.

References

[1] Biloiu I A, Sun X and Scime E 2006 *Rev. Sci. Instrum.* **77** 10F301-1-3
[2] Skiff F 2002 *IEEE Trans. Plas. Sci.* **30** 26-27
[3] Skiff F, Uzun I and Diallo A 2007 *Plasma Phys. Control. Fusion* **49** B259-B265
[4] Honda C, Nishimura K, Maeda M, Muraoka K and Akazaki M 1983 *J. Phys. D: Appl Phys.* **16** 1943-1952
[5] Severn G D, Edrich D A and McWilliams R 1998 *Rev. Sci. Instrum.* **69** 10-15
[6] Skiff F, Good T N, Anderegg F and Paris P J 1989 *Phys. Lett. A* **137** 57-59

Electric field measurements at near-atmospheric pressure by coherent Raman scattering of laser beams

Tsuyohito Ito[1], Kazunobu Kobayashi[2], Sarah Mueller[3], Uwe Czarnetzki[3], and Satoshi Hamaguchi[2]

[1] Frontier Research Base for Young Researchers, Frontier Research Center, Graduate School of Engineering, Osaka University, 2-1 Yamadaoka, Suita, Osaka 565-0871, Japan
[2] Center for Atomic and Molecular Technologies, Graduate School of Engineering, Osaka University, 2-1 Yamadaoka, Suita, Osaka 565-0871, Japan
[3] Institute for Plasma and Atomic Physics, Ruhr-University Bochum, 44780 Bochum, Germany

E-mail: tsuyohito@wakate.frc.eng.osaka-u.ac.jp

Abstract. Electric field measurements at near-atmospheric pressure environments based on electric-field induced Raman scattering are applied to repetitively pulsed nanosecond discharges. The results have revealed that the peak electric field near the centre of the gap is almost independent of the applied voltage. Minimum sustainable voltage measurements suggests that, at each discharge pulse, charged particles that remain from the previous pulse serve as discharge seeds and play an important role for generation of uniform glow-like discharges.

1. Introduction

Recently, non-thermal near-atmospheric (or higher) pressure discharges such as those generated in open air have attracted much attention for use in processing that can not be achieved using low-pressure discharges. Application of repetitive high-voltage nanosecond pulses to electrodes is an efficient method of generating such low-temperature high-pressure discharges. However, not many diagnostic reports describe such high-pressure discharges, probably in part because such plasmas are typically small. For that reason, intrusion of even a small solid probe can alter discharge characteristics. Optical methods are therefore preferred as diagnostic techniques for high-pressure discharges. Successful non-intrusive electric field measurements in high-pressure discharges (corona and surface discharges) were reported [1–3] in high-pressure hydrogen gas where Raman scattering by hydrogen molecules was used for field evaluations. The feasibility of electric field measurements in a nitrogen-containing gas has also been achieved and reported recently [4].

In the present study, laser spectroscopy for electric-field induced coherent Raman scattering by hydrogen molecules is used for direct measurement of electric fields in glow-like repetitively pulsed nanosecond discharges in a high-pressure hydrogen gas. The discharges are generated between parallel plate electrodes and the uniformity over the electrode space and reproducibility are fairly good.

In addition to extremely rapid discharge dynamics, including very rapid formation of electric field profiles, as discussed in Ref. 5, it has been found that the peak electric field (near the centre of the gap) is almost independent of the peak applied voltage. This might be in part related to prior discharge pulses. The repetition rate dependence of the minimum sustainable peak voltage, which also indicates interaction of successive pulses in a discharge, will also be provided.

2. Experiments

2.1. Electric field measurements with parallel plate electrodes

Figure 1 presents normal digital camera images of discharges, in which we have measured electric fields. Nearly uniform discharges between the metal electrodes seem to be generated by negatively biased nanosecond pulses applied to the cathode (the left electrode in Fig. 1) with pulse duration of about 5 ns and a repetition rate of 10 kHz in pure hydrogen at pressure of 0.3 bar, in all voltage conditions tested in this study. One example of the voltage shape and current flowing into the anode is presented in Fig. 2. The electrodes are stainless steel disks with diameters of 17 mm (cathode) and 14 mm (anode/ ground). The discharge gap is 1.15 mm ±0.05 mm. The gas temperature, as estimated by signal strengths of coherent anti-Stokes Raman scattering (436 nm), is 330 K or lower. The electric field is measured by the detection of electric-field induced coherent Raman scattering from hydrogen molecules [1–3,5] by using two pulsed nanosecond laser beams (532 nm and 683 nm) with duration times at full-width at half maximum (FWHM) of 3–5 ns. The reader is referred to ref. 5 for more details on the electric field measurement techniques.

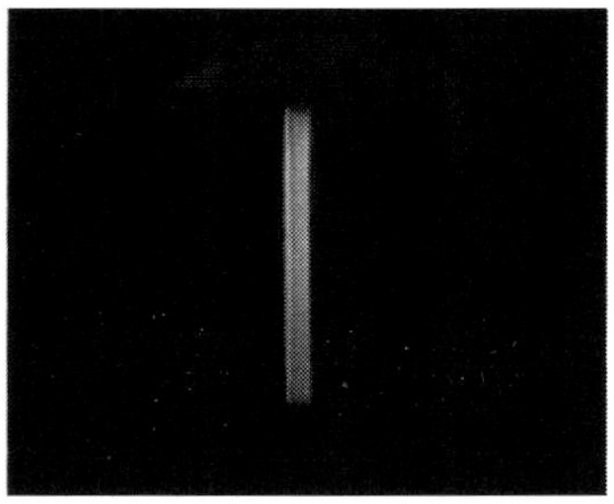

Figure 1. Repetitively pulsed nanosecond discharges generated with parallel plate electrodes in 0.3 atm hydrogen. The gap distance is 1.15 mm; the left side in the picture is the powered electrode (cathode). The peak voltage was about -2.2 kV and the pulse repetition rate was 10 kHz.

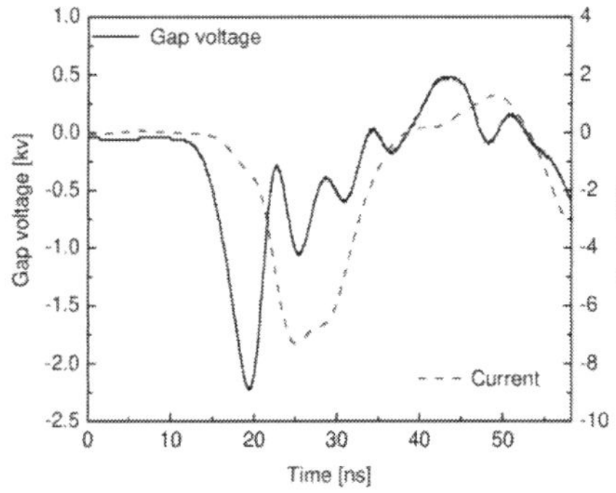

Figure 2. Discharge voltage (solid curve) and current (dashed curve) as functions of time during the discharge generation depicted in Fig. 1.

2.2. Repetition rate dependence of minimum sustainable peak voltage with sphere-plate electrodes

In this repetition rate dependence measurement, we used a slightly different electrode geometry. The cathode is a half sphere with a 6-mm radius; the anode is a disk plate. The gap width is set as 1 mm. The minimum peak voltage for sustaining the discharges as a function of the repetition rate and the environmental pressure has been recorded by slowly decreasing the peak voltage while maintaining the repetition rate and the pressure.

3. Results and discussion

Figure 3 presents the measured electric field near the centre of the discharge gap for three voltage conditions. As discussed in Ref. 5, reasonable agreement is apparent between the nominal electric field (gap voltage/ length) and the measured electric field at the very early stage, when significant ionization has not yet taken place. It is noteworthy that extremely good agreement, despite the temporally averaged signals over a few nanoseconds of laser duration, is explainable by the small timing error, i.e. 0.3 ns, which is likely to exist. Although it is not shown in this paper, the spatial distribution is fairly uniform before showing noticeable disagreement near the centre of the gap. The uniform spatial distribution more directly suggests the negligible net charge effects in the space.

Once the voltage is increased to around -1.2 kV, noticeable differences become apparent between the nominal and measured values. As described above, the measured electric field here is more like the averaged one; the first and second signal points showing disagreement might not directly indicate the significant charge effects at that time. For example, for peak voltage of -2.2 kV, the bright emission starts propagating from the anode to cathode sides at around 18.5–19 ns (see Ref. 5), which we believe is synchronized to the ionization front movement. Consequently, the reduction that is apparent around 17.5–18.5 ns is consistent with the phenomenon by which a considerable amount of ionization first takes place with the ionization front propagation starting at around 18.5–19 ns, which finally produces the thin cathode sheath-like positive net charge layer near the cathode [5]. Since there is little time for newly generated ions to reach the cathode to produce secondary electrons, charged particles that remain from the previous pulse should play an important role for the electron production at the cathode. Ions that remain from the previous pulse and stay near the cathode surface can reach to the cathode in a very short time period and produce secondary electrons at the cathode in a single discharge pulse. The emitted electrons, though they may be small in number initially, generate a significantly large number of electrons at the ionization front due to the high gas density and large electric field. Thus the amount of the ionization events should largely depend on the initial charge densities.

Fig. 3 shows that the peak electric field is almost independent of the magnitude of the peak voltage. This indicates that a discharge with higher power input produces a higher-density plasma that shields the electric field more effectively. The stronger shielding is probably caused by more efficient ionizations near the cathode caused by the higher applied voltage and the higher initial charge densities. Since the voltage is repetitively applied, it indicates that, in each discharge pulse, there is a significant amount of plasma charges that remain from the previous pulse in the discharge chamber.

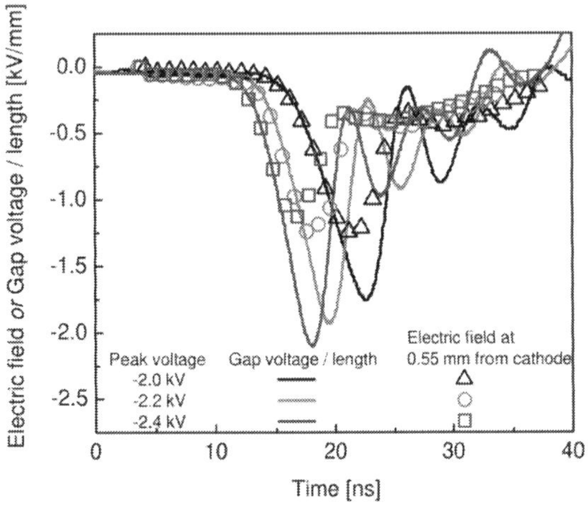

Figure 3. Measured electric field near the centre of the gap and nominal electric field as functions of the peak voltage for three different voltage conditions.

Such effects of the previous discharge pulses are also clearly depicted in Fig. 4, showing the minimum sustainable voltages for three repetition rates. In the repetition rate range tested, the higher the repetition rate is, the lower the required minimum voltage becomes. In all conditions presented in Fig. 4, we observed non-filament-like glow-like discharges. Although not shown here, at very low repetition rates such as 10 Hz, a uniform glow-like discharge cannot be maintained stably. That fact probably indicates that numerous electrons and ions remaining from the previous discharge pulse are required as seed charges for the following discharge pulse to obtain uniform glow-like discharges.

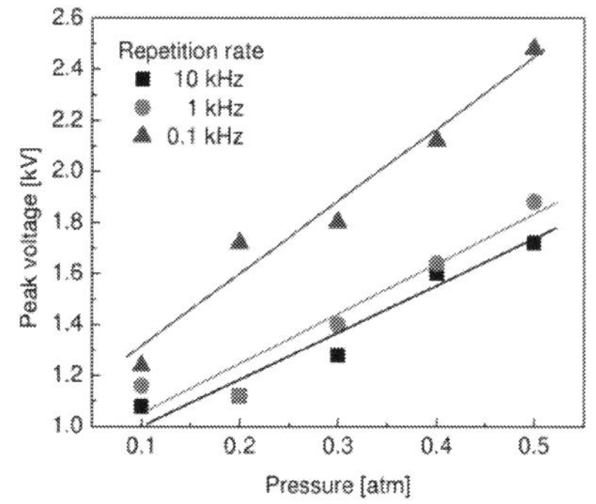

Figure 4. Minimum sustainable peak voltage as functions of the environmental pressure for three pulse repetition rates.

4. Summary

Repetitively pulsed nanosecond discharges were studied using electric field measurements based on electric-field induced Raman scattering. Results show an interesting inverse relation between the applied voltage and the electric field near the centre of the gap, possible because the effects of the previous discharge pulses are provided with supporting results of sustainable voltage as a function of the repetition rate. This kind of repetitively pulsed nanosecond discharges probably requires the initial seeded charges from the previous discharge pulses for such uniform glow-like discharge generation. With the low repetition rate, with intervals too long to retain a considerable amount of seeded charges, discharges likely turn to localized filament-like discharges.

Acknowledgments

This research was performed at the Frontier Research Base for Global Young Researchers, Osaka University, with funding through the program Promotion of Environmental Improvement to Enhance Young Researchers' Independence, Special Coordination Funds for Promoting Science and Technology, Japan Ministry of Education, Culture, Sports, Science and Technology (MEXT). This research has also been funded in part by a Grants-in-Aid for Scientific Research from Japan Society of Promotion of Science (JSPS)/MEXT. Part of the research team (S. M. and U. C.) was supported by the DFG under the framework of Research Group 1123, project TPA3.

References

[1] Evsin O A, Kupryanova E B, Ochkin V N, Savinov S Y and Tskhai S N 1995 *Quant. Elect.* **25** 278
[2] Akimov D A et al. 1999 *JETP Letters* **70** 375
[3] Tskhai S N et al. 2001 *J. Raman Spec.* **32** 177
[4] Ito T, Kobayashi K, Mueller S, Luggenhölscher D, Czarnetzki U and Hamaguchi S 2009 *J. Phys. D: Appl. Phys.* **42** 092003
[5] Ito T, Kobayashi K, Czarnetzki U and Hamaguchi S, *submitted to J. Phys. D: Appl. Phys.*

LIF investigations of O and NO products in air like RF plasma jet

S De Benedictis[1], PF Ambrico and G Dilecce
C.N.R., Istituto di Metodologie Inorganiche e dei Plasmi –sede di Bari
c/o Department of Chemistry, University Campus, via Orabona 4, 70126 Bari, Italy

[1] E-mail: santolo.debenedictis@.cnr.it

Abstract Laser induced fluorescence by one (LIF) and two photon (TALIF) excitation has been employed to characterize NO, O species in the expanded stream of N_2-O_2 air like low pressure plasma jet. The gas, excited in a coaxial RF capacitive discharge at pressure P_1, expands through a de Laval nozzle into a vessel at P_2 =0.25 Torr at expansion ratio P_1/P_2 of about 35. The multiple expansion–compression waves of the jet are traced by laser induced fluorescence of NO and O dissociation products expanding through the nozzle. The quantitative O and NO densities, obtained by in-situ calibration of TALIF and LIF signals are discussed.

1. Introduction

Air like plasma jet expansions as sources of supersonic flows for laboratory experiments on the interaction with material targets used in thermal protection systems of the space shuttle for Earth atmosphere re-entry are still involving aerospace laboratories [1]. Aerospace testing facilities based on plasma jet, employ high enthalpy high Mach number supersonic flows (as an example several tens of MJ/Kg and Mach higher than 7) produced by arc discharge at high pressure (0.1-1 bar). Such huge devices and their operating condition are very complex from the point of view of microscopic, atomic and molecular, diagnostics [2]. On the other hand expanded flows through a supersonic nozzle are intrinsically non-equilibrium flows whatever is the discharge condition [3], this making diagnostics of non equilibrium systems very important [4]. In this context advanced laser diagnostics can be tested on non equilibrium features of small wind tunnel installation and then applied confidently to more hostile expanded flows plasma jet facilities. Here we consider low enthalpy air like plasma jets produced by a capacitively coupled low pressure (~10 Torr) radio-frequency discharge [3]. Such an installation, plasma jet and diagnostics, has been used in previous studies to characterize gas/plasma expansions fed by $N_2/O_2/NO$ gas based mixtures [5-9]. Here we focus our attention on NO and O product species by measuring their axial and radial profiles by calibrated laser induced fluorescence by one (LIF) and two photons (TALIF) as well as on the behavior of some electronic state emissions of N_2 and NO by Optical emission spectroscopy (OES) with the aim to get insights on the presence of some energy carriers in the expanded flow.

2. Experimental

2.1 RF plasma jet installation

The RF plasma jet installation has been described in details in [5]. It is based on a discharge reactor (plasma source) ending with a stainless steel de Laval nozzle. Gas expanding through the nozzle in a low pressure chamber is pumped by root pump. The sketch of the discharge chamber is shown in figure 1. Essentially the discharge is located at the bottom of a stainless steel cylinder (length L= 500 mm and internal diameter d=40 mm) whose jacket is cooled by tap water. The plasma source is a capacitively coupled discharge with a co-axial RF electrode configuration driven under continuous and pulsed regime. The throat diameter of the nozzle is 4 mm while the convergent/divergent section lengths are respectively L_c =15 mm and L_d = 10 mm. The discharge region of plasma source is not accessible for diagnostics so the expansion region is investigated by a spectroscopic setup supporting computer controlled movements of both target and discharge holders.

[1] To whom any correspondence should be addressed.

2.2 Spectroscopic measurements

An air like plasma jet has been investigated by a spectroscopy setup based on OES and LIF, not shown here, and described in detail in [5].

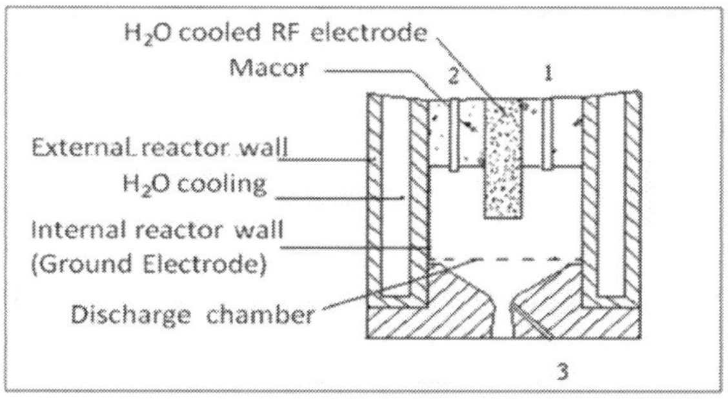

Figure 1. Sketch of reactor chamber. #1 pressure reading; #2 gas inlet; #3 throat pressure reading outlet, here it is sealed.

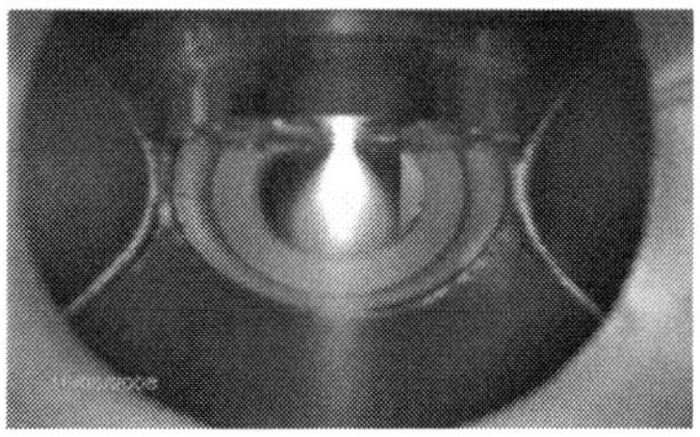

Figure 2. Snap shot of N_2/O_2 air like plasma jet

The plume investigated looks like that in the snapshot of figure 2 that refers to the condition in which the discharge reactor is fed at pressure P_1^{OFF} =7.1 Torr with N_2/O_2 mixture (Φ_{N2}= 780 sccm and Φ_{O2}=220 sccm, i.e. mass flow rate $M_\Phi(N_2\text{-}O_2)$=2.09x10^{-5}Kg/s) that expands into vessel kept at P_2^{OFF}= 0.25 Torr. The nominal power delivered to the reactor is W_0= 215 watt. When the plasma is activated, P_1^{ON} rises at 8.72 Torr while P_2 does not vary (P_2^{ON} =P_2^{OFF}). Under present experimental condition the Mach number is about 3, and the jet enthalpy about 1 MJ/Kg.

3. Results and discussion

3.1 NO Laser Induced Fluorescence

The detection of NO by laser induced Fluorescence has been done by exciting the (0,0) band of NO_γ (A $^2\Sigma_u^+$ -X $^2\Pi$) system and detecting fluorescence of (0,2) band. The laser has been tuned close to λ_{exc}=226.287 nm (vacuum) with the excitation of $^QP_{21}+Q_1$ band head of $^2\Sigma\text{-}^2\Pi_{1/2}$ sub band where the signal is very high and less sensitive to temperature variation. UV excitation photons are produced by third Harmonic generation of dye laser pumped by second harmonic of Nd-YAG laser. The laser line width is large to overlap several J lines of various sub-band of J manifold. LIF calibration has been operated in-situ by flowing a mixture of N_2+0.3% NO in the vessel at 1 Torr.

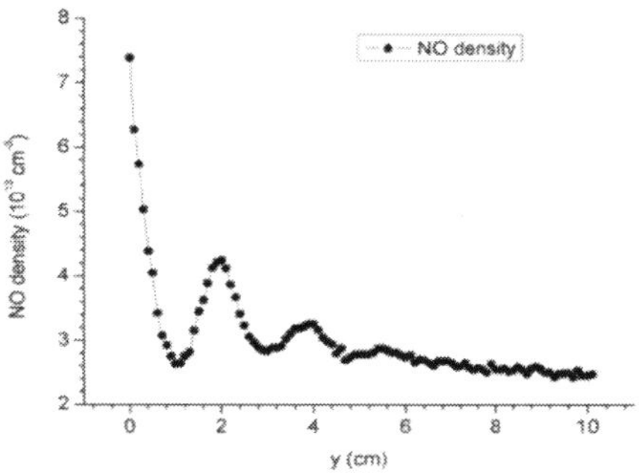

Figure 3. NO Axial scan by LIF

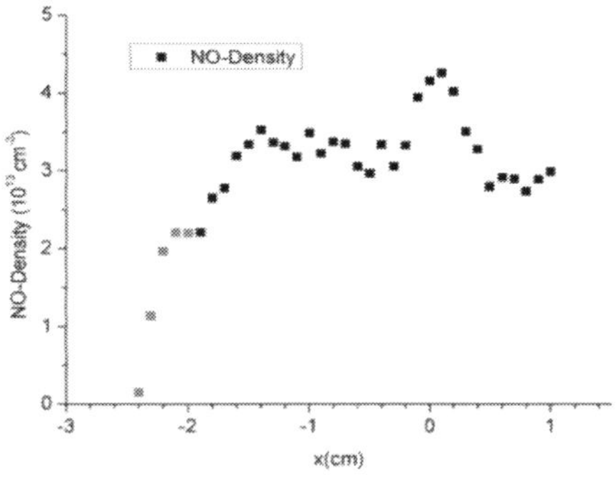

Figure 4. NO Radial scan by LIF at 2 cm distance from nozzle (red points: uncertain background region)

The results allow the following remarks:

i. The axial profile at the distance y=10 cm from nozzle (see figure 3) evidences a clear sequence of expansion-compression waves that are almost fully relaxed. Numerical simulations of O_2/NO cold expansion through this nozzle in [6] showed that at such a distance from nozzle exit a subsonic flow region is reached.

ii. The radial profiles carried out by radial scans between x=1 cm and x=-2.5 cm evidence a centreline minimum at y-positions corresponding to expansion waves and a centreline maximum at y-positions corresponding to compression waves. The figure 4 shows one of such profiles measured at y= 2 cm (first compression wave). The radial hill, whose width is about ±0.5 cm, grows on a noisy broad baseline that declines for x>1.8 cm. The profile indicates that at this distance from nozzle the region of unperturbed gas surrounding jet (background) could be at x >2.5 cm. The x-scan covers a section of the jet of about 2.5 cm radius while the cylindrical vessel walls are at about x= 10 cm. The radial scan at y=10 cm on contrary does not evidence significant structures within the scattering (±15%) of data and NO centerline density at y=10 cm is pretty close to the value at x= -1 cm y= 2 cm. This fact would indicate a quite broad distribution of NO in the jet.

iii. Considering the background NO density, $[NO]_b \sim 1.5 \times 10^{12}$ cm^{-3} , measured at x=-2.4 cm, y=2 cm, representative of NO density in the background region (where $T_{gas} \approx 300$ K, P= 0.27 Torr, $N_{(N_2\text{-}O_2)} \sim 9.5 \times 10^{15}$ cm^{-3}), its fraction in N_2-O_2 mixture would be about 0.014% .

3.2 O-atoms Two Photon Laser Induced Fluorescence (TALIF)

TALIF measurement of $O(^3P)$ atoms and the in-situ calibration in plasma jet are described in details in [5, 6]. The two photon excitation-detection scheme is:

$$O(2p\ ^3P_2) + 2h\nu_L \rightarrow O(3p\ ^3P_2) \rightarrow O(3s\ ^3S_1) + h\nu_E$$

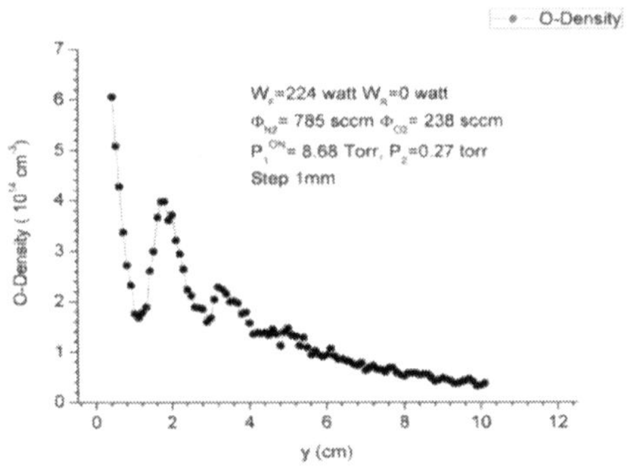

Figure 5. Axial scan of O by TALIF

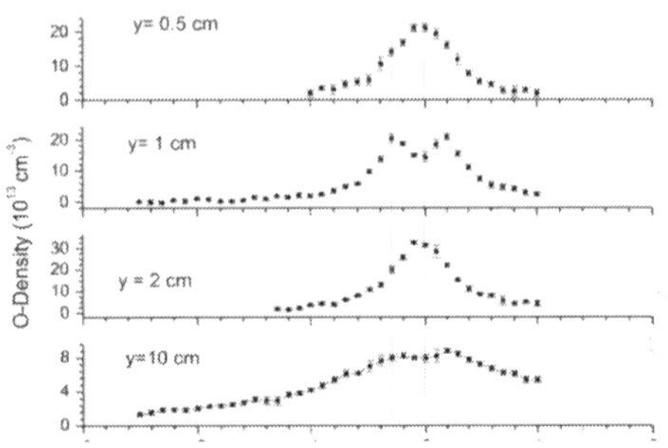

Figure 6. Radial scan of O by TALIF

Where excitation is done at λ_L = 225.58 nm and detection at λ_E = 844.6 nm. UV excitation photons are produced with the same laser configuration used for LIF measurements of NO. Detection is done by red-extended photomultiplier. The axial and radial profiles of O density, obtained upon TALIF calibration carried out as described in [6], shown in figures 5 and 6 offer the following insights:

i. The axial profile of O-atoms evidences the same expansion-compression structures of NO.

ii. The radial scan at various distance y from the nozzle instead reveals a different distribution of atoms in the jet section. A saddle profile with clear double maximum at x ≈ ±0.4 cm is seen in the expansion region and single maximum in the compression wave region. The background is clearly reached at y= 1 cm x= -2 cm: $[O]_b \approx 3 \times 10^{12}$ cm^{-3}. Instead at y= 10 cm, x= -2.5 cm laser beam crosses the jet border : $[O]_b \approx 7 \times 10^{12}$ cm^{-3} .

iii. The fraction of O-atoms in the background gas, $[O]_b / N_{(N_2\text{-}O_2)}$, is about 0.03-0.07 % depending on the uncertainty of background values.

3.3 Optical Emission spectroscopy

Optical Emission Spectroscopy (OES) from plasma plume has been carried out trough a monochromator SPEX 500 using both Photomultiplier and ICCD detectors. The spectral window is set on the band head and the entrance slit of the monochromator is typically 0.2 x 1 mm^2. The recording of photomultiplier signal is by photon counter under computer control.

Figures 7 and 8 report the axial scan of the emission intensity of NO ($A\ ^2\Sigma_u^+$ -$X\ ^2\Pi$) (0.2) band of γ system and of N_2 ($C^3\Pi_u$ -$B^3\Pi_g$) (0.2) band of Second Positive system (SPS) respectively.

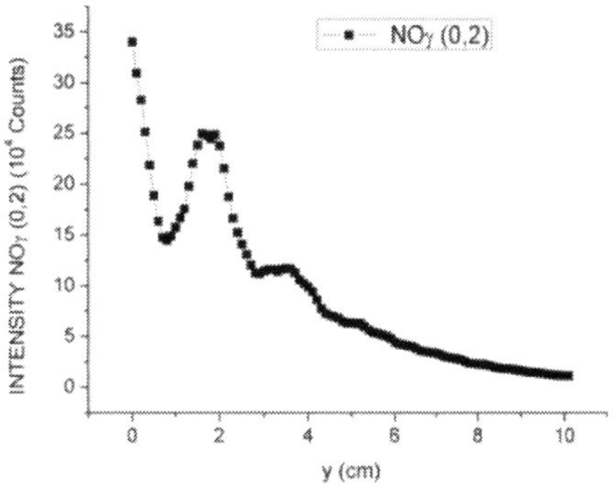

7.

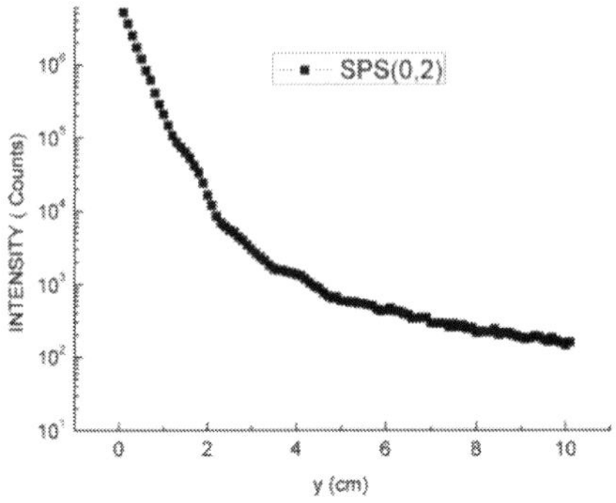

8. 2

The expansion traced by such species shows a significant dynamic range difference. While NOγ shows clearly the structure of multiple expansion-compression waves, the decline of N_2(SPS) by four order of magnitude at 4 cm from nozzle masks them significantly. The positions of maxima and minima coincide with those seen by NO and O species in the ground state. Excited species like N_2*($C^3\Pi_u$ at the excitation threshold $E_{th} \sim 11$ eV) and NO*($A^2\Sigma$ at $E_{th} \sim 5.5$ eV) give insights on the electrons and $N_2(A^3\Sigma_u)$ metastable precursors [3] as well as on nitrogen and oxygen collision processes. In air like mixture it is known that the density of $N_2(A^3\Sigma_u)$ is rather low [10] because of strong quenching by N, O, and O_2. The analysis of the vibrational distribution of $N_2(C^3\Pi_u)$ state confirms that in the plume even at long distance from nozzle (figure 8) N_2 electronic excitation occurs by electron impact from ground state. The excitation of NO_γ, that typically is by $N_2(A^3\Sigma_u)$ energy transfer [11], consequently has to occur by N+O+M recombination excitation.

4. Concluding remarks

The study has evidenced a plasma jet rich of structures observable both by Laser induce fluorescence and Optical Emission Spectroscopy investigations. Apart of the density measurements of NO and O, relevant species of air like plasma jet, by LIF and TALIF, the combined used of LIF and OES diagnostics has been seen useful to get insights on the active species in the jet. In the present RF plasma jet the presence of energetic electrons, likely sustained by plasma potential in the plume, has to be considered non negligible even at distance from nozzle as far as 4 cm. In the core of jet there is a significant density of O and NO products while in the background their fraction, about 0.05% and 0.014% respectively, is relatively low. The composition of gas in the background, due to recirculation of gas from bottom of chamber should represent a lower limit of jet gas composition if deactivation and recombination processes in the jet and in the background are negligible.

Acknowledgments

This work has been supported by Italian Space Agency (project ASI-CAST WP2326).

References

[1] Purpura C, De Filippis F, Barrera P and Mandanici D, 2008 *Acta Astronautica* **62** 410-421
[2] De Filippis F, Purpura C, Viviani A, Acampora L and Fusco L, 47th AIAA Aerospace Sciences Meeting, 2009, Orlando, Florida, 2009 *AIAA* **1596** 1-21
[3] De Benedictis S and Dilecce G 2002 *Pure Appl. Chem.* **74** 317-326.
[4] Schram D C, Plasma 2009 *Sources Sci. Technol.* **18** 014003 1-13
[5] Broc A, De Benedictis S and Dilecce G 2004 *Plasma Source Sci. and Technol.* **13** 505-514.
[6] Dilecce G, Vigliotti M and De Benedictis S 2000 *J. Phys D: Appl. Phys.*, **33** L53-L56.
[7] Ambrico PF, Dilecce G and De Benedictis S 2004 *Czech. J. Phys* **64** suppl. C, 1-6
[8] Dilecce G, Simek M, Vigliotti M and De Benedictis S 2000 *Appl. Spectrosc.* **54** 824-831.
[9] Broc A, De Benedictis S, Dilecce G, Vigliotti M, Sharafutdinov R and Skovorodko P 2004 *J. Fluid Mech.* **500** 211-237.
[10] De Benedictis S and Dilecce G 1997 *J. Chem. Phys.* **107** 6219-6229.
[11] De Benedictis S, Dilecce G and Simek M 1997 *J. Phys. D: Appl. Phys.* **30** 2887-2894.

Journal of Physics: Conference Series **227** (2010) 012020 doi:10.1088/1742-6596/227/1/012020

Investigations on the Generation of Atomic Oxygen Inside a Capacitively Coupled Atmospheric Pressure Plasma Jet

N Knake, D Schröder, J Winter and V Schulz-von der Gathen

Institute for Experimental Physics II, Center for Plasma Science and Technology, Ruhr-Universität Bochum, 44780 Bochum, Germany

E-mail: nikolas.knake@ruhr-uni-bochum.de

Abstract. Inside a miniaturized capacitively coupled cold atmospheric pressure plasma jet operated at a helium base gas flow with a minor molecular oxygen admixture atomic oxygen is created. The build up of atomic oxygen along the discharge channel and its further decay in the effluent is investigated by means of xenon calibrated two photon laser induced fluorescence spectroscopy (TALIF). The longitudinal and the transversal atomic oxygen distribution is measured from the discharge core through the transition area into the effluent. A particular emphasis is set on the influence of collisional quenching at elevated pressures.

1. Introduction

Cold jet-like plasma devices gained tremendous importance, not only in the field of medical application [1]. Reactive oxygen species such as atomic oxygen are supposed to be of major importance regarding wound treatment and germ reduction [2]. Though atomic oxygen densities have been measured in the effluent, investigations of its creation within the discharge are often limited due to small sizes and encapsulation of discharge devices. The miniaturized micro scaled atmospheric pressure plasma jet provides excellent access for active and passive optical diagnostics within the discharge core. Atomic oxygen densities in the free effluent [3] of few 10^{14} cm^{-3} and first independent density measurements in the discharge core [4] - more than one order of magnitude higher - show the importance of the transition area between plasma and charge free effluent. An improved, extended device with a prolongued gas flow channel allows a continuous observation of particle densities from the beginning of the discharge far into the effluent avoiding the creation of vortices and backflow of outer atmosphere.

2. Discharge device and experimental setup

The 13.56 MHz driven jet (figure 1) consists of two rectangular steel electrodes of 40 mm length, 1 mm thickness and 1 mm distance of which one is powered and the other one is grounded. BK7 glass spacers of same size and shape form an electrically isolated continuation of the discharge gap. Quartz glass panes cover the channel from both sides confining the gas flow of 1.5 slm He and 9 sccm O_2 to a channel of 1×1 mm^2 cross section. The discharge is mounted on a three axis stepper motor controlled stage within a steel chamber allowing a precise scanning of the discharge through the fixed focus of the optical setup (figure 2). A frequency tripled Nd:YAG

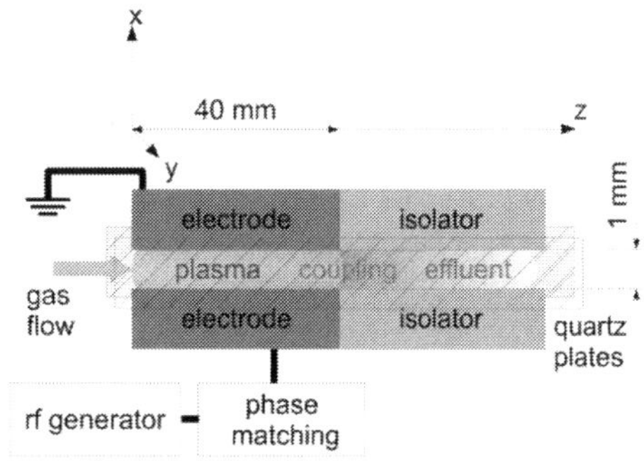

Figure 1. Draft of the augmented micro atmospheric pressure plasma jet.

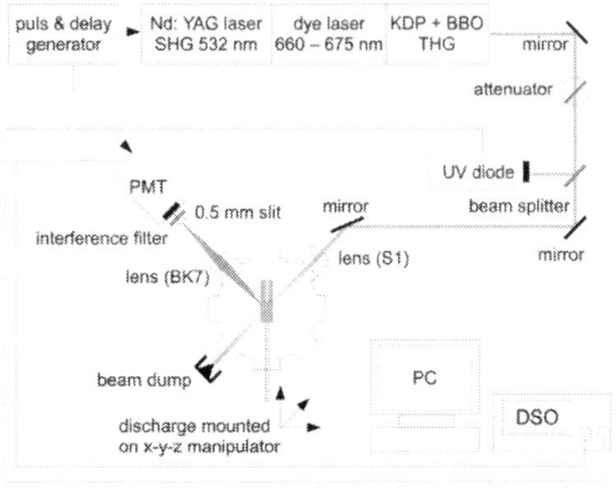

Figure 2. Draft of the experimental setup for TALIF spectroscopy.

laser pumped dye laser is used to generate nanosecond UV laser pulses at around λ=225 nm for two-photon excitation of atomic oxygen ($2p^4P_{2,1,0}-3p^3P_{1,2,0}$). The laser beam is focused into the center of the chamber and fluorescence light at λ=844nm is imaged into a cooled, gated infrared sensitive photo multiplier. Resolution is limited transversally by the focal beam diameter of $< 200\ \mu$m and longitudinally by the 0.5 mm slit in front of the PMT. A fast UV-Diode provides reference to the laser pulse energy and is recorded together with the PMT signal by a digital storage oszilloscope (HP 54510A, 250 MHz, 1 Gs s^{-1}) and instantly evaluated on a PC. Absolute calibration is performed by comparative measurements on xenon reference densities. Details on the calibration process can be found in [5, 6].

3. Results

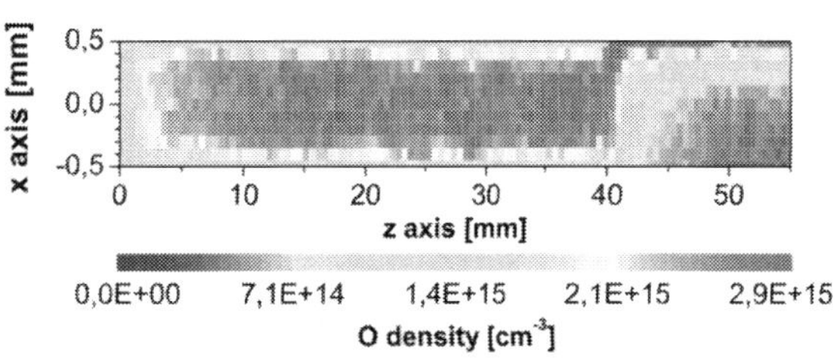

Figure 3. Atomic oxygen density map in cm^{-3}. Geometric shading by electrodes is not compensated. Additional laser reflexions appear at the upper edge of the isolated gap continuation.

A signal map of 1×55 mm^2 ($\Delta x = 0.1$ mm and $\Delta z = 0.5$ mm) can be seen in (figure 3). The jet was operated at 12 W generator power, 1 mm electrode distance, 1.5 slm helium and 9 sccm molecular oxygen admixed. Due to the small cross section of the channel in comparison to its length, the channel has to be precisely aligned to the manipulator axis (Δx and Δy may not shift more than 0.1 mm along the entire channel) to perform on axis measurements. Alternatively, the manipulator position must be adjusted respectively.

3.1. Spatial profiles

Figure 4 shows the normalized signal distribution in transversal direction (between electrodes) at different positions along the discharge channel. A convex profile can be found for each longitudinal position. This is dominantly caused by geometric shadowing of the fluorescence light by the electrodes. For calibration, a uniformly distributed xenon reference density inside

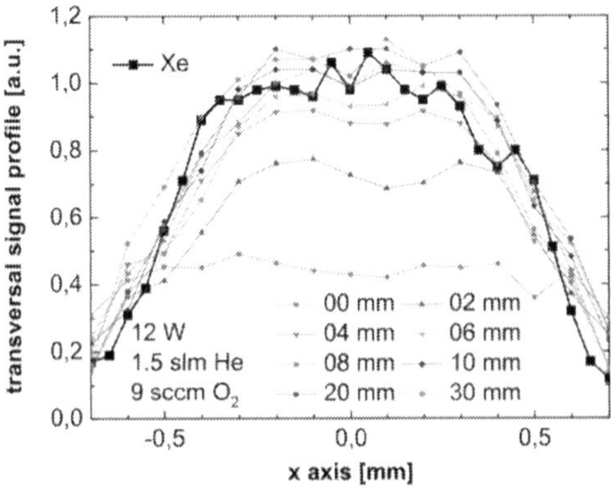

Figure 4. Normalized signal profiles in transversal inter electrode direction for atomic oxygen at different longitudinal positions z along the gas channel compared to the xenon reference profile (black).

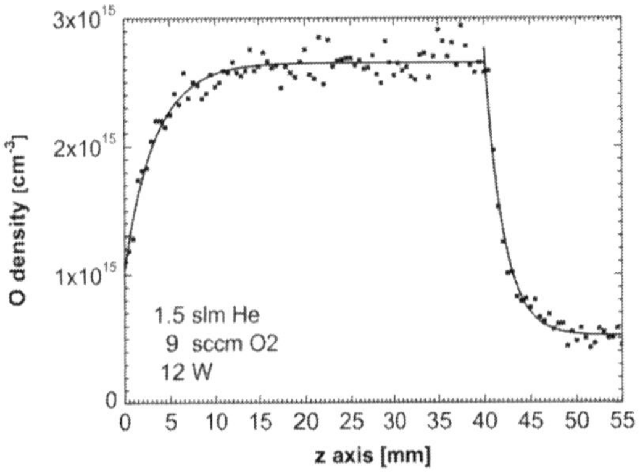

Figure 5. Profile of the atomic oxygen density in the transversal discharge center from beginning of plasma at $z = 0$ mm into the effluent starting at $z = 40$ mm. Ascent and descent show exponential behaviour.

the jet is mapped resulting in a transversal profile of the same shape. Thus the transversal O density distribution can be assumed uniform as well, within the detection limits.

Figure 4 shows the on axis ($x, y = 0$ mm) atomic oxygen density profile along the gas channel. The build up of a plateau can be seen from $z = 0$ to $z = 15$ mm and a decay can be found from the end of the plasma at $z = 40$ mm. The shape of rise and decay can be fitted best by $y = y_0 + A \ \ exp^{-x/\tau}$ functions, with $y_0 = 2,65 \times 10^{15}$ cm^{-3}, $A = -1,62 \times 10^{15}$ cm^{-3} and $\tau = 3,39 \pm 0,24$ mm for the ascent and $y_0 = 5,25 \times 10^{14}$ cm^{-3}, $A = 1,18 \times 10^{24}$ cm^{-3} and $\tau = 1,99 \pm 0,13$ mm for the descent. The overall accuracy of the absolute density values is estimated to be a factor of two. Both, y_0 and τ are important parameters to compare with chemical models for the oxygen chemistry at atmospheric pressure such as [7, 8]. Distances can easily be be transformed into reaction times by means of gas speed and time of flight.

3.2. Collisional de-excitation

An important issue for the applied TALIF calibration technique is the collisional de-exciation of the excited oxygen atoms since the density ratio of atomic oxygen and xenon ist proportional not only to the PMT signal ratio but also to the ratio of effective branching ratio a_{ij} of the observed excited states $< i >$ of oxygen and xenon respectively $a_{ij} = \frac{A_{ij}}{\sum_{k<i} A_{ik} + \sum_q n_q k_q^i}$. Here A_{ij} denotes the transition probability into a lower state, n_q density and k_q^i the quenching rate of species q with that state. At atmospheric pressure this collisional de-excitation can easily exceed photonic de-excitation. Thus, it is of major importance to know its amount for a reliable calibration. For the effluent, usually data from low to mid pressure measurements is extrapolated and only the major colliding species are taken into account, since the available database is still very limited [3, 4, 5, 6]. Since $\frac{1}{\sum_{k<i} A_{ik} + \sum_q n_q k_q^i} = \tau_{eff.}$ equals the effective lifetime of the respective excited state, we investigated the decay behaviour of the PMT signal for each measurement and compared it to the theoretical value with only helium atoms and bi-molecular oxygen in the ground state as colliders. For this we extrapolated mid pressure measurements of $k_{He} = 0.017 \times 10^{-10}$ cm^3 s^{-1} and $k_{O_2} = 9.4 \times 10^{-10}$ cm^3 s^{-1} from [5]. Varying the molecular admixture to the helium base gas flow (figure 6), we find excellent agreement with the assumed theoretical value down

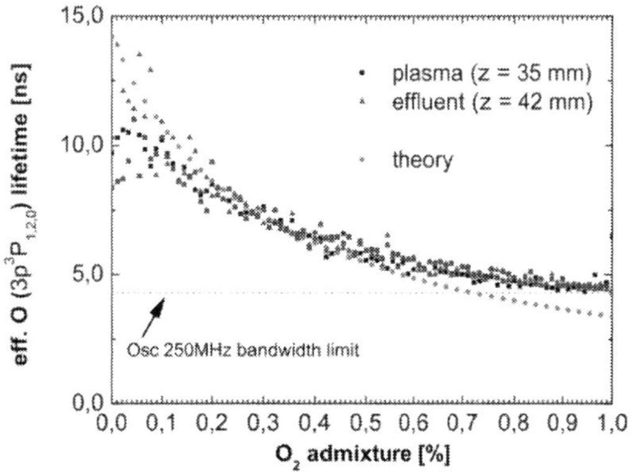

Figure 6. Effective lifetime of the $3p^3P_{1,2,0}$ atomic oxygen state in plasma (z = 35 mm) and effluent ($z = 42$ mm) compared to theoretical data.

to ≈ 5 ns converging to the bandwith limit of the used oscilloscope (250MHz). These life time measurements confirm that under our conditions it is reasonable for the effluent area ($z = 42$ mm) as well as for the plasma core ($z = 35$ mm) to restrict the considered quenching species to only He and O_2 in their ground state and to apply values extrapolated from low or mid pressure data.

4. Concluding remarks

We found that atomic oxygen density inside a miniaturized capacitively coupled cold atmospheric pressure plasma jet builds up converging exponentially into an equilibrium plateau within the plasma itself and decaying also exponentially within the effluent. The corresponding rise and decay constants are an important parameter to compare to theoretical models of the oxygen chemistry in such atmospheric plasma jets. The transversal density distibution can be found to be uniform within the detection limits. Also, asumptions of collisional de-excitation for the atmospheric pressure range could be confirmed within the bandwith limit of the used oscilloscope.

Acknowledgments

We thank J Waskönig, L Graham, K Niemi, St Reuter and T Gans from Queen's University Belfast for valuable discussion. This work was supported by the 'German Research Foundation' (DFG) in the frame of the 'Research Unit FOR 1123 - Physics of Microplasmas' and project 'SCHU-2353/1-1'.

References

[1] Becker K, Koutsospyros A, Yin S M, Christodoulatos C, Abramzon N, Joaquin J C and Brelles-Marino G 2005 *Plasma Phys. Control. Fusion* **47** B513B523
[2] Weltmann K-D, Brandenburg R, von Woedtke T, Ehlbeck J, Foest R, Stieber M and Kindel E 2008 *J. Phys. D: Appl. Phys.* **41** p. 194008
[3] Knake N, Reuter St, Niemi K, Schulz-von der Gathen V and Winter J 2008 *J. Phys. D: Appl. Phys.* **41** 194004
[4] Knake N, Niemi K, Reuter St, Schulz-von der Gathen V and Winter J 2008 *Appl. Phys. Lett.* **93** 131503
[5] Niemi K, Schulz-von der Gathen V and Döbele H F 2005 *Plasma Sources Sci. Technol.* **14** 375
[6] Dilecce G, Vigliotti M and De Benedictis S 2000 *J. Phys. D: Appl. Phys.* **33** L53
[7] Jeong J Y, Park J, Henins I, Babayan S E, Tu V J, Selwyn G S, Ding G, Hicks R F 2000 *J. Phys. Chem. A* **104** 8027
[8] Park G, Lee H, Kim G and Lee J K 2008 *Plasma Process. Polym* **5** 569-576

Journal of Physics: Conference Series **227** (2010) 012021 doi:10.1088/1742-6596/227/1/012021

Observations and modeling of the electron cyclotron emission background in the Levitated Dipole Experiment

P P Woskov[1], J Kesner[1], D T Garnier[2], M E Mauel[2] and S H Nogami[3, 4]

[1]Plasma and Science Fusion Center MIT, Cambridge, MA 02139, USA
[2]Columbia University, New York, NY 10027, USA
[3]University of San Diego, San Diego, CA 92110, USA

E-mail: ppwoskov@mit.edu

Abstract. Hot electron cyclotron emission (ECE) measurements and modeling calculations have been carried out for the Levitated Dipole Experiment (LDX)[5]. Heterodyne radiometers at 110, 137 and 165 GHz have been used to view the plasma from bottom and midplane locations. An intense environment of harmonic ECE of up to 1 keV at 110 GHz and afterglow decay times of up to 6.9 seconds were observed at lower pressures. Interpretation of the emission levels in terms of hot electron parameters requires integrating modeled emission over a large magnetic field range of 0.09 to 3.2 Tesla and a **k** to **B** angular view range of 0 to 90° within the field of view of each receiver as the hot electron density peak is followed around a given flux contour. Hot electron temperatures of the order of 100 keV, depending on knowledge of the electron density peak location, were determined by taking the ratio of radiometer signals at different frequencies. Abrupt radial movements of the highly peaked hot electron density where evident, particularly with ECRH turn on and turn off.

1. Introduction

Hot electrons are a characteristic feature of electron cyclotron resonance heated (ECRH) plasmas [1-3]. The intense harmonic electron cyclotron emission produced can be exploited as a diagnostic of the hot electrons, but also presents a significant background for active probe aided diagnostics in the affected frequency range. Interest in developing such diagnostics is motivated by recent research of the magnetic dipole field configuration, like that observed around planets, as an alternative confinement concept for fusion energy. A magnetic dipole based fusion reactor would use a He^3 catalyzed D-D fuel cycle that would minimize the power by neutrons and would not require external tritium production [4]. Levitated coil remote power refrigeration may be feasible in a reactor size levitated device. Free floating superconducting magnetic dipole experiments are operating at MIT/Columbia, LDX [5], and University of Tokyo, RT-1 [6].

LDX uses a free floating 1.1 MA, 68 cm mean diameter, 560 kg superconducting coil (F-coil) in a large 5 m diameter vacuum chamber [7]. The plasma is confined on the outside of the current carrying coil rather than inside a set of coils as in most other magnetic confinement devices. The F-coil is routinely floated by a for up to 3 hours between cryogenic recoolings by a 280 kA lifting coil (L-coil)

[4] U.S. DOE National Undergraduate Fellowship appointment
[5] http:// www.psfc.mit.edu/ldx/

during which time plasmas are pulsed by ECRH. Up to a total of 17 kW power from four sources: two at 2.45 GHz (2.5 kW and 1.9 kW), one at 6.4 GHz (2.5 kW), and one at 10.5 GHz (10 kW) are pulsed to sustain plasmas for up to 14 s with a slowly decaying afterglow that can be detected for up to 25 s at lower densities after all plasma sustaining sources are off. Line averaged electron densities of up to 2 x 10^{11} cm^{-3} with highly peaked profiles (~$1/r^4$) are diagnosed with a diagnostic set that includes a 4-channel 60 GHz interferometer.

2. Analytical Basis

It is assumed that the hot electron ECE can be represented by blackbody intensity I_{BB} at a single temperature even if the electron distribution is not Maxwellian in which case the temperature is then specific to the ECE measurement parameters. If the radiation transport is also optically thin, then emission intensity I_ω detected by each radiometer with frequency ω over path length s can be expressed by [8]

$$I_\omega \simeq I_{BB}\alpha_\omega s \qquad (1)$$

where α_ω is the ECE absorption coefficient that is a function of frequency, magnetic field vector amplitude and direction, and electron density and temperature. Here we use the expression for α_ω given by Nassri and Heindler equation (2) [9]. In interpreting ECE in the present experiment it is necessary to integrated over a wide range of harmonics and view angles observed simultaneously by each fixed frequency radiometer. This is different from previous ECE diagnostic experience where typically only one harmonic at one view angle is observed at a time. The viewing geometries used in LDX are illustrated in figure 1 showing the diffraction limited $1/e^3$ Gaussian beam antenna views at 137 GHz superimposed on the flux contours. A highly peaked plasma, limiting s to a few cm, follows along an inner flux contour going through the outer midplane at < 80 cm horizontal. Along this flux line the **B** field strength can vary from as low as 0.09 on the outside to as high as 3.2 Tesla on the inside. The angle of **B** to the view **k** varies from being parallel to perpendicular. A partial obstruction of the vertical view by the bottom F-coil catcher is not shown [7].

Figure 2 shows a calculated ECE harmonic spectrum for the bottom view at 137 GHz summed over both x and o components and weighted by the Gaussian receiver antenna pattern. The lower harmonics dominate the emission even when they occur in the outer fringe region of the radiometer field of view. This is particularly true at the midplane where the upper and lower edges of the view contribute most.

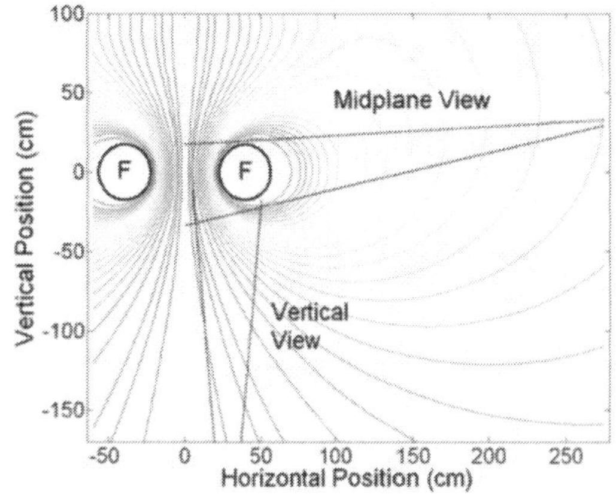

Figure 1. 137 GHz millimeter-wave radiometer fields of views in LDX from the bottom and midplane (F: F-coil).

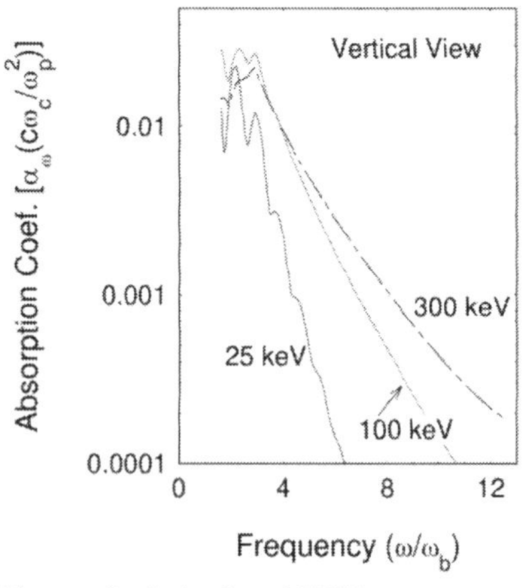

Figure 2. Calculated ECE spectrum at 137 GHz viewing from the bottom in LDX for several electron temperatures.

3. Experimental Setup

The ECE background in LDX was explored over several plasma campaigns using heterodyne radiometers at 110, 137, and 165 GHz singly and in combined pairs for the two viewing geometries shown in figure 1. The receivers had double sideband noise temperatures (DSB) of 2000 to 3000 K and the same 3 GHz DSB IF electronics. Small signal calibration was accomplished with liquid nitrogen and room temperature blackbody sources (eccosorb) and the IF detector and dc amplifier calibrated response was extended to a 40 dB dynamic range using a microwave source that was measured by an Agilent E4418B Power Meter. Figure 3 shows how pairs of radiometers were combined on one view port to have an identical field of view within the limit of diffraction. The scalar horn output of the receivers was combined in wide bandwidth 20 mm diameter circular corrugated waveguide with a quartz beamsplitter. Then the waveguide diameter was expanded to either 29 mm or 35 mm for launching the HE_{11} mode view into LDX through calibrated fused quartz windows over a free space distance of the about 2.5 m to the plasma. The plasma does not fill the radiometer views so the reported temperatures here are lower limits of the local emission.

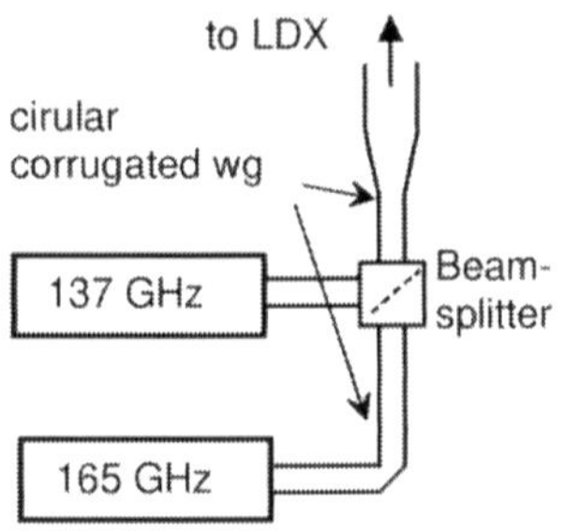

Figure 3. Radiometer pairs combined into single view with corrugated waveguide and beam splitter.

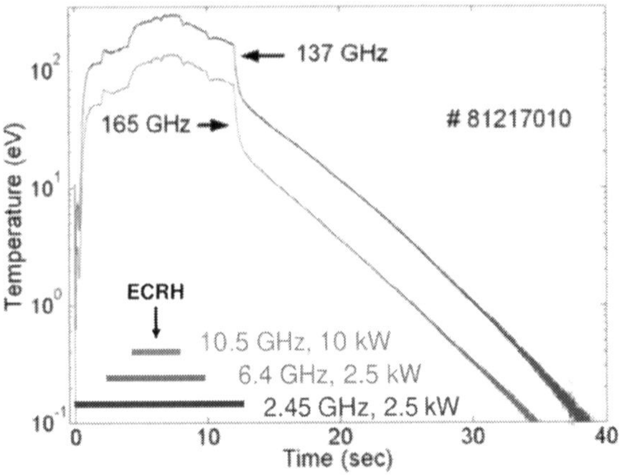

Figure 4. Observed LDX ECE from bottom view at 137 and 165 GHz plotted on log scale showing long afterglow.

4. Results

An example of the ECE observed vertically from below LDX at a horizontal radius of 28 cm in a D_2 plasma with a line averaged peak plasma density of 0.8 x 10^{11} cm^{-3} is shown in figure 4. A total of 15 kW of ECRH was applied as indicated by the bars in the lower left corner. The ECE increases with ECRH power and peaked at radiometer detected temperatures of 300/135 eV for 137/165 GHz, respectively. A long afterglow after all sustaining power is turned off was detected for over 25 s with a 1/e decay time of 6.9 s. Background pressure was 3.6 x 10^{-7} Torr. This afterglow decay is similar to that observed in the levitated spherator [1] and also depends exponentially on background pressure in LDX.

The ECE was observed to peak at 900/330 eV for 110/137 GHz, respectively for both radiometers paired at the midplane view as shown in figure 5 for a D_2 plasma with a line average density of 1.6 x 10^{11} cm^{-3}. In this case it was possible to determine the hot electron temperature by taking the ratio of the two radiometer signals to cancel out the dependence on electron density and path length *s*. However, knowledge on which flux contour the density peaks is required. Two assumptions are shown in lower figure 5. The absolute temperature is sensitive to location, but the transient features do not change. For the plotted inner position, the hot electron density initially increases to about 130 keV in the first second and then deceases as the density builds up with more ECRH power. Transient displacements of the radial position of the hot electron peak are evident, particularly with ECRH switch on and off, supporting the assumption that the hot electron profile is highly peaked. In the afterglow the apparent increase in hot electron temperature to 200 keV could be explained by both the preferential loss of cold electrons and the inward radial displacement of the density peak.

The contribution of internal reflections to the measured signals is an important question that can affect the interpretation. Measurements made by two near identical 137 GHz radiometers from below and at the midplane are shown in figure 6. The ratio of the calibrated signals from these two radiometers is in not 1 as would be expected for uniform reflective emission distribution. The contrast goes as high as a factor of 5 during plasma transients. The vertical view radiometer sees the inner dipole region, which is blocked from the midplane view. The increasing ratio at turn on is due to the warming up of the electrons. At turn off, the much higher ratio increase may be an indication of an additional transient imbalance in the electron distribution between inner and outer regions.

This investigation of the harmonic ECE in LDX shows that an intense background is present in levitated magnetic dipole plasmas, and that useful localized interpretation of the hot electron parameters are possible despite the large magnetic field gradient and orientation variation present. Active millimeter-wave and terahertz probe diagnostics would need to consider this radiation.

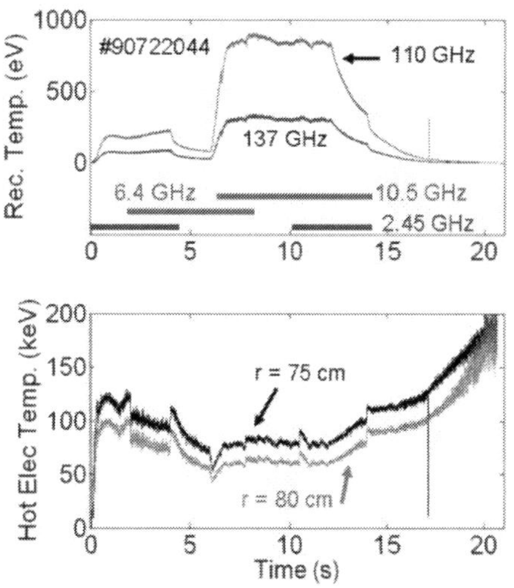

Figure 5. Top: ECE for midplane view with 110 and 137 GHz radiometer pair. Bottom: Hot electron temperature for two assumed peak density locations.

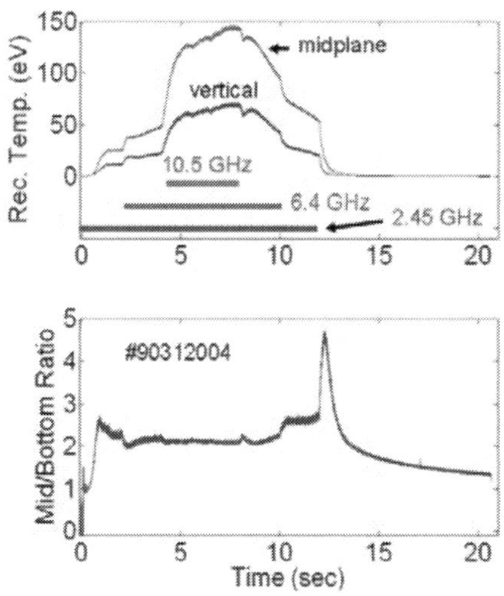

Figure 6. Top: Simultaneous 137 GHz ECE from midplane and vertical views. Bottom: Ratio of the two radiometer signals.

References

[1] Okabayashi M, Voitsenya V, Ripin B, Schmit J and Yoshikawa S 1973 Hot electron plasma in a levitated spherator, *The Phys. Fluids* **16** 1337-48

[2] Hokin S A, Post R S and Smatlak 1989 Constance B *Phys. Fluids B* **1** 862-73

[3] Barué C, Briand P, Girard A, Melin G and Briffod G 1992 Hot electrons studies in Minimaflios ECR ion source *Rev. Sci. Instrum.* **63** 2844-6

[4] Kesner J, Garnier D T, Hansen A, Mauel M E and Bromberg L 2004 Helium3 catalysed D-D fusion in a levitated dipole *Nuclear Fusion* **44** 193-203

[5] Garnier D T, Boxer A C, Ellsworth J L, Kesner J and Mauel 2009 M E *Nuc. Fusion* **49** 055023

[6] Ogawa Y, Yoshida Z, Morikawa J, Saito H, Wantanabe S, Yano Y, Mizumaki S and Tosaka T 2009 *Plasma and Fusion and Rresearch* **4** 020

[7] Zhukovsky A *etal* 2005 *Fusion Enginnering and Design* **75-79** 29-32

[8] Bekefi G 1966 *Radiation Process in Plasmas* (New York: John Wiley and Sons)

[9] Nassri A and Heindler M 1988 *Phys. Fluids* **31** 95-8

Journal of Physics: Conference Series **227** (2010) 012022 doi:10.1088/1742-6596/227/1/012022

Recent Progress in Two-Color Laser Diagnostics

K. Kawahata[1], T. Akiyama[1], K. Tanaka[1], S. Okajima[2], K. Nakayama[2],

[1] National Institute for Fusion Science, Toki, 509-5292, Japan
[2] Chubu University, Kasugai, 487-8501, Japan

E-mail: Kawahata@lhd.nifs.ac.jp

Abstract. This paper describes an innovative laser diagnostic instrument, a two color FIR laser interferometer/polarimeter, under development at the National Institute for Fusion Science, aiming for the establishment of reliable electron density/current density profile measurement techniques in the next step magnetically confined fusion devices such as ITER. In the process of searching for new laser oscillation lines, we have successfully achieved the powerful laser lines simultaneously oscillating at 57.2 and 47.7 μm, which lead to the construction of a new type of two color laser interferometer. So far we have been developing a two color laser interferometer, and confirmed its original function, vibration subtraction, in a test stand. In this conference, development of laser diagnostics which is a combined system of a two color laser interferometer and a polarimeter including hardware development necessary for future fusion devices will be presented.

1. Introduction

Measurements of the refractive index of plasmas by using electromagnetic waves are a well-established technique for measuring electron density profiles in high temperature plasmas. In the present large fusion devices such as JET, JT-60U and LHD, far-infrared (FIR) laser diagnostics have widely been used for density profile measurement. The wavelength of the FIR lasers used in these devices is around 100 to 200 μm, which is optimized from the viewpoints of the plasma refraction and mechanical vibration effects for target plasmas. Recent progress in particle fuelling technique combined with particle pumping system makes it possible to produce high density plasmas [1] of more than 10^{21} m^{-3} in the Large Helical Device (LHD). In these high density plasmas, a steep density gradient is formed leading to the fringe jumps on the density traces measured by FIR laser interferometers. To overcome this problem we have been developing short wavelength FIR laser diagnostics around 50 μm, and achieved a powerful 57.2 μm CH_3OD laser [2] pumped by a continuous wave 9R(8) CO_2 laser. After that, by optimizing lasing conditions we have successfully achieved high power two color laser oscillations [3] at 57.2 and 47.7 μm. The total output power is over 2.4 W, which is the highest cw laser output power in the far-infrared wavelength regime. This new two color laser oscillations enabled us to develop a new two color FIR laser interferometer [4], which has several benefits compared with conventional two color FIR laser interferometers as follows: (i) low refractive effects in high-density plasma; (ii) no additional laser interferometer for vibration compensation; (iii) no optical path difference between two color laser interferometers; (iv) one detector simultaneously

1 K. Kawahata, National Institute for Fusion Science. 322-6, Oroshi-cho, Toki, 509-5292, Japan
2 S. Okajima, Chubu University, 1200 Matsumoto-cho, Kasugai, 487-8501, Japan

Journal of Physics: Conference Series **227** (2010) 012022 doi:10.1088/1742-6596/227/1/012022

detects the beat signals of two color laser oscillation lines. We already demonstrated [5,6] usefulness of this two color laser diagnostic for mechanical vibration compensation in a test stand.

For application to an ITER laser diagnostic, a poloidal polarimeter for current density profile measurement, we have been developing a polarimeter with silicon photo-elastic modulators (PEMs) for the 57.2 and 47.7 μm laser. Since there were no PEMs for the FIR wavelength range so far, a new PEM for the wavelengths was developed [7] with HINDS instruments. Good linearity between actual and evaluated polarization angles is achieved with an angular resolution of 0.01 (0.025) degrees and a temporal resolution of 1 (0.1) ms [8], which satisfy requirements from ITER. In this symposium, recent progress in the development of two color laser diagnostics will be presented

2. Optimization of laser wavelength

For future large fusion devices such as ITER, a poloidal polarimeter system [9] is being designed for measuring the current density profile. A 118.8-μm CH_3OH laser has been proposed as a probing light source for the polarimeter, since the 118.8-μm CH_3OH laser line was the shortest far infrared laser oscillation line with the highest power. But in this case the beam bending effect due to plasma density gradient is large. If we consider shorter wavelength lines, the next candidate is the 10.6-μm CO_2 laser line. In this case, the Faraday rotation angle is less than one degree for a typical poloidal chord on ITER. When a linearly polarized electromagnetic wave, whose electric field vector is parallel to the toroidal magnetic field, transverses a poloidal plasma cross section perpendicular to the plasma current, the phase of the wave is changed by φ and its plane of polarization rotates by angle Ω due to the plasma as follows:

$$\varphi[\mathrm{rad}] = 2.82\times10^{-15}\lambda[\mathrm{m}]\int n_e(z)[\mathrm{m}^{-3}]dz \tag{1)}$$

$$\Omega[\mathrm{rad}] = 2.62\times10^{-13}\lambda[\mathrm{m}]^2\int n_e(z)[\mathrm{m}^{-3}]B_{\parallel}[\mathrm{T}]dz \tag{2)}$$

where λ is the wavelength of probing laser light, n_e is the electron density, $B_{\parallel}$ is the component of the poloidal magnetic field parallel to the laser light. In order to obtain the current density profile from the values of Ω, the electron density profile should be measured by the interferometry or by using the Cotton-Mouton effect. The ellipticity ε due to the Cotton-Mouton effect is given by

$$\varepsilon[\mathrm{rad}] = 2.45\times10^{-11}\lambda[\mathrm{m}]^3\int n_e(z)[\mathrm{m}^{-3}]B_t[\mathrm{T}]^2 dz \tag{3)}$$

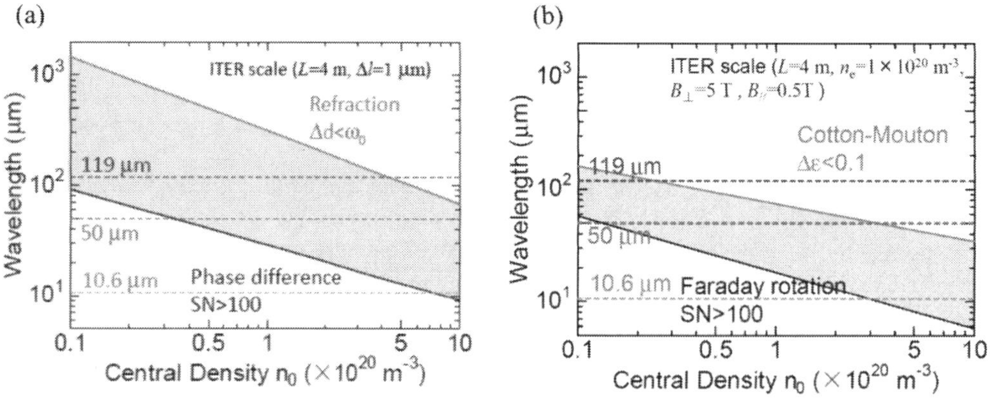

Figure 1. Optimum wavelength of (a) interferometer, (b) polarimeter for ITER scale device

The upper and lower limits of the optimum wavelength for an interferometer are determined by beam refraction and conditions of signal to noise ratio (S/N) of phase measurement, respectively. Figure 1 (a) shows the optimum wavelength space in the case of ITER scale device: optical path length in a plasma L=4 m, $n_e = n_0(1-\rho^2) = 1\times10^{20}$ $(1-\rho^2)$ m^{-3}, changes in the optical path length by mechanical vibrations Δl=1 μm. It is supposed that the S/N is larger than 100 and beam displacement Δd due to the refraction in the plasma is less than the beam radius. On the other hand, optimum wavelength for a polarimeter is determined by the condition to suppress the Cotton-Mouton effect and S/N of measurement of the Faraday rotation angle. Figure 1(b) shows the optimum wavelength space, supposing that the uniform electron density of 1×10^{20} m^{-3} and the uniform perpendicular and parallel magnetic field $B_\perp$, $B_{//}$ are 5 and 0.5 T. Here, the Cotton-Mouton effect is less than 0.1 rad. and the S/N of the Faraday rotation angle is larger than 100; the angle resolution is 0.01 deg. From these two figures, the applicable density range of 50 μm is widest among wavelengths of 10.6, 50, and 119 μm.

3. Combined system of interferometry/polarimetry

So far, we have been developing interferometry and polarimetry independently, and confirmed that the both diagnostics satisfy ITER diagnostics requirements [9]. The achieved angular resolution of the polarimeter is 0.01 degrees with a time constant of 1 ms [8]. Then, the system is being upgraded to the simultaneous measurements of a phase shift and a rotation angle. One of the key issues for simultaneous measurements is the development of a high quality filter to select one wavelength for polarimetry. Then, we have developed a Fabry-Perrot filter operating in THz wavelength regime. Figure 2 shows a constructed tunable Fabry-Perot filter (a) and its power transmittance (b) as a function of cavity length. A metal mesh was applied as a reflector. The number of mesh wires is 1000/inch, and a wire width is 7.37 μm with the thickness of 5 μm. The calculated value of reflectance is 82 % and the measured one is 86 %.

(a)

(b)

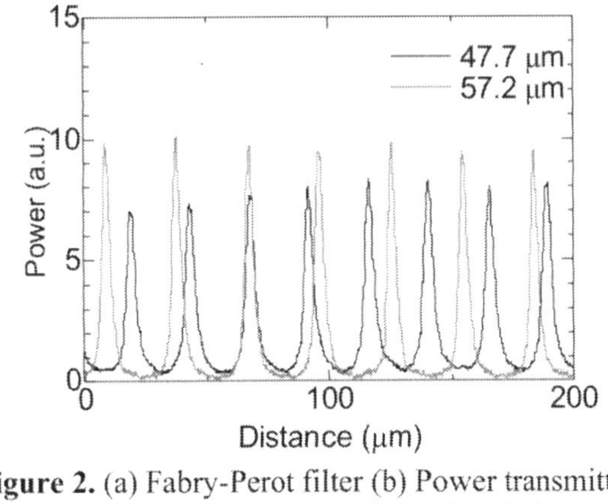

Figure 2. (a) Fabry-Perot filter (b) Power transmittance as a function of a cavity length. Horizontal scale is measured from the initial position of the reflector.

In the polarimeter part, one wavelength has to be selected for the detection of a rotation angle as is shown in Fig. 3(a). As a filter to select one wavelength, a Fabry-Perot interferometer filter is used. Figure 3 (b) shows an effect of the usage of the Fabry-Perot interferometer filter on the angle resolution. Compared with that in the case that 57.2 μm is selected with a polarizer, the angle resolution slightly deteriorates although the required resolution from q-profile measurement in ITER is still satisfied. The deterioration is due to contamination of other wavelength, 47.7 μm, whose added optical retardation by the PEM is different. This would be improved by optimization of the Fabry-Perot interferometer filter, since larger cavity length would provide good elimination of unwanted component. In the real application to ITER diagnostics, we need to take into account the Faraday effect on the laser beam propagating through the vacuum windows. The rotation angle of θ is proportional to the beam path l_w through the window and to the magnetic field strength parallel to the beam propagation at the window H, that is $\theta = V l_w H$, where the constant V is well known as the Verdet constant. By the use of the two color polarimeter, the Faraday rotation component at the vacuum window is expected to be eliminated [10].

(a) (b)

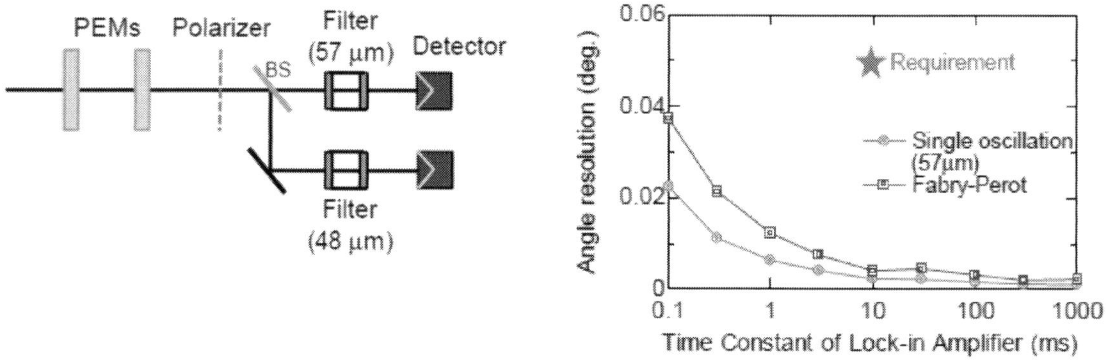

Figure 3. (a) a polarimeter part of the optical system, (b) angular resolution with and without the Fabry-Perot interferometer.

4. Summary

We have been developing a two color laser interferometer/polarimeter using developed two color laser oscillation lines at 57.2 μm and at 47.7 μm. Laser oscillation lines around 50 μm are found to be optimum for the large fusion devices such as the ITER considering beam bending effect and expected signal to noise ratio for phase shift and Faraday rotation measurements. We found that both laser diagnostics, interferometry and polarimetry, satisfy the ITER diagnostics requirements. The achieved angular resolution is 0.01 degrees with the time constant of 1 ms.

Acknowledgments

The authors would like to thank Professor A. Komori for his encouragement. This work was supported in part by a Grant-in-Aid for Science Research from the Japanese Ministry of Education, Culture, Sports, Science and Technology, "Priority area of Advanced Burning Plasma Diagnostics" (16082208) and also by the NIFS budget codes of NIFS09ULHH501 and NIFS08KCHH021.

References

[1] Ohyabu N, Morisaki T, Masuzaki S, Sakamoto R, et al. 2006 *Phys. Rev. Lett.* **97** 055002
[2] Okajima S, Nakayama K, Tazawa H, Kawahata K, et al. 2001 *Rev. Sci. Instrum.* **72** 1094
[3] Nakayama K, Tazawa H, Okajima S, Kawahata K, et al. 2004 *Rev. Sci. Instrum.* **75** 329
[4] Kawahata K, Tanaka K, Tokuzawa T, Akiyama T, et al. 2004 *Rev. Sci. Instrum.* **75** 3508
[5] Kawahata K, Akiyama T, Pavlichenko R, Tanaka K, et al. 2006 *Rev. Sci. Instrum.* **77** 10F132
[6] Kawahata K, Akiyama T, Tanaka K, Nakayama K, et al. 2008 *Rev. Sci. Instrum.* **79** 10E707
[7] Akiyama T, Kawahata K, Okajima S, Nakayama K, et al. 2007 *J. Plasma Fusion Res.* **2** S1113
[8] Akiyama T, Kawahata K, Tanaka K, et al. 2008 *Rev. Sci. Instrum.* **79** 10E720
[9] Donne A J H, Graswinckel M F, et al. 2004 *Rev. Sci. Instrum.* **75** 4694
[10] Kawano Y, Chiba S, Shirai H, Inoue A. Nagashima A, 1999 *Rev. Sci. Instrum.* **70** 714

Pulse-burst operation of standard Nd:YAG lasers

D J Den Hartog,[1,2] J R Ambuel,[3] M T Borchardt,[1] J A Reusch,[1] P E Robl,[3] and Y M Yang[1]

[1] Department of Physics, University of Wisconsin–Madison, Madison, Wisconsin 53706 USA
[2] Center for Magnetic Self-Organization in Laboratory and Astrophysical Plasmas, University of Wisconsin–Madison, Madison, Wisconsin 53706 USA
[3] Physical Sciences Laboratory, University of Wisconsin–Madison, Stoughton, Wisconsin 53589 USA

E-mail: djdenhar@wisc.edu

Abstract. Two standard commercial flashlamp-pumped Nd:YAG lasers have been upgraded to "pulse-burst" capability. Each laser produces a burst of up to fifteen 2 J *Q*-switched pulses (1064 nm) at repetition rates 1–12.5 kHz. Variable pulse-width drive (0.15–0.39 ms) of the flashlamps is accomplished by IGBT (insulated gate bipolar transistor) switching of electrolytic capacitor banks. Direct control of the laser Pockels cell drive enables optimal pulse energy extraction, and up to four 2 J laser pulses during one flashlamp pulse. These lasers are used in the Thomson scattering plasma diagnostic system on the MST reversed-field pinch to study the dynamic evolution of the electron temperature.

1. Laser configuration and standard operation

The Thomson scattering plasma diagnostic system on the Madison Symmetric Torus (MST) reversed-field pinch (RFP) has been successfully operated for several years with two standard commercial Nd:YAG laser systems (Spectron SL858) [1]. This type of flashlamp-pumped laser system is common, and similar systems are available from several manufacturers. As delivered by the manufacturer, our lasers were capable of ≥2 J single pulse, or ~1 J pulses at 50 Hz rep rate. This paper describes the upgrades made to enable each laser to produce a burst of up to fifteen ~2 J pulses, at rep rates up to 12.5 kHz, with the ability to balance burst length and rep rate to best suit experimental needs. The optical head of each laser was left unchanged, except for the addition of programmable Pockels cell drive. The major change was replacement of the flashlamp power supplies with a flexible modular system that controls all aspects of flashlamp operation. This paper begins with a brief description of the laser system as originally configured and operated, then describes the upgrades made and the new operational capabilities.

The optical head of each laser consists of an oscillator with Pockels cell *Q*-switch, pre-amplifier, and double-rod amplifier. Each rod is pumped by a single flashlamp, thus the optical head contains four individual linear flashlamps. As delivered by Spectron, the flashlamps were driven with a standard critically damped inductor/capacitor single mesh pulse-forming network; in other words, a charged capacitor bank was discharged by a fast switch (a thyristor) into an inductor in series with the flashlamp [2]. The capacitance and inductance were approximately matched to the impedance of the flashlamp arc to produce a fixed-width damped sine pulse with a FWHM of about 100 μs. The Pockels cell Q-switch was opened past the peak of the flashlamp pulse to produce a single 2 J, 9 ns laser pulse. This was repeated every few minutes to coincide with the plasma production cycle of the MST device. Thus, with two lasers, it was possible to obtain two Thomson scattering measurements every few minutes.

While this data was valuable, it was extremely difficult to measure the evolution of, or fluctuations in, the plasma electron temperature during the approximately 20 ms equilibrium period of a single MST plasma discharge. Such data could only be acquired by taking large ensembles of measurements from many MST discharges. This was costly and tedious, and was compromised by the imperfect reproducibility of MST discharges. Thus there was substantial motivation to modify our laser system to produce a short burst of pulses at a high repetition rate, but with burst production required only every few minutes. Fortunately, this combination of requirements did not require investment in an expensive custom laser system, but could be met by two relatively straightforward upgrades of the existing commercial lasers.

2. Laser upgrades and modes of operation

The first upgrade made to the lasers was replacement of the original manufacturer's Pockels cell drive circuit with a Pockels cell driver manufactured by Bergmann Messgeräte Entwicklung KG. This driver enables precise on/off control of the Pockels cell (Cleveland Crystals Impact 10), and is able to repeat the on/off sequence every ~30 μs. This upgraded driver provided two immediate advantages. First, it dramatically reduced the pulse of high frequency electrical noise previously recorded during a laser pulse by all sensitive detectors in the vicinity of the laser. Second, it enabled the production of two Q-switched laser pulses from a single flashlamp pulse. The first pulse was produced by switching the Pockels cell slightly prior to the peak of the flashlamp pump pulse, then switching again 100 μs later to produce the second pulse. By raising the capacitor voltage in the flashlamp pulse-forming network pulse energy was maintained, and two ~2 J pulses were produced per laser. This doubled the data acquisition rate for the Thomson scattering diagnostic, but was still short of continuous data acquisition during the 20 ms equilibrium period of an MST discharge.

The second upgrade made to the lasers was larger in scope and entailed replacement of the original flashlamp drive system. While a simple inductor/capacitor pulse-forming network is an efficient way to convert electrical energy into light, it does not allow easy variation of the flashlamp pulse width, pulse energy, or pulse repetition rate. As part of a larger laser development project, we have designed and fabricated a modular flashlamp drive system based on IGBT (insulated gate bipolar transistor) switching of large electrolytic capacitor banks [3]. The system is described in the reference, so only those details relevant to operation of this laser system will be covered here. In essence, this flashlamp drive system applies an approximately square voltage pulse (adjustable up to 900 V capacitor charge) of variable width (0.15–0.39 ms) at a repetition rate up to 1 kHz. Each flashlamp is driven with a separate power supply containing 32 mF of capacitance. The most typical mode of operation is to drive the laser flashlamps with a burst of fifteen 0.15 ms pulses at a repetition rate of 1 kHz (figure 1). The Pockels cell is switched near the end of each flashlamp pulse to produce a train of fifteen ~2 J pulses from each laser (figure 2). Usually the pulse trains from each of the two lasers in the MST Thomson scattering system are interleaved to produce a burst of thirty pulses at an effective 2 kHz repetition rate. This sequence is repeated every few minutes, coincident with the duty cycle of MST plasma production. Since the burst production duty cycle is low (~0.01 Hz), the laser rods and pumping chambers are completely cooled between each burst. This means that each burst starts with a cold cavity with no thermal gradients in the laser rods. The thermal diffusion time constants for the rods range from one to several seconds, therefore any thermal gradients that develop in the rod during the 15 ms burst should be negligible. The modeling results reported in [4] also indicate that the index gradient effects on pulse energy and beam divergence should be small for these short bursts. Some of the drop in laser pulse energy as the burst progresses (figure 2) could be an index gradient effect, but most of the drop is probably due to changes in flashlamp pump light output during the burst.

Another typical mode of operation is to drive the laser flashlamps with a burst of five 0.31 ms pulses at a repetition rate of 1 kHz. The Pockels cell is switched three times: 0.145, 0.225, and 0.305 ms after the start of each flashlamp pulse. The 0.08 ms between Pockels cell switches allows the laser rod to re-pump, and thus produces a burst of three ~2 J pulses at a repetition rate of 12.5 kHz (figure 3). By operating both lasers in this mode and interleaving the pulses, the system produces five bursts of six ~2 J pulses; the pulse repetition rate within a burst is 25 kHz, and the burst repetition rate is 1 kHz. This mode of operation is useful for recording electron temperature fluctuations with a bandwidth up to about 10 kHz.

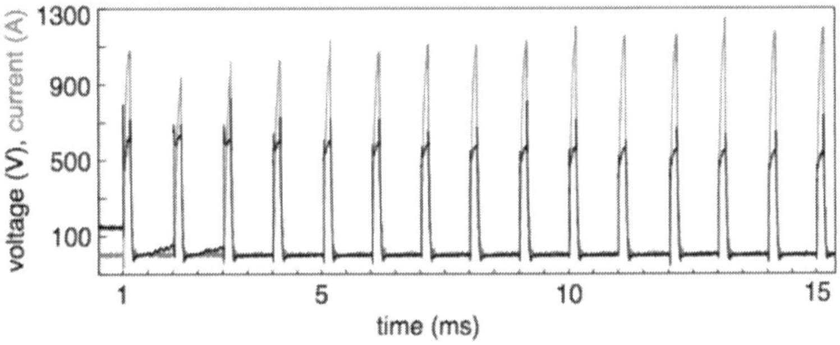

Figure 1. Voltage (black trace) and current (red trace) applied to the largest flashlamp in the system to produce a burst of fifteen 0.15 ms pulses. Each pulse delivers ~70 J of electrical energy to the flashlamp.

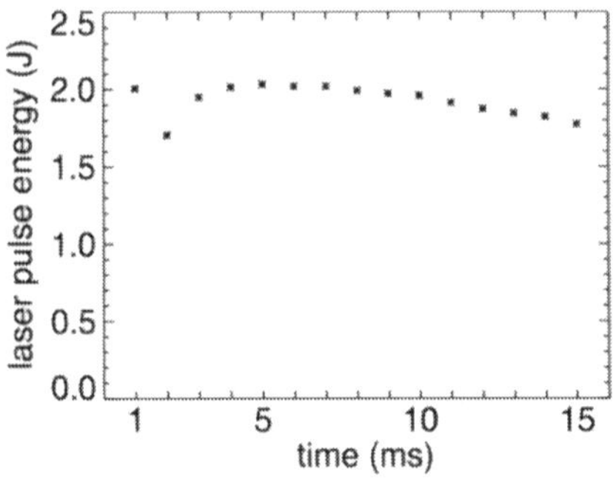

Figure 2. Pulse energies from a burst of fifteen laser pulses produced at 1 kHz repetition rate.

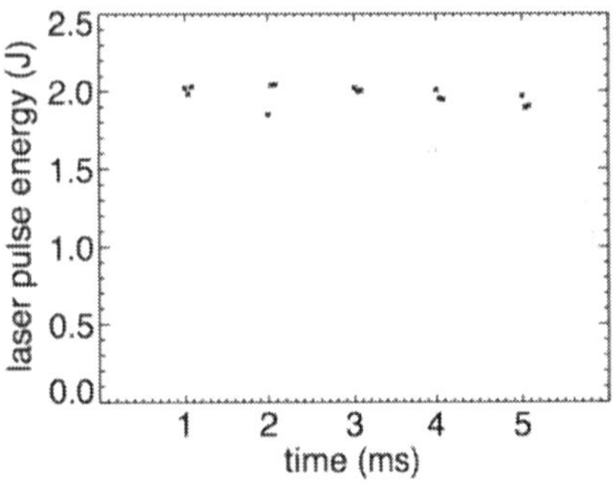

Figure 3. Pulse energies from five bursts of triple laser pulses.

The operational limits of this laser system are set by the explosion energy and wall loading limits of the flashlamps [5]. The explosion energy characterizes the electrical energy applied to the flashlamp that is likely to cause single-pulse catastrophic failure. A single-pulse explosion constant is generally supplied in flashlamp data sheets; the explosion energy then scales as the square root of the pulse width. The flashlamps in the system described in this paper are operated at a maximum of approximately 15% of the explosion energy (table 1). At this limit we expect a lifetime of $>10^5$ pulses. The need to produce ~2 J laser pulses sets the required flashlamp power input (figure 1), thus the practical effect of this explosion energy operational maximum is to limit a single flashlamp pulse to a width of 0.39 ms.

Flashlamp manufacturers typically suggest a time-averaged wall loading of 200 W/cm^2 for liquid-cooled flashlamps with clear fused quartz walls of 1 mm thickness [5]. Application of this limit to pulse-burst operation is not immediately apparent since the power loading within a burst is large, but then the time between bursts is long enough to ensure that all heat is removed from the flashlamps. As a starting point, the flashlamps have been operated such that the burst-integrated heat load is kept well below 200 J/cm^2 (table 1). This operational maximum thus limits the number of flashlamp pulses in a burst; the longer a flashlamp pulse, the fewer a burst can contain. Several months of operational experience producing >10^4 pulses suggests that the operational parameters in table 1 are conservative; in particular it appears that the number of flashlamp pulses in a burst can be increased.

Table 1. Laser system operational modes and example operational parameters for the a) largest, and b) smallest flashlamps in the system. The largest flashlamp pumps the pre-amplifier, the smallest pumps one of the rods in the double-rod amplifer. Both are krypton flashlamps, one with 7 mm bore and 93 mm arc length, the other 4 mm bore and 53 mm arc length.

Number of laser pulses extracted from each flash-lamp pulse	Number of flash-lamp pulses in 1 kHz burst	Flashlamp pulse width (ms)	Fraction of explo-sion energy in single pulse		Burst-integrated wall loading (J/cm^2)	
			a)	b)	a)	b)
1	15	0.15	6%	8%	50	68
2	7	0.23	8%	10%	40	49
3	5	0.31	10%	14%	41	54
4	3	0.39	12%	15%	32	38

Besides actually pulsing the flashlamps, the drive system is required to perform another key function: starting and sustaining a simmer discharge in the flashlamps [5]. If the simmer discharge is not present when the IGBT applies the drive system voltage to the flashlamp electrodes, the flashlamp arc discharge will not start. Initiation of the simmer discharge requires application of a short (~1 μs) high-voltage (≥10 kV) trigger pulse to the flashlamp. In the initial design of the drive system the primary of the trigger transformer was driven with a pulse of fixed width and voltage. Unfortunately the resulting high-voltage pulse was insufficient to reliably initiate the simmer discharge. The solution was to produce an adjustable width input pulse with a small IGBT, and optimize the pulse width to match the characteristics of the trigger transformer as loaded by the flashlamp.

3. Summary

In addition to greatly extending the capability of the MST Thomson scattering diagnostic, a major advantage of the system described in this paper is the relative simplicity of implementation and operation. A number of solid-state pulse-burst laser systems have been previously built (see [3] for brief descriptions), but they have all been custom systems, with limited operational flexibility. In contrast, for the system described in this paper, *Q*-switch and flashlamp parameters are independent and programmable, and enable production of a burst of pulses tailored to experimental needs. In addition, any of a large variety of commercial solid-state laser optical heads could be used as the basis of such a laser system, substantially reducing development and construction effort.

Acknowledgements
D Holly and H Mattison of the Physical Sciences Laboratory at the University of Wisconsin–Madison assisted with the design and fabrication of the flashlamp power supplies. A Falkowski, R O'Connell, E Parke, and H Stephens helped implement this extension of the Thomson scattering system on MST. This work is supported by the United States Department of Energy and National Science Foundation.

References

[1] Reusch J A, Borchardt M T, Den Hartog D J, Falkowski A F, Holly D J, O'Connell R and Stephens H D 2008 *Rev. Sci. Instrum.* **79** 10E733
[2] Koechner W 2006 *Solid-State Laser Engineering* (New York: Springer)
[3] Den Hartog D J, Jiang N and Lempert W R 2008 *Rev. Sci. Instrum.* **79** 10E736
[4] Stone D H and Rotondaro M D 1992 *Appl. Opt.* **31** 1314
[5] PerkinElmer *High Performance Flash and Arc Lamps* (www.perkinelmer.com/opto), Heraeus Noblelight *The lamp book* (www.heraeus-noblelight.com)

Journal of Physics: Conference Series **227** (2010) 012024 doi:10.1088/1742-6596/227/1/012024

LIF measurements for validation of collisional-radiative modelling of atomic helium in the edge of a fusion plasma

M. Krychowiak[1], Ph. Mertens[2], R. König[1], B. Schweer[2], S. Brezinsek[2], O. Schmitz[2], M. Brix[3], U. Samm[2], R. Wolf[1] and T. Klinger[1]

[1] Max-Planck-Institut für Plasmaphysik, EURATOM Association, Greifswald, Germany, e-mail: maciej.krychowiak@ipp.mpg.de

[2] Institut für Energieforschung (Plasmaphysik), FZ Jülich, EURATOM Association, TEC, Jülich, Germany

[3] UKAEA, JET-Experimental Department, EURATOM Association, Culham Science Centre, Abingdon, Oxfordshire, OX14 3DB, UK

Abstract. Local values of the electron density and temperature in the edge of a fusion plasma can be derived with high space and time resolution by the use of line radiation of atomic helium beams. The accuracy of this method is mainly limited by the uncertainties in the collisional-radiative (CR) model which is needed in order to obtain both plasma parameters from the measured relative intensities of atomic helium lines. Laser-induced fluorescence spectroscopy on a thermal helium beam in the edge plasma of the tokamak TEXTOR in Jülich was applied to validate the CR model of helium. By use of a high-power, pulsed laser system (a dye laser pumped by an excimer laser) several laser excitation schemes starting from the n=2 levels have been tried. The fluorescence light was observed at the laser wavelength and elsewhere in the spectrum providing information on population densities of initial levels as well as on collisional population transfer between excited levels. This paper summarises the results of the measurements, showing principal limits and possible improvements of this experimental validation method of the CR model of the diagnostic helium beam.

1. Introduction

Line radiation of atomic helium is widely used for the determination of spatially and temporally resolved electron density and temperature in the plasma. One way of diagnosing the edge region of fusion plasmas is to use radial gas injection into the plasma in the form of thermal [1] or supersonic beams [2][3][4]. The method is based on the intensity measurement of three spectral lines of atomic helium. Electron density and temperature are obtained by comparing two line intensity ratios, with the main dependence on either n_e or T_e, with those calculated using a collisional–radiative (CR) model for helium. Hence, a reliable model is crucial for the accuracy of derived plasma parameters. The model contains a large number of rate coefficients for electron collisional excitation and ionization [2][5][6][7][10] which are in most cases known only from calculations. An earlier work of Brix showing some deviations between the calculated line intensities and corresponding measurements performed on the tokamak TEXTOR in Jülich [2] suggested the need of further improvement of the CR model. Similar conclusion is drawn in a recent study by Schmitz [8].

In this paper we describe our investigations towards a validation of some parameters of the CR model for electron temperatures $T_e \approx 50$ eV, as relevant for the edge of fusion plasmas. The laser-induced

Journal of Physics: Conference Series **227** (2010) 012024 doi:10.1088/1742-6596/227/1/012024

fluorescence (LIF) spectroscopy was applied on a thermal helium beam on TEXTOR using a dye laser pumped by an excimer laser. In [9] we describe the details of the experimental set-up (including the laser system, the gas injection and calibration issues) as well as refer to earlier work in this field. In [9] we also derive particle densities in a selected quantum level from LIF signals. In this paper we concentrate on the determination of rate coefficients for collisional population transfer between levels of similar excitation energy from the shells n=3 and n=4. The rate coefficients of our standard CR model we refer to throughout the paper are taken from the ADAS 2002 preferred data set based on the compilation of Ralchenko et al. [10].

2. Rate coefficients for population transfer between levels of the shells n=3 and n=4

Laser excitation between a level of the shells n=2 and n=3-4 of atomic helium is possible in the visible and near UV spectral range. The (saturated) excitation results in a quick population increase of the upper level and thus an enhanced radiation from this level. Due to high enough electron densities, collision-induced population transfer from the upper level to other levels of the shells n=3-4 is in competition to the radiative decay making the detection of collision-induced fluorescence signals originating from those levels possible. Analysis of all those signals provides access to the rate coefficients relevant for the transfer of the excess population to the other levels, some of them being of direct importance for the accuracy of n_e and T_e derivation from intensity ratios.

In Figure 1 the most successful pumping schema is shown with laser excitation at λ = 388.9 nm ($2^3S \rightarrow 3^3P^o$) and fluorescence detection at the laser wavelength and three other ones resulting from collision-induced population transfer. Example time traces of signals originating from the levels of the shell n=3 (at T_e = 50 eV) are shown in Figure 2. In the following, three possible methods of derivation of rate coefficients are described, including experimental and theoretical difficulties affecting the uncertainty of resulting rate coefficients.

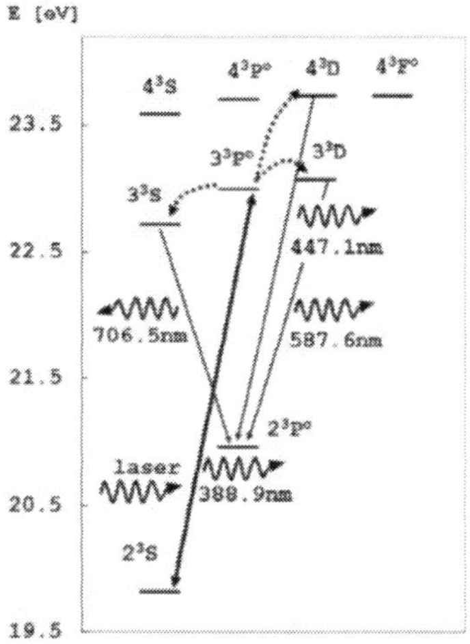

Figure 1: Example pumping channel in triplet helium.

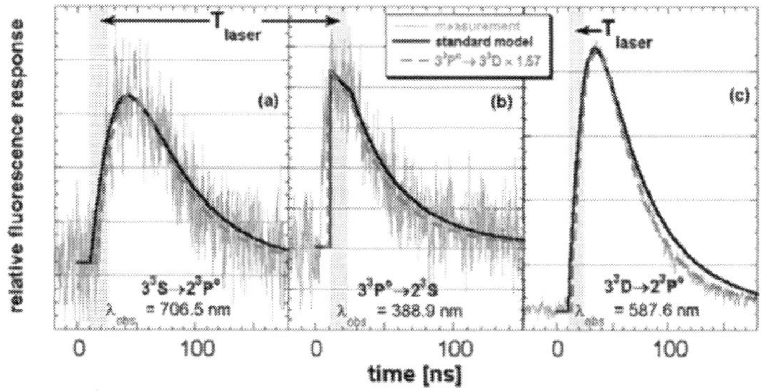

Figure 2: Time traces of fluorescence signals induced by laser excitation at λ = 388.9 nm originating from three different n=3 levels.

2.1. Derivation of a rate coefficient from a time trace of collision-induced fluorescence

Consider Figure 2 c) with the purest signal. The shape of the decaying signal depends mostly on the well known radiative transition probabilities and on the collisional population transfer rate from the upper level to the level 3^3D. As we see, the best fit is achieved upon enlarging the rate coefficient $3^3P^o \rightarrow 3^3D$ by a factor of 1.57. The rate coefficient in question can be derived in this way with an accuracy of 15% which could be further improved by averaging over more laser pulses (in the case of

TEXTOR discharges this means repetition of plasma discharges of same parameters). Since the only quantity, the rate coefficient is retrieved from, is the shape of the decaying part of the signal, the (absolute) calibration of the detection system, the properties of the laser and of the helium beam are not of concern. The only measures needed are a linear detector response as well as a good knowledge of T_e and mostly of n_e since the collisional transition rate is the product of n_e and the rate coefficient. In our experiments we relied on temperatures and densities obtained with a diagnostic thermal helium beam located at a different position (at the equatorial plane at the low field side) complemented by density measurements of a thermal lithium beam at the same position. The accuracy of the plasma parameters obtained in this way is believed to amount to 10% (n_e) and 30% (T_e) [8]. However, the plasma parameters obtained at a different position in the plasma must be mapped to our position being a source of additional uncertainty. Unresolved scattering of the same rate coefficients obtained from different plasma discharges or from different methods suggest a problem with correct mapping of n_e and T_e to our measurement position. To resolve this problem, plasma parameters have to be sampled at the position of the LIF measurements.

2.2. Derivation of a ratio of rate coefficients from a ratio of collision-induced fluorescence

A possibility to obtain ratios of rate coefficients is given by analysis of ratios of signal maxima, e.g. the signals in Figure 2 a) and c) for $<\sigma v>(3^3P^o \rightarrow 3^3D)/<\sigma v>(3^3P^o \rightarrow 3^3S)$, as this signal ratio mostly depends on the ratio of the rate coefficients. The main source of uncertainty in this measurement is the (relative) intensity calibration of the detection system at both wavelengths: application of different gain factors of the photomultipliers in calibration and TEXTOR measurements as well as progressive transmission losses due to contamination of the first, in-vessel mirror limited the measurement accuracy to 20-50%.

2.3. Derivation of a rate coefficient from a ratio of collision- and laser-induced fluorescence

The ratio of signal maxima in Figure 2 c) and b) can be used to determine the rate coefficient of population transfer between both radiating levels. This method, providing rate coefficients and not their ratios, is, however, suffering from several additional uncertainties. We have to cope with the strong laser stray light when detecting the fluorescence from the upper level (the fluorescence signals amounted in many cases only to a fraction of a few percent of the total signal). This deteriorates the signal-to-noise ratio and can lead to a non-linear detector response. Furthermore, very low light intensity of the Ulbricht sphere for wavelengths $\lambda < 400$ nm (detection of the fluorescence at $\lambda = 388.9$ nm) enlarged the error bar of the intensity calibration by up to 20%. An additional issue is the dependence of the pumping efficiency on the collisional population redistribution between the magnetic sublevels of the upper level: it influences the collision-induced fluorescence (e.g. from the level 3^3D) stronger as the fluorescence from the upper level. In the extreme case of very rapid (on the time scale of the laser pulse length) redistribution this would mean in some cases an overestimation of the rate coefficient by up to 50%. This issue could be examined experimentally by measurement of time-resolved traces of both π and σ components of the resonant fluorescence light in the case of exciting e.g. only a π component of the Zeeman-split line.

3. Results and suggestions for future experiments

In triplet helium we deduced the rate coefficient $<\sigma v>(3^3P^o \rightarrow 3^3D)$ from different discharges at the laser excitation at $\lambda_L = 338.9$ nm and $\lambda_L = 587.6$ nm at $T_e \approx 50$ eV to be larger by a factor 1.5 than in our standard model with a relative uncertainty of 40% (unresolved level of scattering of the data). For the rate coefficient $<\sigma v>(3^3P^o \rightarrow 3^3S)$ we obtain a factor 1.3 with relative error of 50% and for $<\sigma v>(3^3P^o \rightarrow 4^3D)$, obtained as a relative value to $<\sigma v>(3^3P^o \rightarrow 3^3D)$, the rate coefficient of our standard model was confirmed with a relative error of 50%. In singlet helium, being a much higher challenge due to very quick radiative decays to the ground state, we were only able to detect collision-induced fluorescence at $\lambda = 667.8$ nm ($3^1D \rightarrow 2^1P^o$, laser excitation at $\lambda_L = 501.6$ nm, according to $2^1S \rightarrow 3^1P^o$) and at $\lambda = 492.2$ nm ($4^1D \rightarrow 2^1P^o$, laser excitation at $\lambda_L = 396.5$ nm, according to $2^1S \rightarrow 4^1P^o$). Owing to the lack of signal detection at the laser wavelengths we are not able to give final statements about the rate coefficients. However, if we assume an overestimation of the model prediction of the 2^1S population by 30% (which could be suggested by our measurements of absolute population densities of the triplet n=2 levels [9]), we can conclude the rate coefficient $\langle\sigma v\rangle(3^1P^o \rightarrow 3^1D)$ to be higher by a factor 1.9 (relative error 50%) at T_e = 45 eV than the one used in our standard model and to match the one obtained by Brix with the ATOM code using the Born approximation [2]. A further conclusion of the assumption is the confirmation of the rate coefficient $\langle\sigma v\rangle(4^1P^o \rightarrow 4^1D)$ as used in our standard model (relative error 50%).

The enhanced values of the triplet rate coefficients affect only the line intensity ratio for the T_e derivation: its application (at T_e = 50 eV) in our model would lower T_e values by up to 7%, depending on n_e. On the other hand, the enhanced value of the singlet rate coefficient affects only the density sensitive line intensity ratio and results in lowering of derived n_e values by 10-20%, depending on n_e. However, we were not able to get any insights to one more rate coefficient ($\langle\sigma v\rangle(3^1P^o \rightarrow 3^1S)$) of a similar importance for both the n_e and T_e derivation, which would complement our investigations. In this sense several improvements are suggested for future experiments, which, applied collectively would allow to suppress the uncertainties by up to a factor of two.

Laser stray radiation: suppressing it by avoiding to hit the nozzle by the laser beam is of utmost importance for better signal-to-noise ratios, avoiding signal distortion by onset of nonlinear response of the photomultipliers as well as to allow signal detection at the laser wavelength in singlet helium (at least for laser excitation of the level 2^1S at λ_L = 501.6 nm). Reproducibility: signal fluctuations of unknown source enlarged the error bars in some cases by up to a factor of two - more experimental time could help to reach satisfactory signal reproducibility. Machine access: more flexible access to the machine to check transmission losses due to build up of coatings on in-vessel optics and for more frequent calibrations. Simultaneous measurement at all relevant wavelengths: getting rid of differing n_e and T_e values at measurement position, more robust intensity calibration. In-situ n_e and T_e measurement: avoiding of mapping uncertainties from other positions in plasma. Use of supersonic helium beam: extending the measurement range towards smaller minor radii, i.e. higher n_e and T_e values. Investigation of collisional population redistribution between the magnetic sublevels of the n=3 levels: more accurate values of the pumping efficiency. In case of using a thermal helium beam: systematic study of the density distribution in the beam by spatial scans of laser-induced fluorescence signals over the beam and by density measurements using a neutral gas manometer for more accurate measurements of absolute densities of the n=2 levels.

References

[1] Schweer B, Mank G, Pospieszczyk A, Brosda B and Pohlmeyer B, *J. Nucl. Mater.* **196–198** (1992) 174–8.

[2] Brix M, *Jülicher Report* **Jül-3638** (1998).

[3] Hidalgo A, Tafalla D, Brañas B and Tabarés F L, *Rev. Sci. Instrum.* **75** (2004) 3478–80.

[4] Andruczyk D, Namba S, James B W, Takiyama K and Oda T, *Plasma Devices Oper.* **14** (2006) 81–9.

[5] Fujimoto T, *J. Quant.Spectrosc. Radiat. Transfer* **21** (1979) 439–55.

[6] Goto M and Fujimoto T, Report NIFS-DATA-43, NIFS, Nagoya (1997).

[7] Brix M and Schweer B, *Proc. 24th EPS Conf. on Controlled Fusion and Plasma Physics* (Berchtesgarden) **21A** (1997) 1837–40.

[8] O Schmitz, I L Beigman, L A Vainshtein, B Schweer, M Kantor, A Pospieszczyk, Y Xu, M Krychowiak, M Lehnen, U Samm, B Unterberg and the TEXTOR team, *Plasma Phys. Control. Fusion* **50** (2008) 115004.

[9] M Krychowiak, Ph Mertens, R König, B Schweer, S Brezinsek, O Schmitz, M Brix, U Samm and T Klinger, *Plasma Phys. Control. Fusion* **50** (2008) 065015.

[10] Yu.V. Ralchenko et. al., Report NIFS-DATA-59, NIFS, Nagoya (2000).

14th International Symposium on Laser-Aided Plasma Diagnostics (LAPD14) IOP Publishing
Journal of Physics: Conference Series **227** (2010) 012025 doi:10.1088/1742-6596/227/1/012025

Laser induced fluorescence measurements on W- and Ba atoms eroded from fluorescent lamp electrodes

J Ehlbeck[1], K Rackow[1], F Sigeneger[1], D Uhrlandt[1], K-D Weltmann[1], S Hadrath[2] and G Lieder[2]

[1] Leibniz Institute for Plasma Science and Technology, Felix-Hausdorff Str. 2, 17489 Greifswald, Germany

[2] RLS-M, OSRAM GmbH, Munich, Germany

ehlbeck@inp-greifswald.de

Abstract. The method of laser induced fluorescence (LIF) is applied to fluorescent lamps (FL) in order to investigate processes of electrode erosion in the vicinity of the electrodes. The life time of FLs which are ignited by instant start is mainly limited by sputtering of the coil electrodes and in final breaking. This sputtering of tungsten mainly occurs during the ignition in the glow discharge phase. Therefore, the density of W atoms is measured in the electrode region during ignition. Temporal and spatial resolved profiles were measured by LIF which has been combined with fast imaging. The life time of FLs which are started with preheated coils is also caused mainly by electrode failures. But the reason differs from the instant start case because here the loss is caused mainly by evaporation. End-of-lamp life is reached if the emitter material which is deposited at the coil to reduce the work function of the coil is lost completely. LIF is used to measure the density of the eroded emitter material, namely Barium atoms. First result of phase resolved absolute Ba atoms densities are presented.

1. Introduction

Today, the lifetime of fluorescent lamps (FL) is mainly limited by the electrode durability. The electrode is the junction between plasma and solid and is therefore responsible for driving the lamp current. In general it consists of a tungsten coil coated with a work function reducing emitter mix of alkali oxides, such as BaO, SrO and CaO. During lamp life such an electrode is damaged by two essential processes. In case of an instant start ballast the electrode is heated up by ion bombardment before transitioning into the arc-mode (glow-to-arc-transition) where the so called hot-spot builds up. During this time electrode material such as emitter material and also tungsten is eroded by sputtering. This process limits the number of switching cycles. After transition into the arc-mode, during steady state operation, the emitter material (primary Barium) erodes from the hot-spot. In case of a good adaption of the electrode to the discharge conditions, this erosion is in general evaporation.

For a detailed investigation of these processes a local and very sensitive diagnostic is needed. To quantify the erosion this method should be able to measure absolute densities. The laser induced fluorescence (LIF) fulfils all this requirements [1-2]. The good temporal resolution (the temporal resolution is limited by the lifetime of the excited level and the laser pulse length) also enables the investigation of the dynamic of such processes.

2. Laser induced fluorescence

The LIF technique allows the detection of ground state or long-lived, excited-state atoms, molecules or radicals. The LIF process can be described as a two-step process [3]. First, a specific species in its ground state $|1\rangle$ (energy E_1) is excited by resonant absorption of laser photons ($h\nu_{laser}$) to a higher electronic energy level $|2\rangle$ of energy E_2 with $h\nu_{12} = E_2 - E_1$. After laser excitation, this excited state relaxes by spontaneous emission to a lower state $|3\rangle$ of energy E_3 by emitting fluorescence photons with the energy $h\nu_{LIF} = E_2 - E_3$. This fluorescence radiation is emitted nearly isotropic [4]. For a spectral width of the laser $\Delta\nu_{laser}$ that is much broader than the line width of the exited transition $\Delta\nu_{12}$ a rate equations model could be used to describe the excitation and relaxation processes [5-8].

In the first approximation the time integrated fluorescent signal S_{LIF} is proportional to the ground state density n_1:

$$S_{LIF} = \int I_{LIF}(t)dt \propto S(\lambda)\cdot\frac{\Delta\Omega}{4\pi}\frac{A_{23}}{A_2}\cdot\sigma_{LIF}\cdot\frac{E_{laser}}{F_{laser}h\nu_{laser}}\cdot V\cdot n_1 \quad , \tag{1}$$

where $S(\lambda)$ is the sensitivity of the detection system at the fluorescent wavelength, $\Delta\Omega$ is the detection angle, the ratio A_{23}/A_2 of the Einstein transition probabilities is the relation between the detected and all possible transitions to lower energy levels, σ_{LIF} is the corresponding fluorescence cross section, F_{laser} is the cross sectional area of the laser beam and V is the observation volume.

2.1. *Absolute density measurements*

For measurements of absolute densities the specific atomic data (such as Einstein coefficients), the laser energy and its spectral profile, the exact observed volume and angle, as well as the sensitivity of the detection system is needed. The main difficulties are the determination of the geometry and the spectral response of the optical imaging and detection. We therefore use the Rayleigh scattering calibration technique to overcome the determination of sensitivity, detection angle and observation volume. This commonly used method is described in text books and literature e.g. by Hadrath [4] in the case of tungsten atoms.

2.2. *Experimental setup*

For the generation of the laser beam a Nd:YAG pumped dye laser operating at 10 Hz is used to pumped the dye laser at 532 nm (figure 1). In case of Barium atoms, the dye laser provides radiation at 700.22 nm and is frequency doubled by a KDP crystal to the desired 350.11 nm. In case of tungsten, the radiation of 575.98 nm is frequency doubled to 287.94 nm. The resulting beam is directed through a 1 mm aperture to get a defined beam diameter.

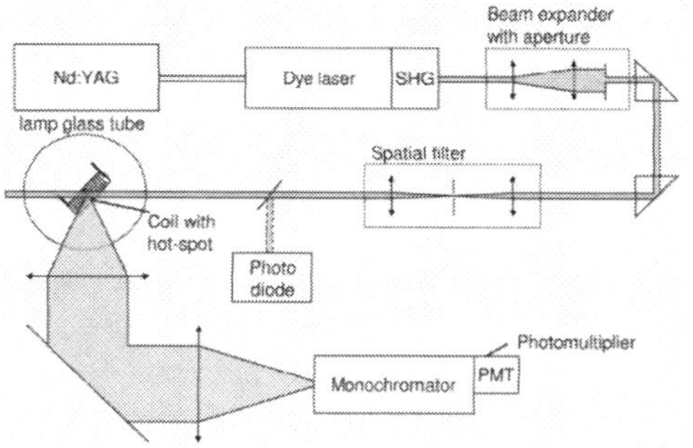

Figure 1. Experimental setup for measurement of Barium ground state densities.

A spatial filter (consist of two lenses (f = 100 mm each) in a distance of 200 mm with a 20 µm pinhole located at the common focal point) is used to reshape the resulting degraded beam profile to a nearly Gaussian shape. This beam is directed through the lamp (perpendicular through the glass surface) directly in front of the electrode. The fluorescent radiation at 582.63 nm is observed perpendicular to the beam. The region of interest is imaged (1 : 3) to the entrance slit of a 0.5 m monochromator (Acton Research SpectraPRO-500I) with a 1800 grooves mm^{-1} grating. The observed volume has a dimension of approximately 0.7 x 0.7 x 1.0 mm^3. For measuring the fluorescent radiation a photomultiplier (Hamamatsu Photonics, R-1477) is used, whereas the laser intensity is measured by a photodiode.

3. Results

3.1. Tungsten erosion

For investigation of tungsten erosion processes during instant start a commercial FL is driven by an electronic control gear with a frequency of around 44 kHz. A few milliseconds after power on (t = 0), the lamp ignites in the glow-mode. After a certain time which is typical in the range between 15 and 25 ms and has a stochastic nature the discharge changes into the arc-mode which is called the glow-to-arc transition. Due to the low laser repetition rate of 10 Hz there is only one laser pulse per ignition available to excite eroded tungsten atoms. Therefore the temporal evolution of the sputtered tungsten density is reconstructed on the base of subsequent lamp starts.

In figure 2 the investigated section of the coil is presented and the location where the discharge attaches is marked. In contrast to the modeling results of Haverlag et al [9] the discharge attaches not only at the bare tungsten ends of the coil but also at the emitter covered part. One possible reason could be that the electrode is covered inhomogeneous by the emitter. LIF-measurements were performed at five equidistant positions 2 mm in front of the coil and with distances of 0.5 mm between them.

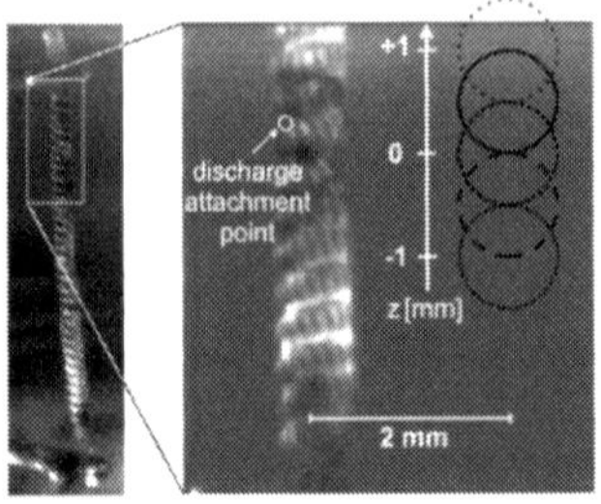

Figure 2. Diagram of the location of the hot spot (discharge attachment point) and the LIF measurements Laser beam (black cycles) at a distance of 2 mm.

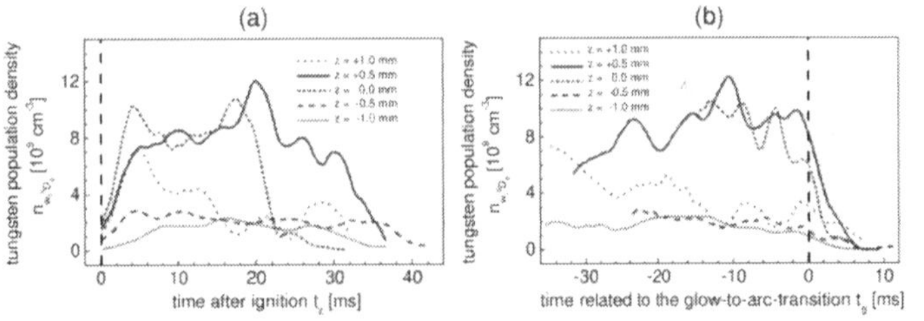

Figure 3. Two reconstructions of the same tungsten population density measurements, according to the time after ignition (*a*) and the time related to the glow-to-arc-transition (*b*) at different positions near the coil. Due to the neglect of the transmission of the bulb the density will be about four times higher.

By measurement of the lamp voltage and the discharge current we could determine the moment when the ignition starts and the moment when the glow-to-arc-transition takes place. The reconstructed tungsten density profiles are presented in figure 3. In figure 3(*a*) one can see the temporal evolution of the tungsten population density in relation to the moment of the ignition of the lamp (Δt_Z) and in (*b*) the same densities in relation to the moment of the glow-to-arc-transition (Δt_g).

Due to the large scatter of the time interval between the ignition and the glow-to-arc-transition for different instant starts the left figure better describes the tungsten densities directly after the ignition, and the right one better describes the behaviour during the glow-to-arc-transition.

The eroded tungsten density starts immediately with the ignition, reaches a maximum where the discharge contracts at the end of the glow mode and decreases some milliseconds before the glow-to-arc-transition takes place.

3.2. Barium Erosion

For investigation of Barium erosion in steady state operation a current controlled electronic ballast working at 25 kHz is used. The commercial FL is driven with its nominal rms-current of 320 mA. As the laser has a low repetition rate of 10 Hz the measured time dependence results from the reconstruction of several periods, where each point is the average of 600 laser pulses. As the Barium densities are very low, they are measured at a distance of 0.5 mm from the hot-spot (of the coil).

During the anode phase the lamp voltage as well as the current are defined to be positive (figure (4*b*)). The Barium density strongly depends on the electrode and plasma processes (figure (4*a*)). During the cathode phase the electrode needs to sustain the lamp current. To emit the electrons a very thin plasma sheet with a strong potential drop builds up at the electrode surface, the so called cathode fall. The emitted electrons pass this sheet and are accelerate by the field to a kinetic energy in the range of the cathode fall (approx. 15 eV). These high energetic electrons ionize and excite the rare gas as well as the Barium atoms in the electrode region. At the beginning of the cathode phase the lamp current is very low and therefore the Barium atom ground state density decreases with a short delay (related to the beginning of the cathode phase). With the end of the cathode phase this excitation and ionisation processes are reduced. During the anode phase these processes are more or less insignificant and therefore the Barium density builds up again.

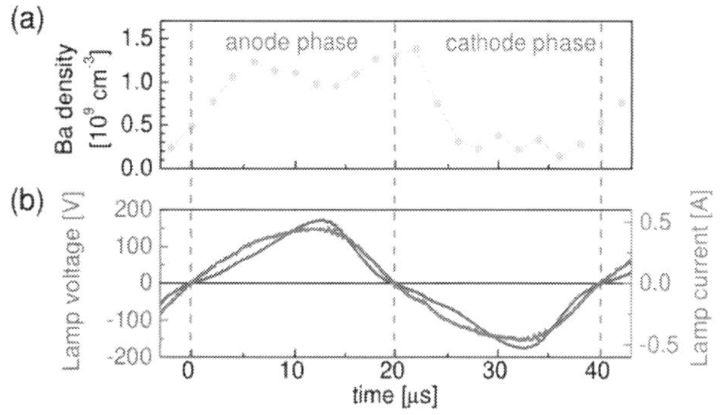

Figure 4. The time dependent Barium density 0.5 mm in front of the hot-spot (*a*) and the regarding lamp current and voltage (*b*).

4. Conclusion

The durability of electrodes mainly influences the lifetime of FL. In the case of instant start the electrodes will be destroyed due to coil fracture at the emitter-free ends, whereas in the case of preheating the Barium erosion during steady state operation is the lifetime limiting factor.

We concentrated on a commercial FL and presented our reconstructions of the temporal evolution of the absolute tungsten and barium population density of the ground state. The temporal behaviour of tungsten erosion during instant start was presented, where constant tungsten erosion during the whole glow-phase was found.

We also presented the temporal Barium densities during steady state operation. Here a strong decay of the density during the cathode phase was found. An evidence for sputtering of barium could not be found.

References

[1] Bhattacharya A K 1989 *J. Appl. Phys.* **65** 4603–7
[2] Bhattacharya A K 1989 *J. Appl. Phys.* **65** 4595–602
[3] Amorim J, Baravian G and Jolly J 2000 *J. Phys. D: Appl.Phys.* **33** R51–R65
[4] Hadrath S, Ehlbeck J, Lieder G and Sigeneger F 2005 *J. Phys. D: Appl. Phys.* **38** 3285–3295
[5] Niemi K, Schulz-von der Gathen V and Dobele HF 2001 *J. Phys. D: Appl. Phys.* **34** 2330–2335
[6] Dullni E 1984 *PhD Thesis* Ruhr-University Bochum
[7] Schnabel R and Kock M 2000 *Phys. Rev. A* **61** 062506
[8] Schnabel R and Kock M 2001 *Phys. Rev. A* **63** 012519
[9] Haverlag M et al 2002 *J. Phys. D: Appl. Phys.* **35** 1695–1701

Journal of Physics: Conference Series **227** (2010) 012026 doi:10.1088/1742-6596/227/1/012026

Application of NIR – CRDS for state selective study of recombination of para and ortho H_3^+ ions with electrons in low temperature plasma

J. Varju, Š. Roučka, T. Kotrík, R. Plašil and J. Glosík

Charles University, Faculty of Mathematics and Physics, Department of Surface and Plasma Science, V Holešovičkách 2, Prague 8, Czech Republic

E-mail: Juraj.Glosik@mff.cuni.cz

Abstract. We present a study of H_3^+ recombination performed at 77 K on the two lowest rotational levels of this ion, which belong to its two different nuclear spin states of the studied ion. A near infrared cavity ring-down spectrometer (~1381 nm, CRDS arrangement) has been used to obtain the time evolution of concentration of both states. From the overall ion density decay during the afterglow we obtained the binary recombination rate coefficient α_{bin} (77 K) = 1.2×10^{-7} cm^3s^{-1}. We have also observed ternary helium assisted recombination of both para and ortho H_3^+. The process is very slow (at 77 K) and the obtained ternary recombination rate coefficient is in contradiction with the theoretical prediction. It is the first time that the binary and ternary H_3^+ recombination rate coefficient was measured at a known population of para and ortho H_3^+ ions in decaying plasma.

1. Introduction

The H_3^+ ion plays an important role in many environments ranging from laboratory plasma through atmospheres of various planets to star forming regions in interstellar space. Recombination of H_3^+ ions with electrons still represents a fundamental problem of contemporary plasma physics, astronomy and quantum chemistry [1,2,3]. Only very recently the dissociative recombination process of H_3^+ ions has been successfully described for energies bellow 1 eV, when the non-Born-Oppenheimer Jahn-Teller coupling was included into theoretical calculations [4,5,6]. This brought the theory to an agreement with storage ring experiments [3,7,8]. The agreement has to be considered as "a first approximation" because the theory predicts large difference in recombination of the H_3^+ ion in the para and ortho state.

The situation in interpretation and understanding of plasmatic experiments is not so straightforward (see e.g. Refs. [2,3]). In plasmatic experiments hydrogen is added to helium to form H_3^+ dominated plasma. The formed H_3^+ ions therefore can have multiple collisions with H_2 and He prior to recombination with an electron. This can influence the overall recombination process in plasma. The improvement can bring state selective monitoring of densities of para-H_3^+ and ortho-H_3^+ in recombination dominated plasma. At temperatures below 300 K the situation is further complicated by the large difference of the recombination rate coefficients of different H_3^+ nuclear spin states (${}^{P}\alpha_{DR}$ and ${}^{O}\alpha_{DR}$ for para-$H_3^+$and ortho-H_3^+, respectively) [6,9]. The theory predicts that at 10 K ${}^{P}\alpha_{DR}(10\text{ K})/{}^{O}\alpha_{DR}(10\text{ K}) \sim 8$). Therefore at low temperatures additional care has to be taken and the ortho/para ratio has to be considered. Changing of the nuclear spin state happens mainly in reactions with hydrogen molecules [10]. For low temperature plasma (e.g. in an afterglow in FALP experiment), where we can exclude processes of H_3^+ formation, the main processes determining the density of para-H_3^+ and ortho-H_3^+ are:

$$\begin{aligned} {}_{\mathrm{O}}\mathrm{H}_3^+ \; + \; \mathrm{e}^- &\xrightarrow{{}^{\mathrm{O}}\alpha} \textit{neutral products} \\ {}_{\mathrm{P}}\mathrm{H}_3^+ \; + \; \mathrm{e}^- &\xrightarrow{{}^{\mathrm{P}}\alpha} \textit{neutral products} \quad , \\ {}_{\mathrm{P}}\mathrm{H}_3^+ \; + \; \mathrm{H}_2 &\underset{k_{\mathrm{OP}}}{\overset{k_{\mathrm{PO}}}{\longleftrightarrow}} {}_{\mathrm{O}}\mathrm{H}_3^+ \; + \; \mathrm{H}_2 \end{aligned} \qquad (1)$$

where ${}^{P}\alpha$ and ${}^{O}\alpha$ are the rate coefficients of the overall recombination of para-H_3^+ and ortho-H_3^+, respectively. The overall recombination includes binary and the helium assisted ternary recombination (recently observed very fast process [11,12]). The rate coefficients k_{PO} and k_{OP} correspond to para→ortho and reverse transition in collisions with H_2, respectively. The balance equations for para-H_3^+ and ortho-H_3^+ can be written:

$$\frac{\mathrm{d}[{}_{\mathrm{O}}\mathrm{H}_3^+]}{\mathrm{d}t} = -{}^{\mathrm{O}}\alpha[{}_{\mathrm{O}}\mathrm{H}_3^+]n_{\mathrm{e}} - \frac{[{}_{\mathrm{O}}\mathrm{H}_3^+]}{\tau_{\mathrm{D}}} + k_{\mathrm{PO}}[{}_{\mathrm{P}}\mathrm{H}_3^+][\mathrm{H}_2] - k_{\mathrm{OP}}[{}_{\mathrm{O}}\mathrm{H}_3^+][\mathrm{H}_2], \qquad (2)$$

$$\frac{\mathrm{d}[{}_{\mathrm{P}}\mathrm{H}_3^+]}{\mathrm{d}t} = -{}^{\mathrm{P}}\alpha[{}_{\mathrm{P}}\mathrm{H}_3^+]n_{\mathrm{e}} - \frac{[{}_{\mathrm{P}}\mathrm{H}_3^+]}{\tau_{\mathrm{D}}} - k_{\mathrm{PO}}[{}_{\mathrm{P}}\mathrm{H}_3^+][\mathrm{H}_2] + k_{\mathrm{OP}}[{}_{\mathrm{O}}\mathrm{H}_3^+][\mathrm{H}_2], \qquad (3)$$

where, τ_D is a time constant given by the ambipolar diffusion, n_e, $[{}_PH_3^+]$ and $[{}_OH_3^+]$ are electron, para-H_3^+ and ortho-H_3^+ densities, respectively. From the equations (2) and (3) we see that to obtain the recombination rate coefficient from the density decay of one particular nuclear spin state we need to have information about the ortho/para balance in the decaying plasma. It is not easy to obtain recombination rate coefficients for para-H_3^+ and ortho-H_3^+ because the densities decays are coupled in the presence of H_2. But if we add equations (2) and (3) together the state changing terms cancel each other out and the resulting balance equation, under the assumption that H_3^+ is the dominant ion, basically describes the decay of electron density in the plasma with $\alpha_{\mathrm{eff}} = {}^{O}\alpha{}^{O}f + {}^{P}\alpha{}^{P}f$, where ${}^{O}f$ and ${}^{P}f$ denotes the fraction of H_3^+ ions in the ortho and para state. The observed recombination rate coefficient α_{eff} depends on population of para-H_3^+ and ortho-H_3^+ and on particular recombination rate coefficients ${}^{P}\alpha$ and ${}^{O}\alpha$. To obtain α_{eff} from spectroscopic measurements, decays of both, para-H_3^+ and ortho-H_3^+, densities at the same conditions have to be measured and then added together. With the assumption of quasineutrality we obtain the electron density decay. From the fit of the resulting electron density decay α_{eff} can be obtained.

2. Experimental apparatus

A cavity ring down spectrometer (CRDS) has been used to obtain time evolution of H_3^+ concentration during a microwave pulsed discharge and afterglow [13,14], see figure 1. The H_3^+ ions were created in discharge tube (cooled by liquid nitrogen) by a microwave discharge in a gas mixture He/Ar/H_2 at 77 K. The concentrations of used gases He: (3–8) × 10^{17} cm^{-3}, H_2: (0.7–8) × 10^{14} cm^{-3}, Ar: (1–5) × 10^{14} cm^{-3}. Transitions R(1,1)^u $03^1 \leftarrow 00^0$ (7237.285 cm^{-1}) and R(1,0) $03^1 \leftarrow 00^0$ (7241.245 cm^{-1}) were used to monitor para-H_3^+ and ortho-H_3^+, respectively. The temperature of ions was determined from the Doppler broadening of the observed absorption lines.

3. Results

First the Doppler broadened line profiles have been recorded for both lines (see figure 3.) and from them the temperature of the studied ions was determined (T = (80±20) K). Next the evolution of absorption at line center at various conditions was measured for both transitions. Those were then recalculated to concentrations under the assumption that para-H_3^+ and ortho-H_3^+ are two separate mutually non-interacting gases in thermal equilibrium. From the measured concentration evolution of both the lowest para and ortho state we can see that the ratio between them is constant during the whole discharge-afterglow cycle (see figure 2). At ~80 K the majority of H_3^+ ions are in the two observed lowest rotational states.

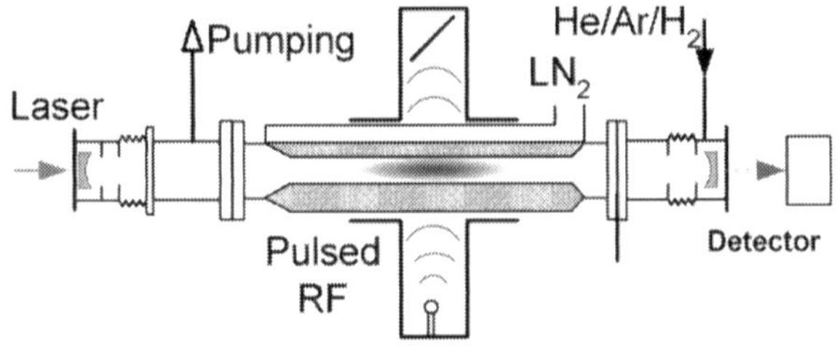

Figure 1. Schematic drawing of the used apparatus. Laser enters from the left, bouncing back and forth between highly reflective mirrors. Light exiting the optical cavity is detected.

The concentration evolutions of para-H_3^+ and ortho-H_3^+ (at the same conditions) were added together and this sum, corresponding to electron density decay, was then fitted to obtain the overall recombination rate coefficient (effective recombination coefficient, α_{eff}). Each CRDS data point in figure 3 was obtained this way. For a comparison of the obtained results with previous measurements (Cryo - FALP [12,13]) see figure 3. To obtain the rate coefficients of the binary and ternary processes we carried out measurements at different helium pressures. The resulting dependence α_{eff} on helium density was fitted with a linear fit ($\alpha_{eff} = \alpha_{bin} + K_{He}[He]$, for details see refs. [11,12]), giving the desired rate coefficients $\alpha_{bin}(77\ K) = 1.2 \times 10^{-7}\ cm^3s^{-1}$ and $K_{He}(77\ K) = 0.7 \times 10^{-25}\ cm^6s^{-1}$.

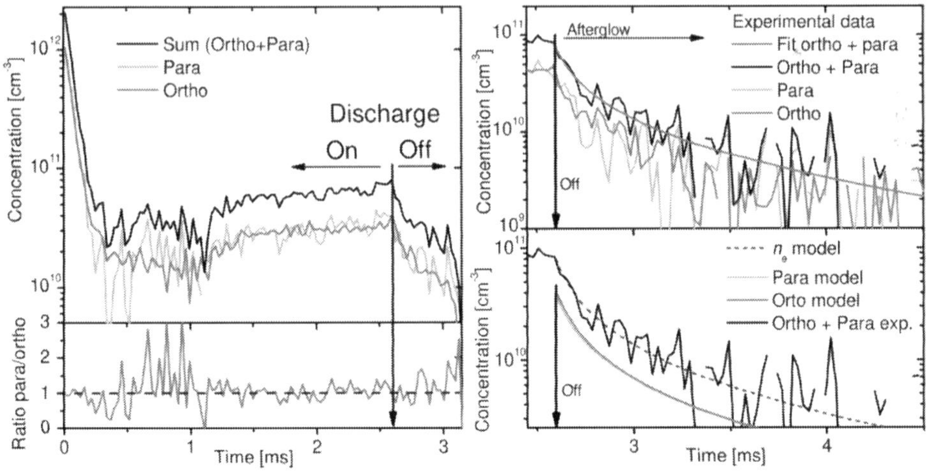

Figure 2. Example of the measured concentration evolutions of both nuclear spin states of H_3^+ at same conditions. In the left lower panel ratio of para-H_3^+ and ortho-H_3^+ densities is plotted. In the right upper panel an example of measured decays during the afterglow is plotted. In the right lower panel the measured and calculated electron density decays are compared. We used $k_{PO} = k_{OP} = 2 \times 10^{-10}\ cm^3s^{-1}$ [9], $[H_2] = 2 \times 10^{14}\ cm^{-3}$. Note that as in the experiment, the calculated para-H_3^+ and ortho-H_3^+ densities are in equilibrium during the afterglow (their densities are equal).

4. Conclusion

The rate coefficient of H_3^+ recombination at 77 K has been measured using time resolved CRDS. Obtained values for the binary and ternary rate coefficient are in agreement with previous measurements made with FALP technique, but without identification of recombining ions [11,12]. The binary rate coefficient also agrees well with theoretical predictions. For the first time ever the para-H_3^+ and ortho-H_3^+ composition of the recombining H_3^+ dominated plasma was measured in situ.

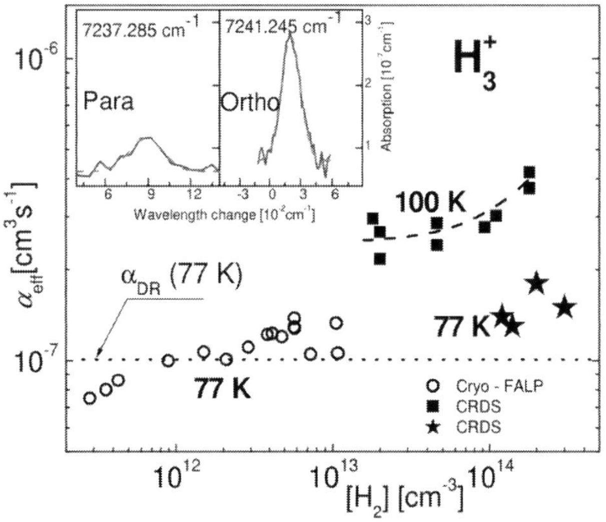

Figure 3. The effective recombination rate coefficients of H_3^+. CRDS and FALP [12] data at 77 K. For comparison also data measured by CRDS at 100 K (using ortho transition) are also plotted (for details see ref. [15]). Horizontal line indicates the theoretical value of α_{DR} at 77 K [6]. The measured absorption lines are plotted in the insert, from the Doppler broadening the ion temperature was obtained.

Acknowledgements

This work is a part of the research plan MSM 0021620834 financed by the Ministry of Education of the Czech Republic and was partly supported by GACR (202/07/0495, 205/09/1183, 202/09/0642, 202/08/H057) by GAUK 53607, GAUK 124707, GAUK 86908 and GAUK 25709.

References

[1] Plašil R, Glosík J, Poterya V, Kudrna P, Rusz J, Tichy M and Pysanenko A 2002 *Int. J. Mass Spectrom.* **218** 105
[2] Johnsen R 2005 *J. Phys.: Conf. Ser.* **4** 83
[3] Larsson M and Orel A E 2008 *Dissociative Recombination of Molecular Ions* (Cambridge: Cambridge University Press)
[4] Kokoouline V, Greene C H and Esry B 2001 *Nature* **412** 891
[5] Kokoouline V and Greene C H 2003 *Phys. Rev.* A **68** 012703
[6] dos Santos S, Kokoouline V and Greene C H 2007 *J. Chem. Phys.* **127** 124309
[7] Kreckel H, Motsch M, Mikosch J, Glosík J, Plašil R, Altevogt1 S, Andrianarijaona V, Buhr H, Hoffmann J, Lammich L, Lestinsky M, Nevo I, Novotny S, Orlov D A, Pedersen H B, Sprenger F, Terekhov A S, Toker J, Wester R, Gerlich D, Schwalm D, Wolf A and Zajfman D 2005 *Phys. Rev. Lett.* **95** 263201
[8] Larsson M, McCall B J, Orel A E , 2008 *Chem. Phys. Let.* **462** 145–151
[9] Pagani L, Vastel C, Hugo E, Kokoouline V, Greene C H, Bacmann A, Bayet E, Ceccarelli C, Peng R and Schlemmer S 2009 *Astron. Astrophys.* **494** 623
[10] Gerlich D, Windisch F, Hlavenka P, Plašil R and Glosik J 2006 *Philos. Trans. R. Soc. London*, Ser. A **364** 3007
[11] Glosik J, Korolov I, Plasil R, Novotny O, Kotrik T, Hlavenka P, Varju J, Mikhaylov I A, Kokoouline V and Greene C H 2008 *J. Phys. B: At. Mol. Opt. Phys.* **41** 191001
[12] Glosík J, Plašil R, Korolov I, Kotrík T, Novotný O, Hlavenka P, Dohnal P, Kokoouline V and Greene C H 2009 *Phys. Rev.* A **79** 052707
[13] Macko P, Báno G, Hlavenka P, Plašil R, Poterya V, Pysanenko A, Votava O, Johnsen R and Glosík J 2004 *Int. J. Mass Spect.* **233** 299
[14] Hlavenka P, Plašil R, Báno G, Korolov I, Gerlich D, Ramanlal J, Tennyson J and Glosík J 2006 *Int. J. Mass Spect.* **255** 170–176
[15] Plasil R, Hlavenka P, Macko P, Bano G, Pysanenko A and Glosik J 2005 *Journal of Physics: Conference Series* **4** 118–25

Reconstruction activities and first results from the Thomson scattering diagnostic on the TCABR tokamak

M P Alonso[1], A C A Figueiredo[1], F O Borges[2], J Leonardo[2], J I Elizondo[2], J H F Severo[2], O C Usuriaga[2], E K Sanada[2], R M O Galvão[2], L A Berni[3] and M Machida[4]

E-mail: alonso@ipfn.ist.utl.pt

[1] Associação EURATOM/IST, Instituto de Plasmas e Fusão Nuclear - Laboratório Associado, Instituto Superior Técnico, P-1049-001 Lisboa, Portugal
[2] Instituto de Física, Universidade de São Paulo, SP, Brazil
[3] Laboratório Associado de Plasma, INPE, São José dos Campos, SP, Brazil
[4] Universidade Estadual de Campinas, Campinas, SP, Brazil

Abstract. An incoherent and infrared Thomson scattering diagnostic (ITS) was transferred from ISTTOK (Lisboa) and reconstructed on TCABR (S. Paulo). In the first phase of this international collaboration, the diagnostic uses a Neodymium:Glass laser with up to 10 Joules per laser pulse and a first generation polychromator with three pairs of interference filters and avalanche photodiodes. It measures 90° scattered radiation in a single volume of observation with a single laser pulse to obtain the instant plasma electron temperature. This paper reports the reconstruction activities already carried out and presents the first experimental results. These activities include: new data model performance, laser refurbishing, new laser delivery system, stray-light reduction in the vacuum vessel, new collection lens and relative diagnostic calibration. A long run of experiments with this diagnostic shows consistency and coherence with the other TCABR diagnostics and gives indications to be able to contribute effectively to the Alfven heating program of this tokamak.

1. Introduction

In a typical Thomson scattering system the discrete volume of the plasma to be analyzed is occupied by electrons that are illuminated by a laser pulse light.

The incoherent and infrared Thomson scattering diagnostic (ITS) developed for the tokamak ISTTOK in Lisboa was transported and reassembled on the tokamak TCABR in São Paulo, following an agreement where different local teams and hardware merged in the denominated infrared based ITS diagnostic installed on TCABR (ITS-TCABR).

This paper includes a brief overview of the ITS-TCABR hardware in Section 2. The Section 3 covers the commissioning activities carried out which will describe the prerequisites and pitfalls of the infrared based ITS diagnostic. Preliminary scattering results are presented in Section 4 along with discussion and commissioning activities still to be completed that are covered in the Section 5.

[1] Author to whom any correspondence should be addressed.

2. ITS-TCABR Hardware

TCABR is a tokamak with 61 cm major radius, 18 cm minor radius, toroidal magnetic field of 1.1 T, and typical electron density and temperature values of 2×10^{19} m^{-3} and 400 eV respectively. The hardware of the Thomson scattering diagnostic is almost the same diagnostic on ISTTOK [1].

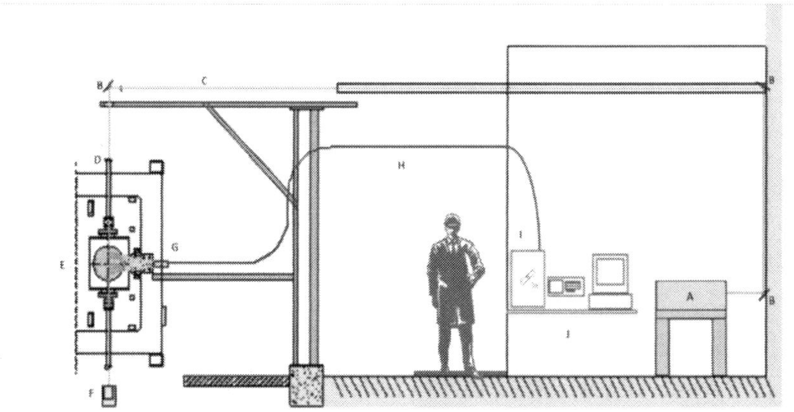

Figure 1. The ITS diagnostic lay-out at the TCABR tokamak with approximate sizes: (A) Nd:Glass laser (Quantel PG38); (B) Dichroic mirrors from CVI; (C) laser delivery system; (D) focusing lens and input window to the vacuum vessel from Thorlabs; (E) poloidal section of the TCABR tokamak; (F) main laser measurement and absorption dump from Scientech; (G) observation window and doublet collecting lenses from Thorlabs; (H) single optical fiber with 1.8 mm diameter; (I) polychromator with 3 pairs of avalanche photodiode and interference filters from former AEA-UK; (J) digital oscilloscope and PC data acquisition. The plasma is represented as a pink disk and the future collection lens is shown.

3. COMMISSIONING ACTIVITIES

3.1. The laser

The Neodymium glass rod in the oscillator was with a small internal optical imperfection and a pair of Neodymium glass rods was purchased from the Chinese enterprise "Shanghai Daheng Optics". Then a final alignment of the laser was made by the Quantel representative in Brazil. To decrease the risk of damage of the glass rod, and to avoid the use of two defective high energy capacitors, the maximum energy of the laser pulse (30 J) was reduced to 10 Joules.

3.2. Laser beam delivery system and collecting lenses

Inside the 2 m long laser box, we introduced the first high power dichroic mirror. This mirror deflects the laser beam by 90° sending it to the adjacent wall at about 2 m distance. In this wall there are two other dichroic mirrors that bend the laser beam back to the tokamak. This simple setup ensures mechanically stable positions and avoids significant temperature effects. Between the tokamak and the laser, we installed a 3 m high steel pillar. This pillar has two perpendicular branches: one branch upward, with 2 meter long, is for holding the last mirror and the focusing lens with f=1 m and the lower branch is for the collecting lens with a F# 10.2 and making a 18 mm spot size on the laser cord. The total laser beam path length is 12 m.

3.3. Polychromator

The ISTTOK polychromator is the first generation of these devices made in the Culham Laboratory (UK). It was adapted on-site to the new environment of the TCABR laboratory.

Internal optical alignment as well as its electronics has been tested using a fast infrared photodiode with a rise time of 10 nsec. The interference filters were replaced and tested for optical transmission. The polychromator was then recalibrated for electron temperature measurement.

3.4. Stray-light reduction

Stray-light reduction is an important step in any ITS diagnostic. Then, from the start of this diagnostic activity we simulate the ITS-TCABR using optical ray tracing with the program Zemax. We made a comprehensive preliminary study of spurious light trying to identify the most critical parts of the system in an early phase.

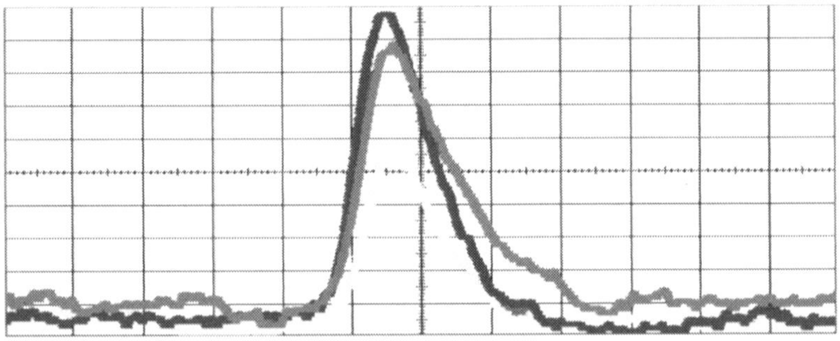

Figure 2. Experimental ITS-TCABR data. The graph shows the 3 pulses from the 3 pair of photodiodes and interference filter. Horizontal axis is time at 20 nsec/div. Vertical axis is voltage at 10 mV div.

4. Preliminary scattering results

A successful intensive experimental campaign of two weeks (in May 2009) was dedicated to ensure that the ITS-TCABR diagnostic was able to work in regular mode without the need of changing any diagnostic parameters or making any change in the optical alignments.

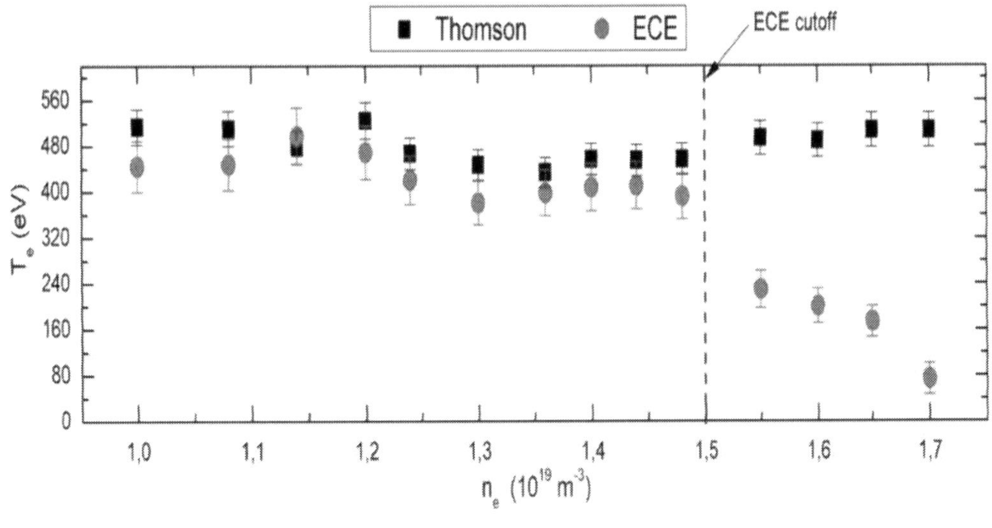

Figure 3. Electron temperature obtained by the ITS-TCABR compared with the values from the ECE diagnostic in the same shots. The right side shows the electron temperature results obtained from the TCABR plasma center with Alfven heating and with a consequent added increase in the plasma density.

In this experimental campaign electron temperatures from 50 eV up to 600 eV have been measured which are in good agreement with these provided by ECE, especially in ohmic discharge. In discharges with additional Alfven heating the disagreement observed is most probably due to errors in the ECE measurements and these are related with the value of the electron density cutoff.

5. Conclusions

The ITS-TCABR observation volume has 18 cm length and 2 mm diameter. The data analysis model was completed. The Nd:glass Laser was installed in place, was tested and refurbished. The general stray-light deduced by optical ray-tracing gave satisfactory and practical results. The laser pulse delivery system was concluded and is stable. The polychromator was tested, calibrated and commissioned in site. The ITS-TCABR diagnostic is working satisfactorily and in a routine mode. All the digitalized raw data are stored in a local hard disk of a personal computer.

This paper describes briefly the commissioning activities highlighting the prerequisites and pitfalls of the ITS-TCABR diagnostic. This system fires a single laser pulse with 5 Joules (40 ns) per plasma shot, measuring the electron temperature in the center of the plasma using interference filters and photo-diodes. It was commissioned in May 2009. It was a success history because we fulfill the decided scientific parameters on time. During the commission experimental campaign, more than 200 T_e measurements were carried out, during two weeks, with having no need for change any diagnostic parameter. The determined electron temperatures from 83 eV (shot# 23913) up to 584 eV (shot# 24031) are in agreement with those provided by the ECE diagnostic.

Acknowledgments

The authors wish to thank to the TCABR's team for the support; to the "Fundação de Amparo à Pesquisa do Estado de São Paulo" FAPESP, for one fellowship that made these first results possible. This work has been conducted under the European Fusion Development Agreement (EFDA), as well as within the framework of the Contract of Association between the European Atomic Energy Community and the "Instituto Superior Técnico" (IST), and it has also received financial support from the "Fundação para a Ciência e a Tecnologia" (FCT, Portugal) and the "Fundação de Amparo a Pesquisa do Estado de São Paulo" ((FAPESP, Brazil). The content of the publication is the sole responsibility of the authors and does not necessarily represent the views of the European Commission, the FCT, the IST, or their services.

References

[1] Alonso M P, Wilcock P D and Varandas C A F, 1999 *Rev. Sci. Inst.* **70,** 783
[2] Alonso M P, Berni L, Severo J H, Borges F O, Elizondo J I, Machida M, Varandas C A F and Galvão R M O, 2008 *Plas. Fus. Sci.* **996,** 192
[3] Alonso M P, Figueiredo A C A, Berni L A and Varandas C A F, 2008 *Plas. Sci. IEEE Trans. Plas. Sci.* **36,** n4 p1 1094
[4] Bellintani J V, Elfimov A G, Elizondo J I, Fagundes A N, Fonseca A M M, Galvão R M O, Guidolin L, Kuznetsov Y K, Lerche E A, and Machida M, 2006 *AIP Conf. Proc.* **875,** 350
[5] Berni L A, Alonso M P and Oliveira R M, 2004 *Rev. Sci. Inst.* **75,** 3884

Absolute calibration of LIDAR Thomson scattering systems on large fusion devices

R Scannell, G Naylor, M Walsh

EURATOM-UKAEA Fusion Association, Culham Science Centre, Abingdon, OX14 3DB, UK

E-mail: rory.scannell@ccfe.ac.uk

Abstract. Two lasers of different wavelengths could be used in the ITER LIDAR Thomson scattering system. This raises the possibility that absolute calibration from one laser could be used to calibrate the other laser system. Lower laser wavelengths have the advantage that the Raman cross section increases, $\sigma_{Raman} \propto 1/\lambda_0^4$. However at lower laser wavelength the spectral width of the scattered spectrum decreases $\Delta\lambda_{Raman} \propto \lambda_0^2$ making measurement closer to the laser wavelength nessecary. The choice of calibrating gas presents a number of trade offs. Raman calibration using a hydrogenic molecule produces a broad spectrum, useful since it is not close to the laser wavelength. However, Raman calibration in Nitrogen or Oxygen produces a far greater number of scattered photons for the same calibrating gas pressure. The f-number of scattered light collected by a LIDAR system varies significantly with plasma major radius. The large change in collected solid angle changes the angle of incidence of the scattered light onto the detector and hence the spectral transmission of the optical filter. This change in spectral transmission must be very accurately measured to determine the detected cross section of Raman or Rayleigh scattered light. In the case of Raman scattering, uncertainty in the absolute calibration can be reduced by collecting both Stokes and anti- Stokes lines with a number of different optical filters.

1. Introduction

Calibrating LIDAR Thomson scattering systems to produce reliable electron density profiles presents a number of challenges. The accuracy of the absolute calibration is strongly dependent on the spectral calibration of the spectrometer and the gas temperature of the calibrating molecule. As well as the various permutations of laser wavelength and calibrating gas, engineering constraints must also be considered. While it is possible to put 150mbar of calibrating gas in smaller Tokamaks, this is not always possible on large devices due to interaction with cryo-pumping and cooled components. At low pressure the overall number of collected scattered photons may limit the feasibility of certain calibration techniques. Although only Raman calibration is considered in this paper, Rayleigh calibration is also a possibility. However, Rayleigh calibration is subject to stray laser light and would require an optical filter at the laser wavelength that would not be useful for Thomson scattering.

The design of a core ITER LIDAR Thomson scattering system is detailed in [1]. The performance of one such system based on a 1064nm laser is analysed in [2]. This 1064nm laser system would benefit from an additional laser of a different wavelength. The use of an additional laser was originally proposed by [3] with a focus on spectral calibration. In this paper the absolute calibration of a Thomson scattering system for ITER using both 532nm and 1064nm lasers is discussed. The absolute calibration from an auxilary 532nm laser could be extended to provide absolute calibration of a standard 1064nm laser system.

Journal of Physics: Conference Series **227** (2010) 012028 doi:10.1088/1742-6596/227/1/012028

2. Raman Calibration

2.1. Calibration Laser Wavelength

The proposed Thomson scattering system could be calibrated in Nitrogen, Oxygen, Hydrogen or Deuterium depending on which gases are allowed in the ITER vacuum vessel. Currently the technology for detectors in the visible regions is available [4]. Detectors are also available in the infrared region, but have relatively poor effective quantum efficiency (EQE). The detectors in the visible region will allow measurement of Thomson scattered light from a 1064nm laser for a broad range of electron temperatures, but currently only low EQE detectors can detect Raman scattered light from a 1064nm laser. Frequency doubling a 1064nm laser would allow the high EQE detectors to detect the Raman scattering from the 532nm laser pulse. This Raman scattering from the 532nm pulse could be used to provide an absolute density calibration that could be scaled to the 1064nm laser system.

The Raman cross section scales with laser wavelength λ_0 as:

$$\sigma_{Raman} \propto \frac{1}{\lambda_0^4} \tag{1}$$

and hence there is 16 times more Raman scattered light from a 532nm laser than from a 1064nm laser of equivalent energy. Due to the strong scaling of scattered light with laser wavelength, even if high EQE fast infrared detectors are developed absolute calibration from a 532nm laser will remain an attractive option. A similar scaling of cross section occurs for Rayleigh scattering.

At lower laser wavelength the separation between Raman lines decreases. The wavelength of the Raman line resulting from the j→j-2 transition [5] is in general given by:

$$\lambda_{j\rightarrow j-2} = \frac{1}{\frac{1}{\lambda_0} + B_0(4J-2)} \; for \; j = 2, 3, ... \tag{2}$$

which may be approximated as in [6] to:

$$\lambda_{j\rightarrow j-2} \approx \lambda_0 - \lambda_0^2 B_0(4J-2) \tag{3}$$

where B_0 is the rotational constant[5, 7, 8], which is 198.96m^{-1} in N_2, 143.8m^{-1} in O_2, 2991.05^{-1} in D_2 and 5933.9m^{-1} in H_2. The gap between subsequent Raman lines $\Delta\lambda = \lambda_{j+1\rightarrow j-1} - \lambda_{j\rightarrow j-2}$ is given by:

$$\Delta\lambda \approx -4B_0\lambda_0^2 \tag{4}$$

which for Nitrogen and Oxygen evaluates to $\Delta\lambda = 0.6 - 0.8nm$ for a scattering wavelength of 1064nm and $\Delta\lambda = 0.15 - 0.2nm$ for a scattering wavelength of 532nm. For this reason, the Raman scattered lines from a 532nm laser are quite close to the laser wavelength. Optical filters must have good transmission to within 2-3nm from the laser wavelength to detect this spectrum, while at the same time good blocking at the laser wavelength. The spectral broadening of the laser wavelength should be small, since it will have a significant affect on the Raman cross section in a spectral channel.

2.2. Calibration Gas

Raman calibration in H_2 or D_2 gas overcomes the difficultly of proximity to the laser wavelength and reduces the strong dependency of calibration accuracy on the accuracy of the filter transmission. This is a strong advantage, since the optical filter transmission will vary with collection solid angle across the plasma. A further advantage of calibration using a Hydrogenic molecule is that broadening of Raman lines due to non-monochromatic laser pulses will not affect the cross section since the lines are not at the edge of the filter.

There are fewer spectral lines for Hydrogenic molecules. There are four strong Stokes lines corresponding to transitions j=0→2, j=1→3, j=2→4 and j=3→5. There are two strong anti-Stokes lines corresponding to the transitions j=3→1 and j=2→0. Due to the smaller number lines a much lower

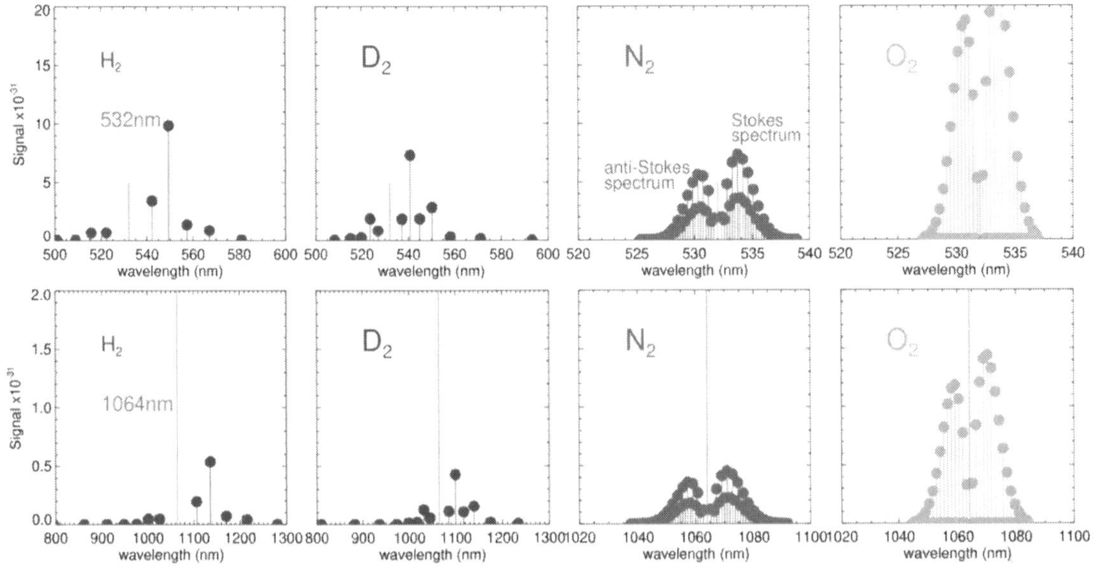

Figure 1. The top plots show the Raman spectrum for scattering from a 532nm laser from gases at 298K. The bottom plots, whose scale a factor of 10 lower, show the spectrum from a 1064nm laser.

Table 1. Number of detected photoelectrons from Raman scattering at a gas pressure of 10mbar from both Stokes and anti-Stokes lines. Optical transmission 10.7%, f/12, scattering length 67mm, EQE 4%.

	H_2	D_2	N_2	O_2
1.25J, $\lambda_0 = 532nm$	249	260	1493	3845
2.5J, $\lambda_0 = 1064nm$	29	31	186	480

number of scattered photons are detected for the equivalent gas pressure relative to Raman scattering from N_2 or O_2. The accuracy of the calibration is sensitive the gas temperature, since this changes the relative population of the different rotational states and hence the intensity of the different lines. A further practical consideration is that Hydrogenic gases are explosive in reaction with air and so may pose a safety risk which translates to a low allowable pressure.

The number of detected photoelectrons from Raman scattering at a gas pressure of 10mbar for a variety of molecules are given in table 1. There are significantly more scattered photons from the high Z elements, however scattering from Hydrogenic molecules would produce detectable results from a 532nm laser. The spectral distribution of the scattered photons is shown in figure 1. Oxygen provides a greater number of scattered photons than Nitrogen, although both elements produce scattered spectra close to the laser wavelength.

2.3. Collecting both Stokes and anti-Stokes lines

The collection f/number of the ITER core lidar system varies from f/6 at the outboard edge of the plasma to f/18 at the inboard edge of the plasma. The variation in the position of the scattered light collected relative to the focal length of the collection lens can cause a variation in angular distribution of light on the optical filters in the spectrometer. The resulting spectral distortion can skew the radial density profile if not correctly calibrated out.

Collecting both Stokes and anti-Stokes lines provides robustness against systematic errors due to

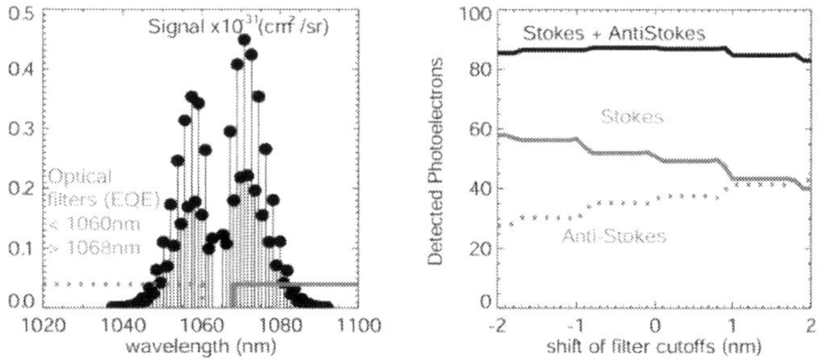

Figure 2. Variation in detected photoelectrons from Raman scattering in Nitrogen for spectral channels on either side of the laser wavelength.

change in collection position within the plasma. This is illustrated in figure 2 for Raman scattering from Nitrogen gas from a fundamental harmonic YAG laser. The results shown in this figure use a shift in the central wavelength of the filter as a simple model for the variation in the angular distribution of light on the optical filters. Applying the same wavelength shift to the filters capturing Stokes lines and anti-Stokes lines causes a change in the number of detected photoelectrons in one filter to compensate for the change in the number of detected photoelectrons in the other filter.

3. Conclusions

Absolute calibration of LIDAR systems can be acchieved using a number of different choices of gas and laser wavelength. Oxygen is not a likely choice of gas since it will pollute the vacuum vessel. The combination of Nitrogen gas and a 1064nm laser would work well and is already widely used on current TS systems. Another option is calibrating using a Hydrogenic gas at low pressure. This has the advantage that it might be considered cleaner for the vacuum vessel and Raman lines are further from the laser wavelength than for Nitrogen. The lower cross section could be paritally compensated by use of a 532nm wavelength laser.

Acknowledgement

This work has been funded jointly by the United Kingdom Engineering and Physical Sciences Research Council and by the European Communities under the contract of association between EURATOM and UKAEA. The views and opinions expressed in this paper do not necessarily reflect those of the European Commission.

References

[1] M. J. Walsh, M. Beurskens, P. G. Carolan, M. Gilbert, M. Loughlin, A. W. Morris, V. Riccardo, Y. Xue and R.B. Huxford and C.I. Walker 2006 *Rev. Sci. Instrum* **77** 10E525 .
[2] M. Beurskens, L. Giudicotti, M. Kempernaars, R. Scannell and M.Walsh 2008 *Rev. Sci. Instrum* **79** 10E727
[3] O. R. P. Smith, C. Gowers, P. Nielsen, and H. Salzmann 1996 *Rev. Sci. Instrum* **68** 725
[4] M. Kempenaars, J. C. Flanagan, L. Giudicotti, M. J. Walsh, M. Beurskens and I. Balboa 2008 *Rev. Sci. Instrum* **79**, 10E728
[5] S. C. McCool, I. L. McCool, R. D. Bengston and P. E. Phillips 1981 *Calibration of Thomson Scattering Density Measurements* (DOE/ET/53042-7, FRCR 234, University of Texas).
[6] M. V. de Sande 2002 *Ph.D. thesis, Technische Universiteit Eindhoven*
[7] C. M. Penney, R. L. S. Peters and M. Lapp 1974 *Journal of the Optical Society of America* **64** 712 .
[8] G. Vaughan, D. P. Wareing, S. J. Pepler, L. Thomas and V. Mitev 1993 *Applied Optics* **32**, 2758

Thomson scattering from laser induced plasma in air

K Dzierżęga[1], A Mendys[1], S Pellerin[2], E Thouin[2], G Travaille[3], B Bousquet[3], L Canioni[3] and B Pokrzywka[4]

[1] Institute of Physics, Jagiellonian University, ul. Reymonta 4, 30-059 Kraków, Poland
[2] GREMI - site de Bourges, Université d'Orléans, rue Gaston Berger BP 4043, 18028 Bourges, France
[3] Centre de Physique Moléculaire Optique et Hertzienne, Université Bordeaux I, 351 Cours de la Libération, 33405 Talence CEDEX, France
[4] Mt. Suhora Observatory, Pedagogical University of Cracow, ul. Podchorążych 2, 30-084 Kraków, Poland

E-mail: krzysztof.dzierzega@uj.edu.pl

Abstract. The laser induced plasma in air produced by 6 ns, 532 nm Nd:YAG pulses with 25 mJ energy was studied using the Thomson scattering method and plasma imaging techniques. Plasma images and Thomson scattered spectra were registered at delay times ranging from 150 ns to 1 μs after the breakdown pulses. The electron density and temperature, as determined in the core of the plasma plume, were found to decrease from 7.4×10^{17} cm^{-3} to about 1.03×10^{17} cm^{-3} and from 100 900 K to 22 700 K. The highly elevated electron temperatures are the result of plasma heating by the second, probe pulse in the Thomson scattering experiments.

1. Introduction

Laser induced breakdown spectroscopy (LIBS) is a well known and commonly used technique for analysis of solid, liquid and gaseous samples [1, 2]. However, it suffers from low sensitivity and poor detection limits when compared to other spectroscopic methods [3]. Increasing the sensitivity of LIBS would also allow one to decrease the pulse energies and minimize destructive effects on the sample. It was recently demonstrated that double pulse (DP) LIBS is an effective way to improve the analytical capabilities of the LIBS method [4]. Higher sensitivity is achieved by enlarged plasma volume, more efficient production of atoms in excited states and longer sustained emission. Although the mechanism resulting in signal enhancement in DP LIBS is not fully recognized, lowering of the laser shielding effect, heating of the target surface and reheating of the formed plasma plume have been invoked as possible explanations of the experimental observations. DP LIBS has been realized in orthogonal and parallel configurations of the laser beams using either the pre-ablation or the re-heating mode [4]. In the orthogonal re-heating configuration, the first pulse irradiates the sample and generates the plasma plume while the second pulse, propagating orthogonally to the sample and to the first laser beam, re-heats the ablated elements resulting in higher plasma emissivity.

In this configuration, the impact of the second laser pulse on the plasma plume can be studied using standard emission spectroscopy, but also by the analyzing its scattering by plasma electrons in the Thomson scattering (TS) experiment. Unlike emission spectroscopy, the principal advantages of TS are high spatial resolution and ease with which the measured data can be interpreted. The standard plasma parameters, the electron temperature and density, can be directly derived from the electron feature of the TS spectrum without any assumptions about the plasma symmetry, equilibrium state or its chemical composition, whereas the composition itself can be deduced from the ion feature. Detailed theory and applications of TS can be found in several monographs [5, 6].

In this work we used TS to investigate double-pulse laser induced plasma in air, since most of LIBS experiments with solid samples are performed in such an environment.

2. Experimental arrangement

The experiments were performed in atmospheric pressure air with two second-harmonic (532nm) Nd:YAG laser beams arranged in orthogonal geometry. The first pulse induced the plasma plume while the second, with much lower fluence and below the breakdown threshold for air, was used simultaneously to reheat the plasma and for observation of Thomson scattering. These beams were delivered to the interaction region and focused to the spots of 30 and 150 μm diameters using lenses of 10 cm and 50 cm focal length, respectively. The energy of the laser pulses was 25 mJ and 50 mJ, respectively and their duration was about 6 ns. This way, the fluences of the laser pulses in the interaction region differed by more than one order of magnitude. The delay between laser pulses was controlled by a delay generator and monitored using a fast photodiode and an oscilloscope.

The image of the induced plasma and the Thomson scattered light were observed in the perpendicular direction to the laser beams (see insert in Fig. 2), by imaging the investigated plasma volume onto the entrance slit of a spectrograph with 1.6 nm/mm reciprocal dispersion. The TS spectra were collected over a wavelength range of about 20 nm, with the slit width set to 50 μm. Then the images of the plasma were recorded at zero order of the spectrograph with the slit fully opened to 3.8 mm. The spectral and spatial resolution were found to be 0.032 nm and 31 μm, respectively. Single-shot TS spectra and plasma images were recorded using a gated two-dimensional intensified charge-coupled device (ICCD) camera and averaged over 1000 shots. The ICCD was synchronized either to the first (plasma imaging) or to the second (TS measurements) laser pulse and its integration time was set to 8 ns and 15 ns, respectively.

3. Results and discussion

Series of images of the plasma induced by the first laser pulse were recorded from 10 ns to 1000 ns after the pulse. Since measured intensities are integrated along the line of sight, the spatial distribution of the total emission coefficient was retrieved by applying the Abel transformation. Typical images of the plasma plume after Abel transformation are presented in Fig. 1a, where the intensity in each individual picture is normalized to the maximum value. The evolution of

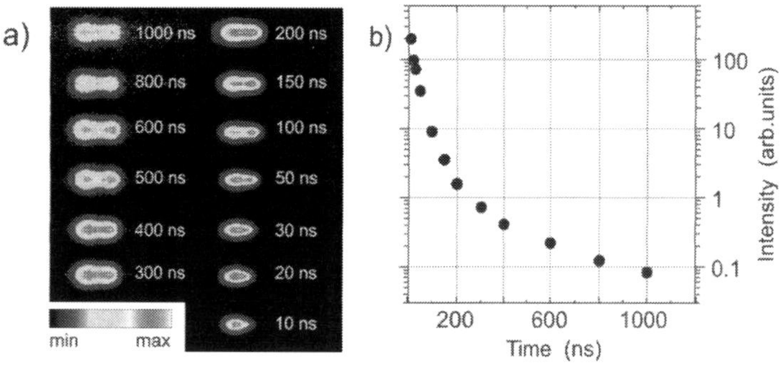

Figure 1. (a) Temporal evolution of Abel inverted images of the plasma plume (an air spark). The timings in the images represent the time after the breakdown pulse. The exposure time was 8 ns. (b) Temporal evolution of plasma emission coefficient taken from the images at the central part ($z = 0$) of the image and at the y position marked by the line.

the intensity of the plasma emissivity for the central part ($z = 0$) of the image and at the y position marked by the line is shown in Fig. 1b. Rapid radial and axial expansion of the plasma

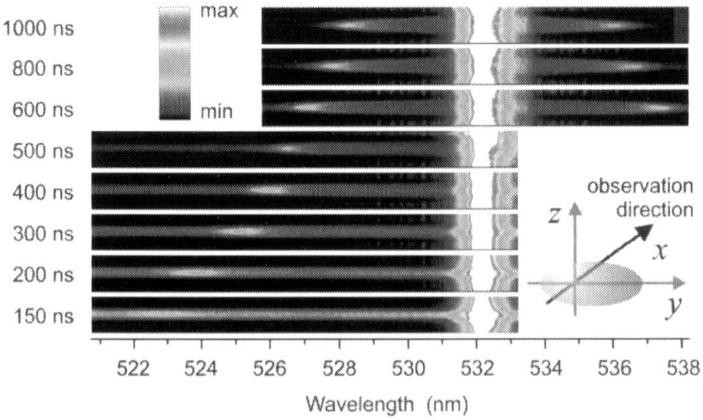

Figure 2. Thomson scattered spectra by laser induced plasma taken for different delays between laser pulses. The ICCD exposure time was 15 ns. The plasma generating beam is applied along the y-axis, the probe beam along z-axis while scattered light is observed in the x direction. Spectra of scattered light were taken from regions around the core of the plasma ($x = y = 0$).

plume occurs during the first 200 ns, while the axial expansion is predominantly in the direction of the propagating laser beam. It can be seen here that the plasma expansion can be related to fast decay of plasma emissivity (mainly due to the bremsstrahlung process) on the axis due to the decrease of the electron density caused in turn by recombination processes. For longer times, spatial evolution becomes more complex and the plasma size remains nearly constant, which corresponds to a smaller decay rate of its emissivity.

Images of the TS spectra as registered by the ICCD camera are shown in Fig. 2. They were obtained by illuminating the central part of plasma plume with the second laser beam and along the line marked in Fig.1: the z axis at $y = 0$. The consecutive images refer to different time delays between laser pulses. The saturated part of the spectra near 532 nm, apart from the TS (ion feature) and plasma light, contains also the Rayleigh scattered light as well as stray light and was discarded from further analysis. On the other hand, the electron feature appears as two peaks symmetrically situated with respect to the laser wavelength indicating the partially collective character of TS. This part of the TS spectrum was observed, with reasonable signal-to-noise ratio, for time delays longer than 150 ns. For shorter delays, the electron feature gets very broad and disappears in the midst of very strong continuous emission of plasma. For time delays shorter than 600 ns, due to large separation between peaks, only one side of the TS spectra could be recorded at the same time. The collected TS spectra extend spatially over a distance which corresponds to plasma diameter. The spectra in Fig. 2 clearly reveal that the

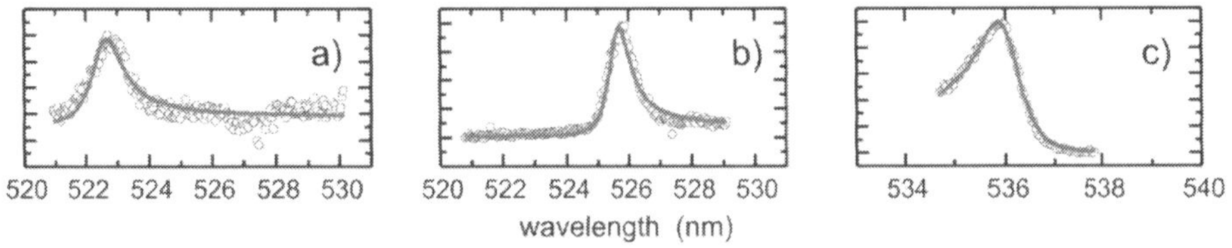

Figure 3. Example of the measured and fitted Thomson scattered spectra. The consecutive figures correspond to 150 (a), 400 (b) and 1000 ns (c) delay time between the breakdown and TS probe laser pulses. The fitted n_e and T_e are 7.44×10^{17} cm^{-3} and 100 900 K (a), 3.35×10^{17} cm^{-3} and 45 900 K (b) and 1.03×10^{17} cm^{-3} and 22 700 K (c).

peaks diverge and become broader approaching the origin of the plasma, meaning that both n_e and T_e are increasing. For more quantitative analysis theoretical TS spectra were fitted to the experimental ones as described in [8]. Such theoretical and experimental spectra are shown in

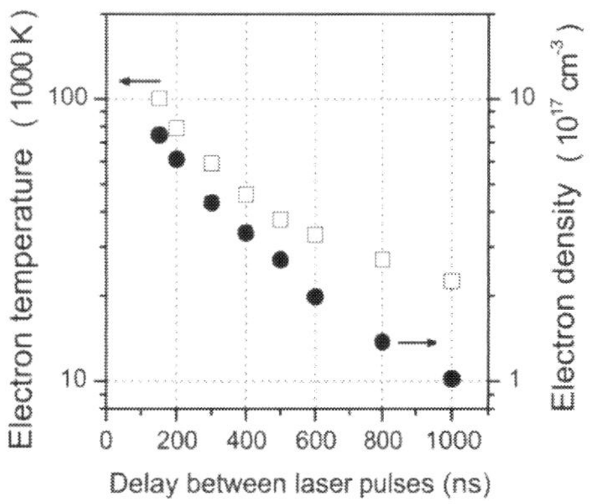

Figure 4. Electron density (○) and temperature (●) depending on delay time between the breakdown and the probe laser pulses derived from the TS spectra.

Fig. 3 for time delays of 150, 400 and 1000 ns. The fitted n_e and T_e depending on time delay between laser pulses are depicted in Fig. 4. The electron density decreases with delay time from about 7.4×10^{17} cm^{-3} to about 1.03×10^{17} cm^{-3}. At the same time the electron temperature drops from 100 900 K to about 22 700 K. It should be emphasized that the TS spectra are integrated over the duration of the second (probe) pulse, which heats the plasma in the inverse bremsstrahlung process. The problem of thermal plasma heating by the laser pulse and its influence on the TS results was recently studied theoretically by Murphy [7] and experimentally by Dzierżęga *et al* [8]. It was shown that application of the laser pulse results (under conditions of typical TS experiments) in the increase of T_e. On the opposite, the change of n_e is insignificant due to the long ionization time scale with respect to the duration of applied laser pulses. Thus, the electron temperature derived from TS spectra can significantly overestimate that just before the probe pulse. On the other hand, the electron density, at the first approximation, can be assumed to be unaffected. This means that derived values of T_e and n_e are close to those at the end of the second pulse, not to the unperturbed ones. Moreover, since the process of inverse bremsstrahlung is proportional to $n_e^2/\sqrt{T_e}$, the increase of T_e is expected to be higher when the plasma is probed by the TS pulse at the early stages of its evolution.

In a future experiment, plasma heating during the probe TS pulse will be studied in order to determine the initial "undisturbed" temperature of the plasma generated by the first pulse.

Acknowledgments

We wish to acknowledge the support of this work by project d'Action Intégreés – Polonium 7836/R09/R10,7835/R09/R10 and partially by the Polish Ministry of Science and Higher Education grant NN202 031136.

References

[1] Cremers D A and Radziemski L J 2006 *Handbook of Laser-Induced Breakdown Spectroscopy* (Chichester, John Wiley & Sons)
[2] Miziolek A W, Palleschi V and Schechter I 2006 *Laser-Induced Breakdown Spectroscopy (LIBS), Fundamentals and Applications* (Cambridge, Cambridge Academic Press)
[3] Fichet P, Tabarant M, Salle B and Gauthier C 2006 *Anal. Bioanal. Chem* **385** 338
[4] Babushok V I, DeLucia Jr F C, Gottfried J L, Munson C A, Miziolek A W 2006 *Spectroch. Acta Part B* **61** 999
[5] Evans O E and Katzenstein J 1969 *Rep. Prog. Phys.* **32** 207
[6] Hutchinson I H 2002 *Principles of Plasma Diagnostics* (Cambridge, Cambridge Academic Press)
[7] Murphy A B 2002 *Phys. Rev. Lett.* **89** 025002; Murphy A B 2004 *Phys. Rev. E* **69** 016408
[8] Dzierżęga K, Zawadzki W, Pokrzywka B and Pellerin S 2006 *Phys. Rev. E* **74** 026404

Edge Thomson Scattering in RFX-mod: operation and first measures

A Alfier, A Fassina, R Pasqualotto, S Dal Bello, V Cervaro, L Lotto and M Tollin

Consorzio RFX, Corso Stati Uniti 4, 35157 Padova

E-mail: alessandro.fassina@igi.cnr.it

Abstract: During the 2009 experimental campaign the Edge Thomson Scattering System started the operations, after a commissioning period. The main modification has been the adoption of interferential filters spectrometers with APD detectors, similar to those used on the main TS. Other improvements were the optimization of the laser input path and of the collection optics, which increased the signal level of about 50 %.
With the present setup, temperature and density measures show good compatibility with those of the main TS diagnostic; density data have been calculated after Rotational Raman Scattering calibration, a method which is not affected by impurity deposition on the collection window. Present operational limits of the diagnostic are discussed as well as short terms planned operations, focusing in particular on laser multi-pulse issues and laser sharing with the impurity injection system by laser blow-off.

1. Components description

The edge Thomson Scattering (TS) diagnostic [1] has recently started to operate on the RFX-mod machine after a commissioning period. The diagnostic will be shortly described, focusing on the laser characteristics, the beam path, the collection optics, the spectrometer and the acquisition system. Mechanical and optical improvements will be highlighted, discussing the temperature and density data with some examples, as well as some coupling issues with the Laser Blow-off system.
The source is a Q-switch ruby laser (λ=694.3 nm) with 10 J beam energy, a pulse length of 30ns at FWHM, and it can produce two pulses in within a lamp discharge (in less than about 1 ms) 5 J each. The laser is located at the same level of the experiment, in a separate room, together with the main TS laser and the acquisition electronics. The beam reaches the input optics in the torus hall with the correct polarization (vertical) by means of a $\lambda/4$ waveplate; it is focused on a 4 mm pinhole placed in a metal pipe which can be remotely inserted in the vacuum vessel. Before entering the plasma, the beam is left-tilted of 30 degrees and the pinhole is imaged at 100 mm after the pipe exiting; the scattered radiation passes through a 83 mm, F#1.2 camera lens and it is collected by a linear bundle of 12 optical fibers (hard clad silica, 1mm core diameter). The collection optics and the pipe axis are on the equatorial level. The fibers bundle goes in the laser room and here is connected to two APD interference filters spectrometers, of the same type of those mounted on main TS [2]. An optical delay system accommodates three scattering volumes (6 fibers) in each spectrometer: 25m difference in fiber length separates the three TS signals of about 120ns; 4 channels waveform digitizers are used to acquire the signal (500 Mhz, 8 bits dynamic range).

2. Improvements

In the old configuration, a single spectrometer was used for all scattering volumes; it was based on four interferential filters in a cascade design but it was coupled to a CCD camera through a gated Image intensifier. All fibers were allocated into a single bundle whose image was repetitively focused

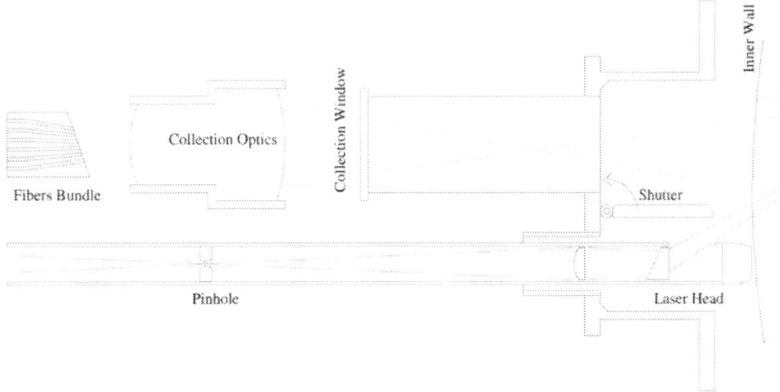

Figure 1: Detail of the laser path and of the collection optics

on four sectors of the camera field, corresponding to the four different spectral channels. A signal synchronized with the laser triggered the image intensifier with a gate whose width was typically 100 ns, optimizing the signal acquired with respect to the plasma light: the TS signal duration, defined as the time in which the signal is higher than the noise, is in fact about 60ns.

The main problem was actually the impossibility to distinguish the contribution of the two sources of background light: plasma thermal and line emission, and diffused laser radiation (stray light). The first contribution could in principle be subtracted via a second CCD acquisition, but a minimum delay time of 15 ms was required by the image intensifier characteristics. This delay time was too long compared to the variations timescale of plasma light. In a similar way, stray light contribution could be subtracted from an acquisition performed during a vacuum shot, but the image intensifier gate times proved to be slightly variable, so that this procedure became inaccurate.

Plasma light detection was solved with APD and fast digitizers; thanks to the new acquisition system, it has been relatively simple to reduce the stray light firing repetitively in vacuum conditions and optimizing the laser path.

To keep the stray light level low, the laser head is inserted very near the plasma boundary: this could have generated thermal stresses on the AR coating of the sapphire wedge which tilts the beam (fig 1); this wedge showed surface damages only on the plasma-facing side – hence the hypothesis of a thermal detriment of the AR coating. To solve the problem, laser energy has been limited to 3 joule and a new set of wedges, without AR coating, has been purchased and mounted.

Being the laser energy limited to 3-3.5 J, the S/N ratio was too low even for average plasma densities, so two fibers per scattering volume have been grouped, in order to measure 6 radial points instead of 12.

Further modifications have been:

-The change in the pinhole size from 3 to 4 mm; a 3 mm pinhole caused the loss of a large fraction of laser power and made the alignment difficult and unstable.

-The rearrangement of the fibers on the objective side; previously fibers were mounted parallel each other; this mean that for the external ones the acceptance angle vignetted by the objective. The fibers now are placed with their optical axis converging in the objective centre, gaining up to 25 % in the signal.

-The shifting of the objective of 50 mm toward the plasma, passing from an F#5 to an F#4.5; this reduced the magnification of the system, and thus the vignetting which affected the outer scattering volumes.

The objective mounting has been modified to hold a single lens: the substitution of the objective (which has an axial transmission of 70 %) by a lens should lead to a further 20 % increment in the signal, and is one of the planned modifications - together with a change in the shutter mechanism to avoid both vignetting and interference with the laser beam.

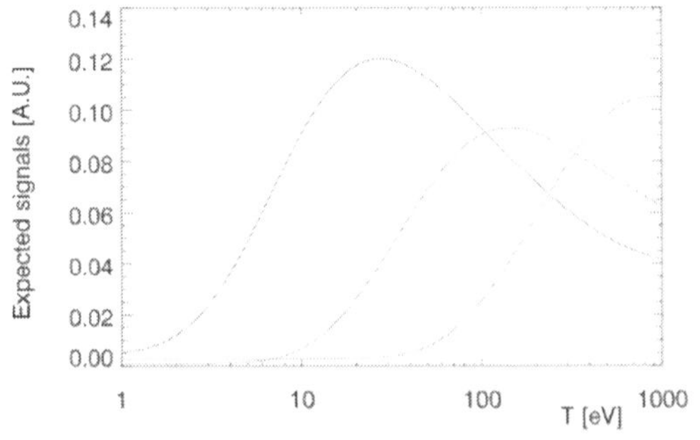

Figure 2: Expected signal on the four spectral channels in one spectrometer as a function of T_e

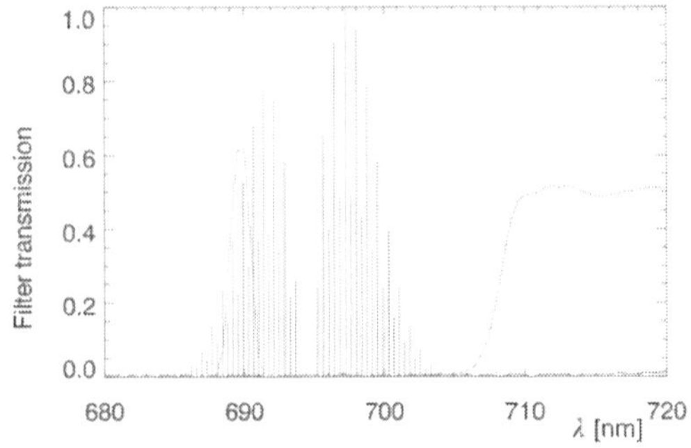

Figure 3: Intensity of nitrogen RRS spectral lines for a ruby laser (a.u.); continuous lines refer to filters transmission functions: the calibration acts only on the first channel.

3. Temperature and density data

Electron temperature is obtained processing the relative heights of the signals to reconstruct the scattered spectrum shape: fig.1 shows the expected relative value of the signal in the four spectral channels, as a function of temperature. Density data, on the other hand, are related to the absolute value of the scattered power and thus require a calibration of the instrument, which has been done via Rotational Raman Scattering in nitrogen (see fig 3). On the core TS system, RRS calibration suffers from material deposition on the windows, and isn't stable during plasma experiments. Edge TS isn't affected significantly by this problem: the pipe where the collection window is mounted is far behind the plasma and, during wall treatments, the shutter is closed. RRS proved also to be useful in checking the laser alignment, maximizing the scattering signal.

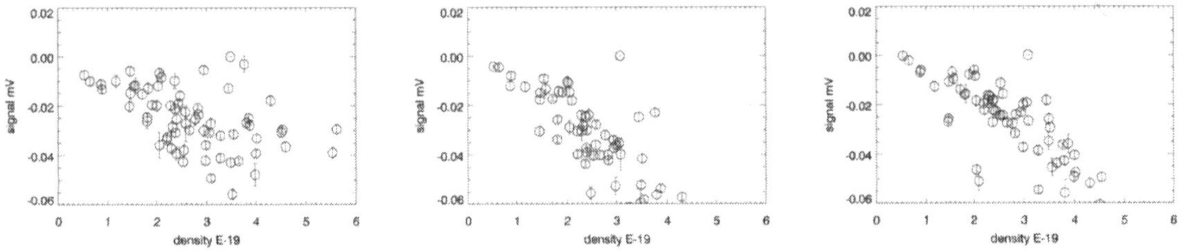

Figure 4: Signal level as a function of plasma density for three spectral channels of a core scattering volume; the lower operating limit of 1.5E19 corresponds to a 10 mV signal and to a S/N ratio of the order of 2.

With the present setup, the system is able to measure particles densities of 3-4 E18 m^{-3} and works properly for core densities over 1.5 E19 m^{-3} (fig 4). Over this value, edge temperature profiles can be compared with those from the Main TS showing the consistency of the two diagnostics in the radial overlap region (fig 5); density data of the internal points are also in good agreement with values given by the interferometer,

4. LBO coupling issues

The edge TS and the LBO system share the same laser beam: during the commissioning, the beam was directed toward the LBO assembly by a mobile mirror, if impurities injection was required during the experiments. The mounting of a mirror required the re-alignment of the LBO system on each use and didn't permit the contemporary operation of the two systems.

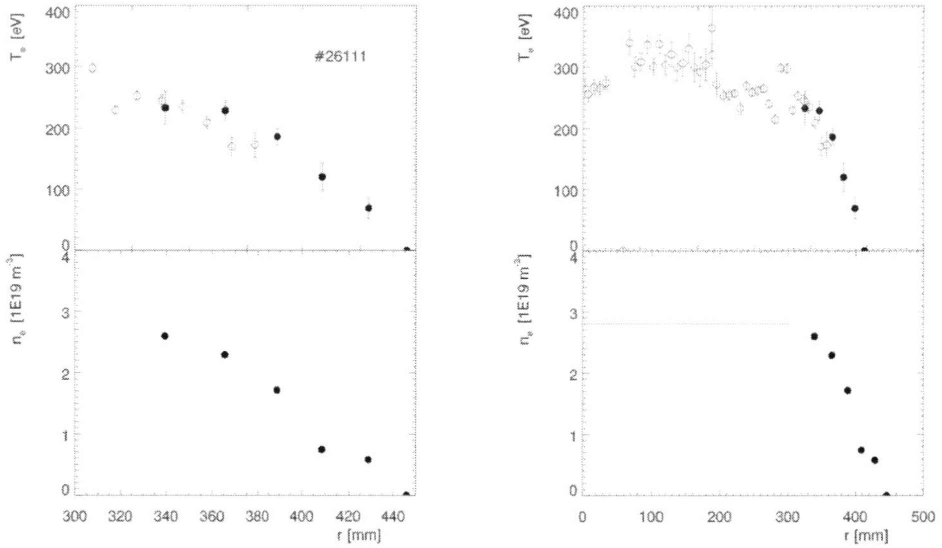

Figure 5: Example of electron temperature and density profiles obtained by the Edge TS, for the edge region (left) and for the whole radius (right) ; temperature data are consistent with the Main TS data (empty dots), while the density values for the inner points agree with the core value measured by interferometer.

Being the energy required by Laser Blow Off and by the Thomson scattering of the order of 3 J, it was decided to mount a 50 % beamsplitter on the laser path with AR coating for λ=694 nm. Simultaneous operation of the two systems has been demonstrated: LBO injection timescale is of the order of few ms and so, even in case of multiple firing, it doesn't perturb the measured temperature profiles.

5. Developments

In the short term, double pulse operation and real time laser triggering are the most interesting possibilities.

The ruby laser can fire up to four times within a lamp discharge, with successive partial openings of the electro-optical shutter (Pokels cell) in the oscillator cavity; despite this, performing more than two pulses represents a risk for optics of the input laser system, due to energy constraint and interpulse balance issues. The total energy of the device is limited to 15 J; multiple firing simply splits the available energy among the pulses. If the present setup is maintained, the presence of the beamsplitter will leave only 5-6J for the Edge TS, sufficient, in normal density conditions, only for two pulses.

Besides, balancing the pulse to pulse energy is quite simple for the oscillator, but, as the pulse are amplified, the process can become unstable and deliver most of the amplifiers energy to a single pulse, resulting in a probable damage of optics. Obviously the issue is more severe as the number of pulses (and thus the energy) increases.

Real time triggering of the laser has been successfully tested for the Main TS: the trigger signal is generated as the dominant mode amplitude exceeds a given threshold. For Edge TS operation, an application can be selective shooting during QSH crashes, to characterize edge transport modifications.

6. References

[1] A. Alfier and R.Pasqualotto 2006 *Rev. Sci. Instrum.* **77** 10E501

[2] A. Alfier and R.Pasqualotto 2007 *Rev. Sci. Instrum.* **78** 01350

Journal of Physics: Conference Series **227** (2010) 012031 doi:10.1088/1742-6596/227/1/012031

First measurements of the multichannel far-infrared polarimeter on RFX-mod

M Brombin[1,2], F Auriemma[1], A Canton[1], L Giudicotti[1,2], P Innocente[1] and E Zilli[1]

[1] Consorzio RFX, Associazione EURATOM-ENEA sulla Fusione, Corso Stati Uniti 4, 35127 Padova, Italy

[2] Padova University, Electrical Engineering Department, Via Gradenigo 6/A, 35131 Padova, Italy

E-mail: matteo.brombin@igi.cnr.it

Abstract. A multichannel far-infrared (FIR) polarimeter has been recently installed and improved in RFX-mod to measure the Faraday rotation angle along vertical chords on a poloidal plasma section. Polarimetric data, associated with measurements of the electron density, permit the reconstruction of the poloidal magnetic field profile, B_θ. The entire diagnostic is described and its main sections outlined. Emphasis is placed on the work performed on the polarimeter to reduce the fluctuations affecting the old diagnostic signals and to increase the S/N ratio. In the recent installation of the polarimeter the optical line was more carefully designed and the mirror holders have been made in insulating material to avoid any interaction with the variable magnetic fields. Moreover all the optics have been fixed on an inertial granite platform. Examples of the first Faraday angle measurements performed on five chords are presented and discussed. The measured Faraday rotation angles are compared to a theoretically calculated value, based on the μ&p model, showing a good agreement between experimental and predicted data in the central region of the plasma. The comparison between experimental and predicted data is reported and discussed in the present work.

1. Introduction

RFX-mod is the largest toroidal fusion experiment where reversed field pinch (RFP) plasmas with current up to 2MA, are studied. Since the magnetic configuration is strongly determined by currents flowing inside the plasma, an accurate measurement of the magnetic field profile is crucial to study the RFP physics. For these reasons a multichannel far infrared (FIR) polarimeter diagnostic has been installed in RFX-mod since the last year. Before this installation, an intensive activity has been undertaken to optimize the design of the new system considering several indications outlined by the previous diagnostic operating in RFX and considering the stronger operative conditions due to high plasma currents up to 2MA. The main problem in the RFX polarimeter was signal fluctuations that degraded the signal over noise ratio. The new setting has healed this problem and the system vibrations have been definitely reduced, allowing better polarimetric measurements.

In this paper, after the description of the main sections of the polarimeter, we present and discuss the experimental results of the first polarimetric measurements performed during the high current plasma campaigns.

2. Experimental set-up

The FIR polarimeter hardware consists mainly of the laser, the beam line, the beam-splitter section, the detection section, the motion electronics for the calibration procedure and the acquisition system.

The source of the laser beam is a CO_2 laser pumped FIR cavity (CH_3OH) which generates the linearly polarized radiation at 118.8 μm, mechanically chopped at 3kHz. The laser system is located in a room far from the torus and it has been recently covered by an air-conditioned box to strongly reduce the humidity in the laser room; without the dry atmosphere the radiation was attenuated up to 70% along the 2 meters long path in air between the FIR output and the optical beam line [1]. This latter guides the beam from the laser room to the beam-splitter box located close to the torus and it has a controlled Nitrogen atmosphere, as well. In the beam line, nine mirrors are displaced to lead the beam from the laser to the beam splitter box, where the FIR radiation is split into five beams by tungsten wire grids. The splitting section is integrated into the beam line and hence air-conditioned. The holders of all these mirrors are made by insulating material, whereas in the past the holder was made by metal and they interacted with the electromagnetic field, causing mirrors vibrations. Moreover, to avoid mechanical vibrations coming from the machine, the mirror holders are placed over a granite platforms.

The insulating material detection section is placed over machine and it is made up of five individual sections, each including a half wave-plate, a wire grid at 45° which divides the beam into two components and two pyroelectric detectors. These are the same as the ones used in the previous system and they have been well described in ref. 2. Each detector is equipped with a preamplifier and the signals are sent, via coaxial cables, to the diagnostic hall, where they are filtered, amplified and acquired. Then, after each plasma pulse they are processed by a software which computes the Faraday rotation angle for each line of sight. The software implements the same operations of the old hardware system [2]. It extracts the amplitude of the component at about 3 kHz (chopping frequency) from the raw signals. For each couple of detectors (named S_A and S_B), related to the same channel, the algorithm computes the sum (S_A+S_B) and the difference ($|S_A-S_B|$). The Faraday rotation angle $\alpha = C\,|S_A-S_B|/(S_A+S_B)$, where C is a constant determined for each chord by the calibration procedure, afterward explained. The time resolution of the output signal is about 3ms.

The automatic calibration procedure, performed before each plasma pulse, computes the zero of the polarimeter output signal without plasma and calibrates the diagnostic before the measurement. For this purpose, each line of sight is equipped by an half wave plate, moved by a step by step motor, that rotates the polarization direction of the radiation, named θ. The pre calibration performs an 80° width angular scan, in step of half degree, looking for the zero position of the output signal. Starting from the zero position of each chords, a fine scan is also performed from -10° to +10° in step of 0.1°, computing the zero position and the constant C for each chords. A diagram of the output signals V_{out} versus θ allows to verify that the polarimeter responds linearly within the angular range of interest. An example of output of the pre calibration is shown in figure 1 on the left, whereas the right part of the figure shows the output of the fine angular scan.

3. First experimental results

The signal to noise ratio (S/N) has been increased by a factor three since the laser box has been conditioned and the FIR output redoubled. The whole improvements of the diagnostic and the lower level of fluctuations in the acquired signals allowed the first polarimetric measurements to be made in RFX-mod. The measurements during the vacuum shot of RFX are good with zero line very well determined and an average error of about 0.2°. In the plasma discharges a high frequency (5 kHz) noise appears on the raw signal of the detectors that are closer to the machine (the B detector of each line of sight), as displayed in figure 2. This noise, due to the electromagnetic coupling of the detectors with the radial magnetic field generated by the saddle coils of the feedback control system, seems not to affect the final result of the measurements. Even if this aspect is still being studied and not yet well understood, the reason is that the processing software extracts the component at the chopping frequency (3 kHz), deleting the high frequency noise. A typical example of measurements performed

Journal of Physics: Conference Series **227** (2010) 012031 doi:10.1088/1742-6596/227/1/012031

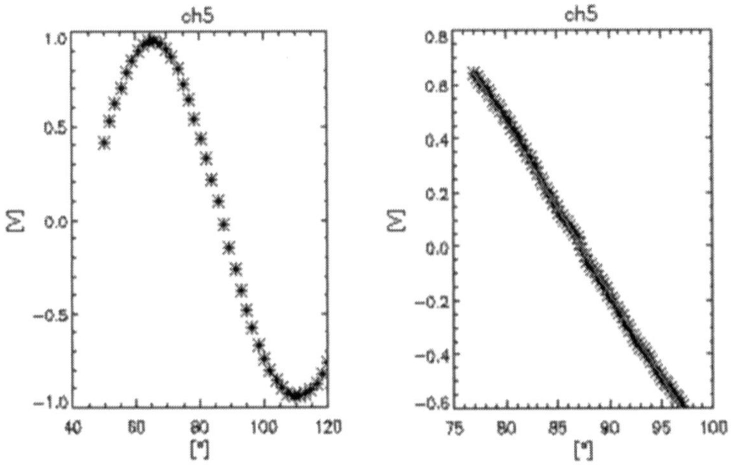

Figure 1. Example of the output of the pre calibration procedure (left side) and of the fine calibration procedure (rigth side) for chord #5.

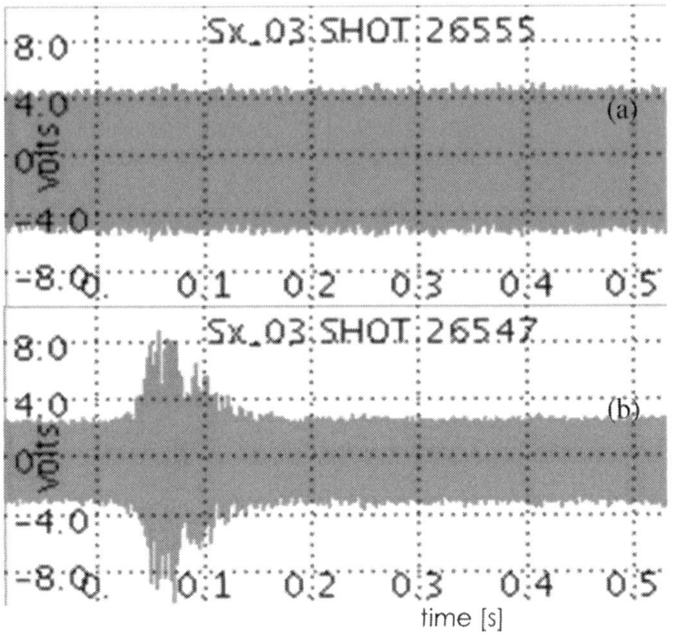

Figure 2. (a) signal from detector B of chord #3 (shot 26555) and (b) with 5 kHz noise (shot 26547)

on five chords is shown in figure 3a, where the time evolution of the polarimeter output signals during a plasma pulse is plotted for the five chords. At the end of shot when the plasma current has finished, some residual fluctuations at about 100 Hz in the polarimeter output are still visible with less impact on the measurements than in the past. The origin of these noise is purely mechanical and so they should be strongly reduced with a better alignment and a higher radiation power which has been achieved after the laser maintenance.

The Faraday angle measured by the polarimeter has been compared to the theoretically calculated values based on the μ&p model [3] and on the interferometric measurements. The magnetic field and the inverted electron density radial profiles allow to compute the theoretical Faraday rotation angle along the minor radius of the torus and its temporal evolution. An example of the comparison of the measured and predicted data is given in figure 3b, where the Faraday rotation profile as a function of the distance of the chords from the centre of the vessel is shown. The measured and the predicted data are in good agreement having the same order of magnitude and following the same general profile shape. The differences increase going from the centre towards the edge highlighting how the μ&p model is not able to describe the actual magnetic profiles especially in the edge of the plasma where the electron density and current density gradient are present. Hence a deeper study of the internal current density profile is currently being performed. The graph also reveals that the experimental values do not pass through zero at the centre of the vessel, performing an experimental determination of the position of the magnetic axis in RFX-mod, that is clearly displaced towards the outer edge of the torus [4]. The shift of the plasma has been investigated in RFX-mod by soft X rays tomography and the preliminary data of the polarimeter are in agreement with the SXR measurements [5].

4. Conclusions and further developments

RFX-mod is now equipped with a multichord FIR polarimeter for measurement of the Faraday rotation angle. Faraday rotation measurements have been obtained in high current (Ip > 1.5 MA) pulses obtaining values from zero up to more then 15°, with a good level of signal over noise ratio. The preliminary results highlighted the shift of the plasma magnetic axis and the measured values

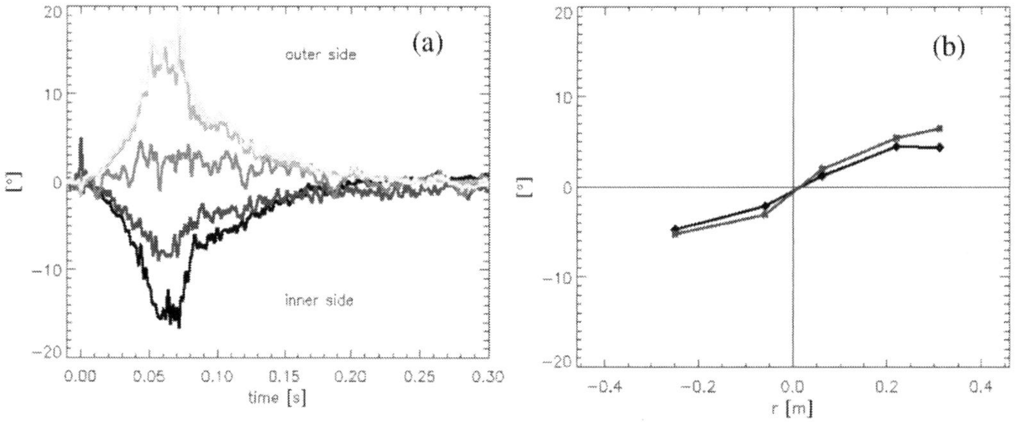

Figure 3. (a) temporal evolution of the Faraday rotation along five chords; (b) radial profile of the Faraday angle: comparison between the measured value (blue line) and the simulated values (black line).

agree with those computed by other diagnostic. The comparison with a theoretical model for the determination of the internal magnetic radial profile shows a general good agreement, whereas the main differences are in the edge region of the plasma. A more suitable description of the internal magnetic profile is being studied. Still some improvement on the diagnostic are needed. In particular, increasing the S/N with a new setup of the beam splitter box and of the detection section will help to obtain a better alignment of the laser beam. A noise on the raw signals of some detectors is present and an electromagnetic screen or a displacement of the detectors is under investigation.

It can be concluded that in the present state the polarimeter diagnostic is capable of giving a valid contribution to the knowledge of the RFX-mod experiments in combination with other diagnostics.

Acknowledgements
Authors would thank V. Cervaro, L. Franchin, M. Maniero, D. Ravarotto and A. Tiso for their technical support during the installation and the first operating time of the FIR polarimeter. This work was supported by the European Communities under the contract of Association between Euratom/ENEA. The views and opinion expressed herein do not necessarily reflect those of European Commission.

References
[1] E Zilli, M O'Gorman, L Giudicotti, F Milani, S L Prunty, A Murari and A Boboc 2000 *Int. J. Infrared Millim. Waves* **21** 1673
[2] E Zilli, F Milani, M O'Gorman, L Giudicotti and S L Prunty 2001 *Rev. Sci. Instrum* **72** 4125
[3] S Ortolani and D D Schnack 1993 *Magnetohydrodynamics of plasma relaxation* (Singapore: World Scientific)
[4] H Soltwisch 1983 *Nucl. Fusion* **23** 1681
[5] P Martin et al. 2000 *Phys. Plasmas* **7** 1984

Journal of Physics: Conference Series 227 (2010) 012032 doi:10.1088/1742-6596/227/1/012032

Analysis of the JET FIR interferometer beam phase changes during plasmas and application for fast fringe jump corrections by electronics

C Gil[1], A Boboc[2], A Barbuti[1], P Pastor[1] and JET EFDA Contributors*

JET-EFDA, Culham Science Centre, Abingdon, Oxon, OX143DB, UK
[1] CEA, IRFM, F-13108 Saint-Paul-lez-Durance, France.
[2] EURATOM-UKAEA Fusion Associations, Culham Science Centre, Abingdon, Oxon, OX14 3DB, UK,
* See Appendix F Romanelli et al., Proceedings of the 22nd IAEA Fusion Energy Conference 2008, Geneva, Switzerland
E-mail : christophe.gil@cea.fr

Abstract The interferometer diagnostics using Far Infrared (FIR) laser beams provide, by phase measurement techniques, precise line-electron density measurements of magnetic fusion plasmas. But the FIR beams still suffer the effect of refraction when traversing the plasma. This can cause, when the density gradients are strong, temporary losses of signal that induce the so called fringe jumps on the estimated phase. On the CEA Tore Supra tokamak, a Field Programmable Get Array (FPGA) electronics has been developed using a time delay method to calculate the phase and correct the fringe jumps at the frequency rate of the probe sinusoidal signal (100 KHz). To test the efficiency of the algorithm on various plasma scenarios, a prototype of the CEA electronics has been installed on the JET tokamak and the data have been compared with those issued from the present JET electronics, which calculates the phase by a different method. Statistical comparisons between the two methods on more than 1500 JET shots are reported in this article and show that the two methods give similar results but that none of them is 100% reliable as still some fringe jumps remain, in particular when Edge Localised Modes (ELMs), pellet injections or disruptions occur. To understand this phenomenon, an analysis of the fast changes of the 100 KHz raw input signals during ELMs and pellet injections has been done with a 1MHz numerical acquisition. The typical durations of signal losses have been found to be few hundreds of micro-seconds. Meanwhile, the line density can increase and then return to its original value. Simulations show that an algorithm that would block the phase calculation during a longer time (i.e. 500 μs) than the disturbed period would help to avoid fringe jumps.

1. Introduction

To diagnose the line-electron density of magnetic fusion hot plasmas, FIR interferometers [1] have been developed using plasma traversing beams in the wavelength range of 100-200 microns for the CEA Tore Supra [2-4] and the JET tokamaks [5]. The diagnostics of the two machines are very similar. They both use the principle of recombination of a 100 KHz frequency shifted beam with the probe beam that has traversed the plasma. The interference produces a 100 KHz signal, of which one can measure the phase change that is proportional to the line density. A reference beam outside the plasma allows a differential phase measurement. As the expected line density can reach $5\ 10^{20}\ m^{-2}$ in the tokamaks, the corresponding FIR beam phase changes are several 2π radians, commonly named fringes.

Thus, one needs to build up a specific electronics that continuously measures the phase, not to be lost in the counting of the fringes. However, at these wavelengths, losses of signal may happen, due to changes of trajectories when the FIR beams are crossing too strong density gradient plasmas. This configuration can occur during ELMs, disruptions or when pellets are injected in the plasma. The loss of the phase information can mislead the phase measurement and generate the so called fringe jumps, each jump corresponding to an error of roughly 10^{19} m^{-2}.
JET and CEA developed different electronics that both enable to reach a very good precision of the order of 5 10^{16} m^{-2}. The JET method is based on a 400 KHz sampling of the 100 KHz signal, which allows reconstructing the sinusoid and the corresponding phase. When the signal amplitude is lower than 50 mV, the phase measurement is stopped [5]. On Tore Supra, the sinusoidal signal is first transformed into a square signal, the ascending fronts of it being generated at every zero crossing of the sinusoid. The electronics thereafter counts the time delay between each front. Recently, the initial analogical electronics has been replaced by a numerical one. The program of the delay counting is now embedded in an FPGA processor. When the amplitude is below a threshold (0.2 volt), the fronts disappear while an algorithm maintains the phase at the most plausible value [6]. This set up has been successfully installed on Tore Supra since 2007 and correctly calculates the phase when the signal is lost as for instance during ICRH break downs, although the measurement is still disturbed by disruptions or sometimes pellet injections.

2. Comparative tests between the JET and CEA electronics

The advantage of the algorithm of the CEA electronics is that it can be easily modified. In order to test this electronics on different plasmas scenarios and possibly improve it, a complete rack of the CEA electronics has been installed in parallel with the JET electronics and connected to the 195 micron vertical traversing channels 3 and 4 during the 2008 spring campaign from shot 76103 to shot 78157.
Figure 1 shows that during a shot the evolutions of the measured phases of the two electronics are similar. One can see (fig 1b) that there is a very slight difference (~3 10^{17} m^{-2}), oscillating at low frequency (0.7Hz). As this oscillation is also seen on the JET temporal derivate (fig 1c), one can assume it is due to the JET non synchronous sampling method that calculates the phase.

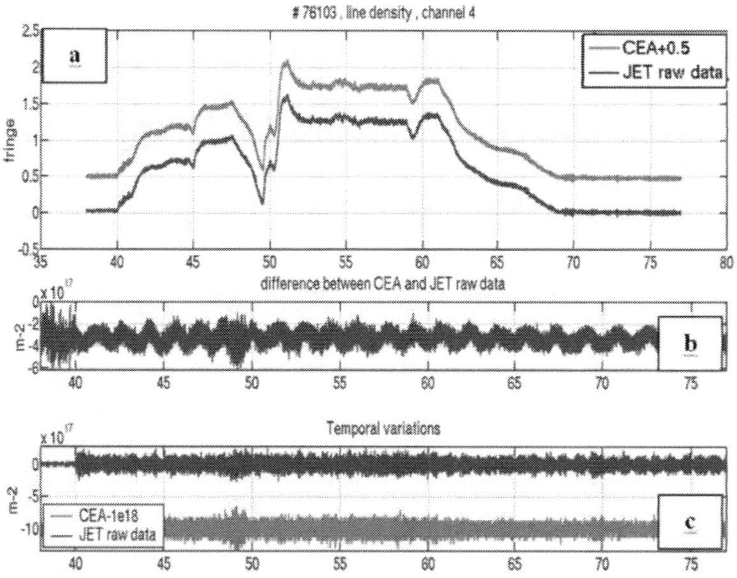

Figure 1. Comparison of the data from the two electronics

A statistical analysis has been performed to compare the fringe jumps that still remain on the two electronics. A good indicator of fringe jump occurrence during a shot is the final value phase, which is zero if no fringe jump occurs, although two fringe jumps of opposite value can erratically cancel out.
Figure 2 shows the statistics for more than 1500 shots. The number of null final values is greater for channel 3 than for channel 4 because the last one is more sensitive to refraction, as it has to specifically go through a very long metal pipe that could cause that part of the refracted beam will

suffer internal reflections that will reduce of beam power. One can see that the two electronics have almost the same behaviour.

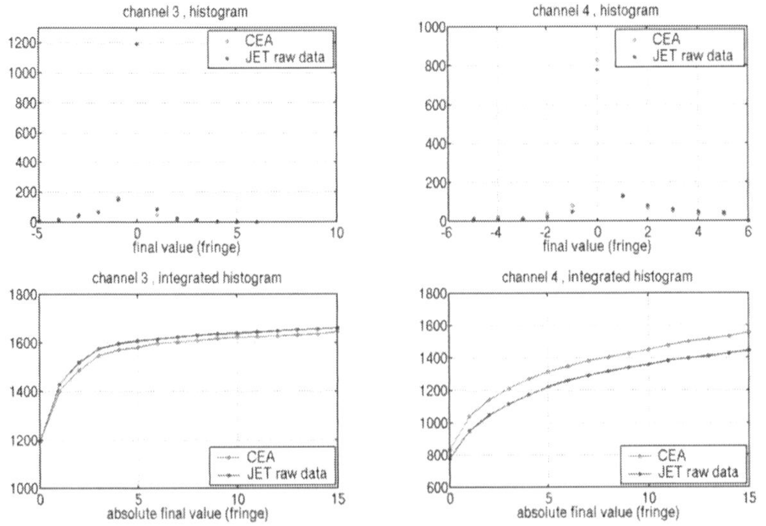

Figure 2. Analysis of the final phase value of shots 76103 to 78157

This is confirmed by table 1, which classifies the successes versus the disturbing events. One can see that the two electronics are confused when disruptions occur. During heating, they are both disturbed because ELMs induce strong density variations. The pellet injection doesn't significantly modify the statistic although some fringe jumps are sporadically observed, especially on channel 4.

Table 1: Stastistical analysis of the final value

Characteristics	Disrupt.	N. of shots	Null final value Channel 3		Null final value Channel 4	
			CEA %	JET %	CEA %	JET %
Mean plasma current>200KA	no	1598	74.3	74.2	50.8	47.9
	yes	130	4.6	6.15	13.1	10.0
Ohmic	no	190	96.3	95.7	96.8	97.9
Heating > 1MW	no	1408	71.3	71.2	44.6	41.1
Pellet and heating	no	233	76.0	76.0	43.8	42.9

3. Analysis of the phase changes during an ELM

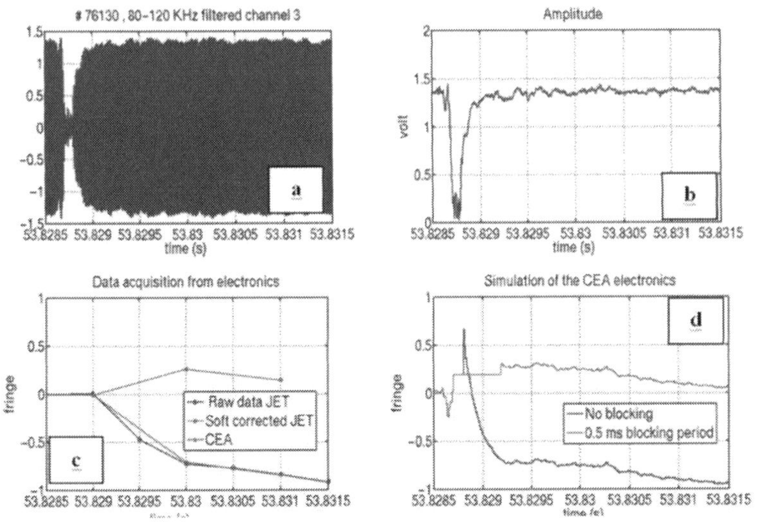

Figure 3. Analysis of a phase change during an ELM

To understand why the two electronics behave similarly, we used the one MHz data acquisition of the 100 KHz interferometric signals, available on JET since 2007. Figure 3a shows the 100 KHz signal change during an ELM. One can see on fig 3b that the amplitude is attenuated during less than 500 μs and is under the 0.2 volt threshold during an even shorter time. But, at this low level, the amplitude is noisy, thus the two electronics are mislead and give wrong changes (fig 3c), compared with the soft post shot corrected value. This can be explained by a simulation of the installed algorithm (fig3d) : The CEA electronics restarts the counting too early in considering that the most plausible scenario is no phase change, even though the ELM induced an important density increase . One can see that an algorithm with a 500µs blocking time fits better the JET soft corrected signal, as it restarts the counting when the amplitude is stable and when the phase has almost returned to its initial value.

4. Analysis of the phase changes during a pellet injection

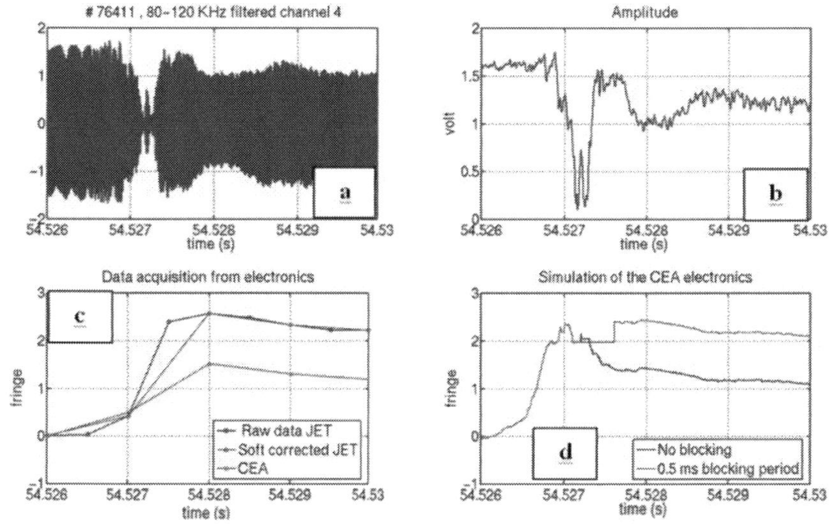

Figure 4. Analysis of the phase change during a pellet injection

Channel 4 amplitude can also be affected by pellets (fig 4). The amplitude is under the threshold at the maximum of the injection peak, but not permanently. Again the CEA present algorithm restarts too early in a very disturbed period and thus is mislead. In the contrary, the algorithm with a blocking period better fits the true phase change.

5. New algorithm and perspectives

One can see how revealing it is to be able to detect in real time the loss of amplitude of the 100 KHz signal and from this time to wait till the line density measurement is less disturbed. Thanks to the FPGA electronics architecture, the algorithm of the CEA electronics has been straightforwardly modified and a version with a 500 µs blocking delay will be tested on JET, in parallel with the initial version. Possibly the blocking period could be adjusted. In case of very long periods of amplitude defaults, the algorithm manages to wait and only restart when the amplitude is again correct.

For the moment, due to lack of information during the disturbed period, the algorithm considers that the most plausible event is that the phase has not changed. But, as during a pellet injection, this assumption can be wrong and it is of interest to explore for the JET tokamak or for future machines, the possibility to estimate by Cotton Mouton effect the line density change after a disturbed period.

Acknowledgement

The authors want to thank D. Mazon and A. Murari for the Task Force D help to organize this task. This work, supported by the European communities under the contract of association between EURATOM and CEA, was carried out within the framework of the European Fusion Development Agreement. The views and opinions expressed herein do not necessarily reflect those of the European Commission.

References:

[1] Veron D 1979 Infrared and millimeter waves (Button ed), **2**, chapter 2
[2] Bruneau JL and Gil C 1989 *Proceed.14th Conf. IRMMW in Wuzburg* 264
[3] Gil C, Elbeze D, Portafaix C 2001 *Fusion Eng. Design*, **56-57** 969
[4] Gil C et al. 2007, *Fusion Eng. Design*, **82** 1238
[5] Innocente P, Mazon D, Joffrin E, Riva M 2003 *Rev. Sci. Instrum.* **74** 3645
[6] Gil C et al. 2008 *Rev. Sci. Instrum.* **79** 10E710

Journal of Physics: Conference Series **227** (2010) 012033 doi:10.1088/1742-6596/227/1/012033

In-vessel Optic System Design of FIR Interferometer/Polarimeter System for KSTAR

Y U Nam[1] and J W Juhn[2]

[1] National Fusion Research Institute, 113 Gwahangno, Daejeon, 305-333, Korea
[2] Seoul National University, 599 Gwanak-ro, Seoul, 151-742, Korea

E-mail: yunam@nfri.re.kr

Abstract. An FIR interferometer/polarimeter system will be constructed in the Korea Superconducting Tokamak Advanced Research (KSTAR) device for electron density profile measurements. Due to the complex inner-wall geometry and long diagnostics ports, a specially designed in-vessel optic system is necessary to ensure a sufficient number of beam lines for inversion of the density profile. A single channel on the vertical center line and seven channels on the tangential plane will be placed using corner-cube reflectors mounted on the vacuum vessel. To avoid interferences with in-vessel structures such as neutral beam protection tiles and diagnostics systems, the positions of these beam lines are designed carefully using an in-vessel beam positioning module placed inside a narrow diagnostics port. Since these in-vessel optic components must be mounted on the vacuum vessel, a vibration compensation system needs to be integrated on the module. Designs of the corner-cube reflectors and the beam positioning modules have been performed for these purposes. A prototype corner-cube reflector will be installed on KSTAR before the 2010 campaign and vibration compensation performance and surface degradation due to plasma will be tested.

1. Introduction

First commissioning operation of Korea Superconducting Tokamak Advanced Research (KSTAR) device was successfully completed last year. During the commission phase, line-integrated electron densities were measured by a 280 GHz millimeter-wave interferometer system [1]. The maximum line-density was about 5×10^{19} m^{-2} which corresponds to 24 fringes with a probing frequency of 280 GHz. The fringe jump effect was compensated with help of a multi-fringe counting circuit and a correction algorithm [2]. For the main operation phase in which a line-density of over 10^{20} m^{-2} is expected, however, these correction techniques will no longer be valid. A 118.8 μm far-infra red (FIR) interferometer/polarimeter system with a triple FIR laser pumped by a CO_2 laser has now been developed for these purposes [3]. It contains a single vertical channel for density feedback control and a total of seven channels of tangential lines for density profile reconstruction. As a superconducting tokamak with a large cryostat, KSTAR has complex inner-wall geometry and long diagnostics ports. A specially designed in-vessel optic system is necessary to ensure a sufficient number of beam lines for inversion of the density profile. The beam line configuration has been optimized and a prototype design of the in-vessel optic systems including vibration compensating in-vessel corner-cube reflectors and an in-vessel beam positioning module using a cassette system have been performed.

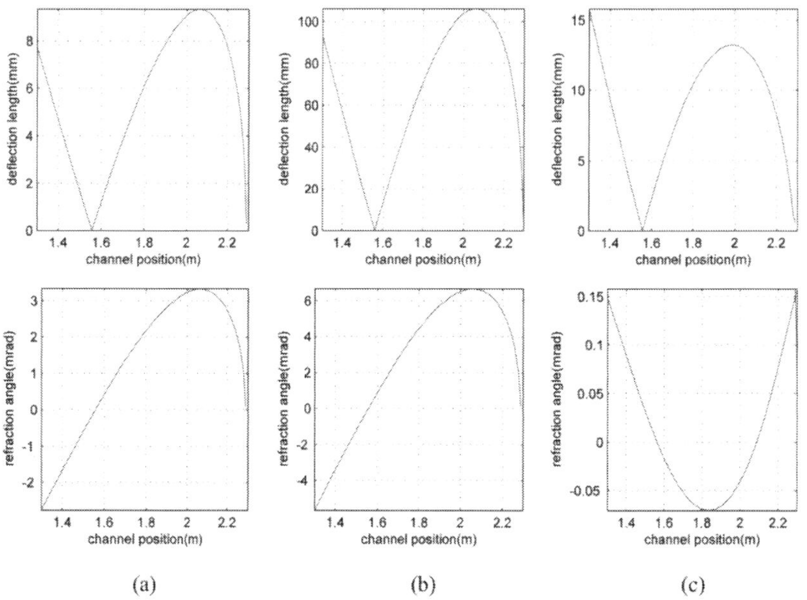

Figure 1. Simulation result of the beam refraction effect on KSTAR. (a) on a reflecting mirror (b) on a detector (plain mirror as a reflecting mirror) (c) on a detector (CCR as a reflecting mirror)

2. In-vessel vibration compensating corner-cube reflector

Heterodyne interferometry is the most reliable diagnostic for electron density measurements. Since it measures phase and not amplitude, it is very insensitive to amplitude modification from power attenuation or electronic noise. However, mechanical vibration can affect the measurement data, especially for a short probing wavelength system. Although two-color interferometry is typically used for vibration compensation, severe vibration could be damped by a mechanical technique [4].

The mechanical vibration during plasma operation mainly comes from the vacuum chamber itself. The best solution is to dismount all optical components from the chamber. Unfortunately, this is not possible on a superconducting tokamak with large cryostat such as KSTAR. A straight beam line through the vacuum chamber is not available. At least one mirror must be mounted on the chamber wall. Since a typical plain mirror can deflect the beam path significantly by small angle tilting due to vibration, a corner-cube reflector (CCR) has been adopted for the in-vessel reflecting mirror. A reflected beam on a CCR always is parallel to the incident beam regardless of the angle of the reflector. A CCR is especially good for plasma measurements since the beam itself is refracted during passage through the plasma due to electron density gradients. Figure 1 shows a simulation result of the beam refraction effect on KSTAR. The peak electron density is set to 10^{20} m^{-3} and a parabolic density profile is assumed. The maximum deflection length on a reflection mirror is below 10 mm, and the maximum refraction angle is below 4 mrad. However, this small refraction angle makes a large deflection for a detector which is 17 m away from the reflection mirror. It can be seen in Figure 1(c) that the refraction angle and the deflection length on a detector can be significantly suppressed by using a CCR as a reflecting mirror,

Although a CCR is insensitive to angle tilting as shown previously, position shift along the beam direction can affect the measurement data. This shift directly affects the beam path length, and results in phase variation. The problem is that the in-vessel environment is significantly unfavorable for vibration damping. Inside the tokamak there is a high vacuum and a high magnetic field. During plasma operation, there are high heat flux and high neutron bombardment on the vessel wall. Typical

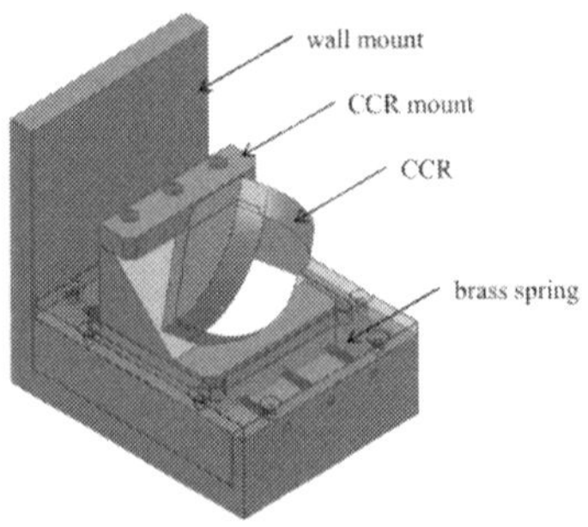

Figure 2. Conceptual design of an in-vessel vibration compensating CCR.

vibration damping techniques use springs, rubber, and pneumatic pressure. Rubber cannot be used inside the vessel and pneumatic methods are difficult to employ. The only solution is to use a simple spring made from a non-magnetic material such as brass. A conceptual design of an in-vessel vibration compensating CCR is presented in figure 2. A prototype corner-cube reflector will be installed in KSTAR before the 2010 campaign and the vibration compensation performance and surface degradation due to plasma will be tested.

3. Beam distribution/positioning module

For a profile reconstruction with line-integrated data, a sufficient number of measurement channels should be ensured. The required number of channels had been calculated using an improved Abel inversion method. The optimized number is seven in the case of KSTAR [5]. Due to the complex inner-wall geometry and long diagnostics ports, designing beam-lines for a total of seven channels is not easy. The positions of these beam lines are designed carefully using an in-vessel beam positioning module placed inside a narrow diagnostics port using a diagnostics cassette to avoid interferences with in-vessel structures such as neutral beam protection tiles and diagnostics systems. Since the optical components in the positioning module cannot be mounted on the vessel wall, the module is supported by a vibration compensation stand placed outside of the vacuum vessel. A beam line configuration and the beam positioning module are presented in figure 3. Positioning of the beam lines was limited by the neutral beam protection armor and the diagnostics cassette. To ensure the beam paths within the narrow diagnostics cassette and to minimize the size of the holes in the armor, the beam lines are stacked in two columns using two plane mirrors. These beams are arranged on the vibration compensation stand and are transmitted to the beam distribution module placed downstairs.

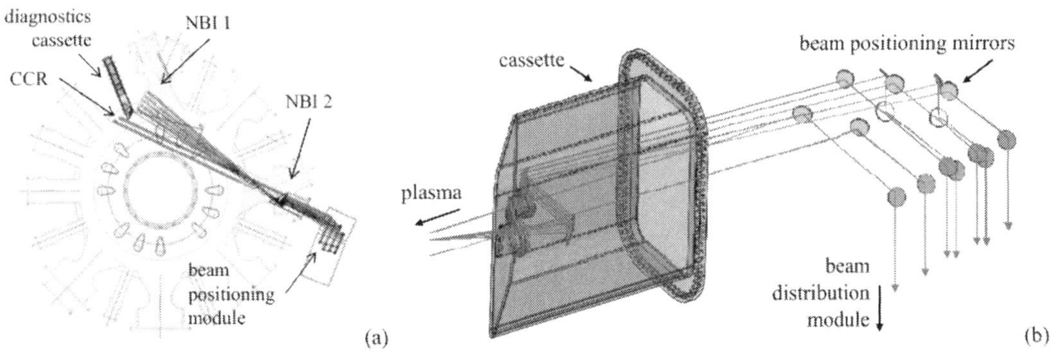

Figure 3. (a) Beam line configuration and (b) beam positioning module.

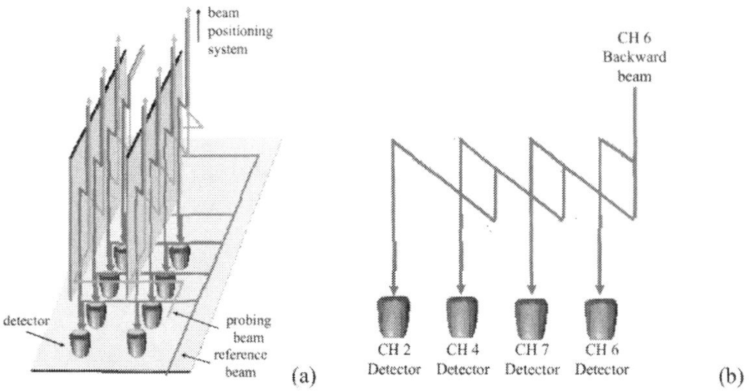

Figure 4. Beam distribution system (a) schematic diagram (b) cross-talk of backward beam

A triple FIR laser system will be placed in the diagnostics room to allow routine maintenance of the laser system. A probing beam and a reference beam are transmitted to the beam distribution module. The beam distribution module is mounted on the same vibration compensation stage as the beam positioning module. Before the beams are split for the probing and the reference detectors, the beam transmission lines do not need to be protected from vibration since the vibration effect is compensated by a phase comparison circuit. A schematic diagram of the beam distribution system is presented in figure 4. One of the important points for designing a beam distribution system is the cross-talk of the backward beam. As it can be seen on figure 4, the main portion of the channel 6 backward beam power is transmitted to the channel 6 detector, however, some portion of beam power can be transmitted to channel 7, channel 4, or even channel 2. Using low transmittance beam splitters is one of the solutions. When the channel 6 backward beam is transmitted to its own detector, only two reflection processes are need. When the channel 6 beam is transmitted to the channel 7 detector, it requires two transmission processes. Therefore, the cross-talk of a backward beam can be reduced using a low transmittance splitter. In the case of 50:50 splitter, the power ratio between its own detector and nearest detector is 4:1. In case of 30:70 splitter and 10:90 splitter, this ratio is reduced to 11:1 and 100:1. A low transmittance splitter also is good for beam power distribution between channels in a cascade type distribution system such as figure 4.

4. Conclusion and future work

The design of the in-vessel optic system of the FIR interferometer/polarimeter system for KSTAR has been performed. In spite of the complex in-vessel geometry on KSTAR, a sufficient number of beam lines for a profile reconstruction can be ensured using a beam positioning module. All in-vessel optics has been designed with consideration of protection from mechanical vibration. A prototype corner-cube reflector will be installed on KSTAR before the 2010 campaign and the vibration compensation performance and surface degradation due to the plasma will be tested. Based upon these results, a single-channel vertical FIR interferometer system will be installed for the 2011 campaign. Finally, a multi-channel tangential FIR interferometer/polarimeter system will be installed for the 2012 campaign.

References

[1] Nam Y U and Lee K D 2008 *Rev. Sci. Instrum.* **79** 10E705
[2] Nam Y U 2009 *Trans. Fus. Sci. Tech.* **55** 180
[3] Cheon M S, Nam Y U, Ha J H and Hwang Y S 2004 *Rev. Sci. Instrum.* **75** 3402
[4] Lee K C, Domier C W, Deng B H, Johnson M, Nathan B R and Luhmann, Jr. N C 2003 *Rev. Sci. Instrum.* **74** 1621
[5] Ha J H, Nam Y U, Cheon M S and Hwang Y S 2004 *Rev. Sci. Instrum.* **75** 3408

Journal of Physics: Conference Series **227** (2010) 012034 doi:10.1088/1742-6596/227/1/012034

Reconstruction method of X-mode ultrashort-pulse reflectometry in LHD

Y. Yokota[1], A. Mase[2], K. Uchino[1], Y. Kogi[3], T. Tokuzawa[4], K. Kawahata[4], Y. Nagayama[4], H. Hojo[5],

[1]Graduate School of Engineering Sciences, Kyushu University, Kasuga 816-8580, Japan
[2]Art Science and Technology Center for Cooperative Research, Kyushu University, Kasuga 816-8580, Japan
[3]Fukuoka Institute of Technology, Fukuoka 811-0295, Japan
[4]National Institute for Fusion Science, Toki 509-5292, Japan
[5]Plasma Research Center, University of Tsukuba, Tsukuba 305-8577, Japan

E-mail: yokotay5@asem.kyushu-u.ac.jp

Abstract. Reflectometry is considered to be one of the key diagnostics to measure density profiles and density fluctuations of fusion oriented plasmas. When an electromagnetic wave is launched into a plasma, the wave is reflected at the corresponding cutoff layer of the ordinary (O) mode or the extraordinary (X) mode. Reflectometry measures the time of flight (TOF) or group delay of the reflected wave. We have applied ultrashort-pulse reflectometry (USPR) to Large Helical Device (LHD) at National Institute for Fusion Science (NIFS). The highspatial analysis method called signal record analysis (SRA) is utilized to reconstruct the density profiles from the TOF signal. Also, it is noted that the remote control system using super science information network (super-SINET) has been introduced to the present USPR system. This remote system is exclusive, and it seems to be quite effective for collaborating experiment of large devices such as ITER.

1. Introduction

Reflectometry is a radar technique utilized to measure plasma density profiles [1-7]. When the electromagnetic wave is launched into the plasma, the wave is reflected at the corresponding cut-off layer. The density profiles are obtained by calculating phase delay or group delay between reflected wave and reference wave, since the density at the cutoff layer is determined by the probing frequency. The advantages of the reflectometry lie in its high spatial resolution, non-invasive nature, and minimal requirement for port access. There are two modes for the waves propagating perpendicular to the magnetic field, that is, the ordinary (O) mode and the extraordinary (X) mode based on the dispersion relation for the electric field is parallel and

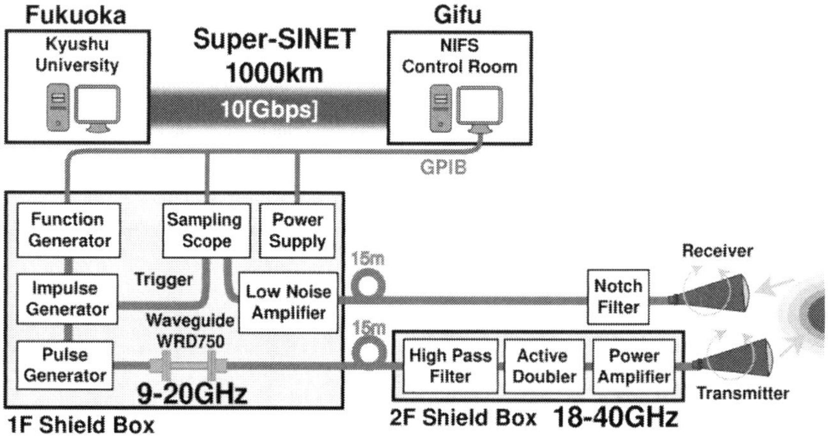

Figure 1. Overview of reflectometry system. Whole system can be controlled from the remote site using super-SINET

paerpendicular to the magnetic field, respectively. The phase difference can be described as

$$\phi_p(f_n) = \frac{4\pi f_n}{c} \int_{r_0}^{r_c} \mu_{o,x}(r, f_n)dr - \frac{\pi}{2}. \tag{1}$$

Where f_n is the frequency, c is the velocity of light, r_0 is the edge of plasma, r_c is the cutoff layer, and $\mu_{o,x}$ is the refractive index of each mode. The O-mode cut-off only depends on the local electron density. On the other hand, the X-mode cut-off depends on the local electron density and the magnetic field strength. For the O-mode propagation, this integral can be analytically solved for the density profile using the Abel inversion. For the X-mode propagation, however, no analytic solution exists and numerical methods have to be used. Though the reconstruction in the X-mode system is more complicated than the one in the O-mode system, the initial point of the density profile is expected to be obtained in the case of the X-mode system. This paper describes the present USPR system (Sec 2), the result of reconstructed density profile obtained by the O-mode system and application of the X-mode system (Sec. 3), and summary (Sec. 4).

2. USPR SYSTEM

Figure 1 shows the schematic of the USPR system applied to LHD at NIFS [8, 9]. The wide frequency band system is required to obtain wide density profile since an incident wave is reflected at the density layer corresponding to its cutoff frequency. Our reflectometry utilizes an ultrashort pulse as a source. Since the bandwidth of an impulse has an inverse relation to the pulse width, we can cover the frequency range of micro- to millimeter waves with a single source. The pulse width, height and the impulse repetition are 22 ps, 3 V and 1 MHz in the present system, respectively. The transmitter and receiver are identical conical horn antennas each having a collimating lens. The antenna gain is 30-34 dB in 18-40 GHz range. We can easily change the propagation from the O-mode to the X-mode by rotating the waveguides. The signal waveform is then digitized by a sampling scope with an equivalent sampling frequency of 250 GHz. We utilize the frequency range of 18-40 GHz in the present system The measurable plasma density is 0.4-2.0 $\times 10^{19}\mathrm{m}^{-3}$.

Remote control system using super-SINET [10] is introduced to the present USPR system. The general-purpose interface bus (GPIB) card is installed in each device. The whole operations

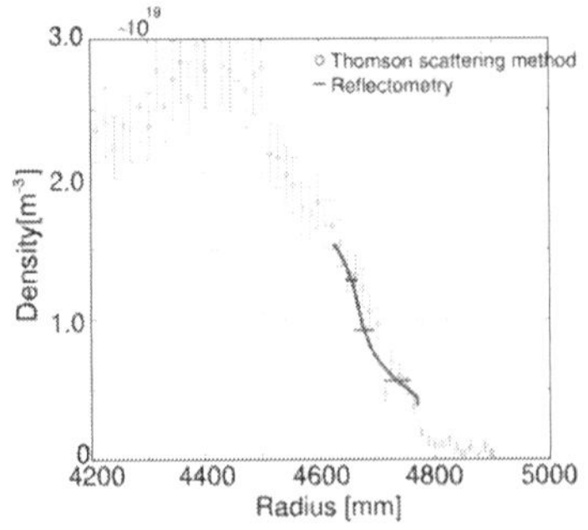

Figure 2. Plasma density profile obtained by the USPR (solid line) together with that obtained by the Thomson scattering method (open circle).

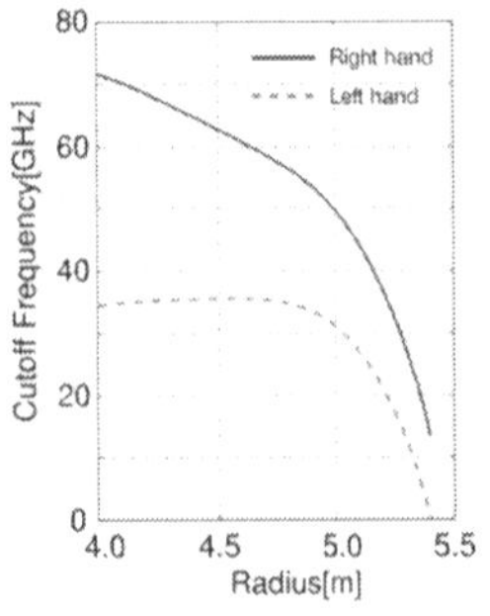

Figure 3. The cutoff frequency of X-mode: magnetic configuration is $B_0 = -1.5$ T. The solid line shows the right hand, the dashed line shows the left hand

such as adjustment of supply voltage fed to the amplifiers and the doubler, the timing control of the impulse, and the data acquisition and monitoring can be performed from the remote site (Kyushu University). It seems to be quite effective for collaborating experiment of large plasma devices such as ITER.

The reflected wave is analyzed with a signal record analysis (SRA) method which is a software handling to calculate the phase difference of reflected waves with and without a plasma as a function of frequency [11]. The conventional system utilizes a filterbank to obtain the phase vs. frequency. Therefore, the measurable points are limited by the number of filters. On the other hand, continuous spectrums can be obtained with the SRA method since the reflected wave is directly recorded to the sampling scope. As a result of that, the reconstruction profiles are expected to have a good spatial resolution.

3. Experimental Result

In the present system, the reflected wave can be obtained in every 0.4 s. In the experiment, the reflected wave can be obtained especially in the case of high density and stable plasmas. This is because the cutoff layer works as a mirror wall in the edge region due to rapid change of density. Figure 2 shows an example of the experimental reconstruction profile for the O-mode system. The result was compared with the one obtained by the Thomson scattering method. It seems to be in good agreement between the profiles obtained by the USPR and the Thomson scattering.

In the X-mode propagation, the cutoff frequency changes by the local electron density and the magnetic field. The cutoff frequency is described as

$$f_x = \pm\frac{f_{ce}}{2} + \frac{1}{2}\sqrt{f_{ce}^2 + 4f_{pe}^2}. \tag{2}$$

where f_{pe} is plasma frequency and f_{ce} is cyclotron frequency. There are the two cutoff layers in the X-mode propagation. The higher frequency is called right hand cut off and the lower one is left hand cut off, respectively. Figure 3 shows the expected cutoff frequency of each position in which magnetic configuration of $B_0 = -1.5$ T, central magnetic field, is used. The density profile of the edge region seems to be obtained for the frequency range of 18-40 GHz in this magnetic configuration. In the X-mode system, the numerical methods is utilized instead of the conventional Abel inversion. The phase difference of the n th frequency is described as the equation (1). For the next frequency f_{n+1}, the measured phase can be represented as

$$\phi_p(f_{n+1}) = \frac{4\pi f_{n+1}}{c}\left(\int_{r_0}^{r_c(f_n)} \mu_x(r, f_{n+1})dr + \frac{1}{2}\Delta r \mu_x(r_n, f_{n+1})\right) \quad (3)$$

where Δr is the distance between the n th and the $(n+1)$ th cutoff layer. All the other quantities, except Δr, are known because the profile for f_n is known. Then Δr is described as

$$\Delta r = \frac{2}{\mu_x(r_n, f_{n+1})}\left(\frac{c}{4\pi f_{n+1}}\phi_p(f_{n+1}) - \int_{r_0}^{r_c(f_n)} \mu_x(r, f_{n+1})dr\right). \quad (4)$$

The density profile can be extended in a step by step manner using the above algorithm. The behaviors of edge plasma and plasma position are important for stability control of magnetically confined plasmas. It is difficult for the other method such as FIR interferometer and Thomson scattering method to measure the density profile of the edge region. The USPR system seems to be useful for this purpose.

4. SUMMARY

In summary, an ultrashort-pulse reflectometry has been installed in LHD for measurement of density profiles in the edge region. Our reflectometer can be operated as the X-mode system as well as the O-mode system by rotating the waveguides. The whole system can be controlled from a remote site using Super-SINET. This remote system is effective for the collaboration experiments. The reflected waves from the O-mode or the X-mode system can be obtained especially in the case of high density and stable plasmas. The reconstructed density profiles obtained by the O-mode system are in good agreement with those obtained by the Thomson scattering system. In the X-mode system, the density profile of the edge plasma seems to be obtained by considering the present magnetic configuration by using the numerical method.

This work is performed as a collaborating research program at National Institute for Fusion Science (NIFS07KCHH014), and is also partly supported by the Grant-in-Aid for Scientific Research, the Ministry of Education, Science, Sports and Culture (No. 16082205 and No. 20360186)

References

[1] Laviron C, Donne A J H, Manso M E, and Sanchez J, *Plasma Phys. Control Fusion.* 1996 **38** 905

[2] Conway G D, Angioni C, Dux R, *et al.*, *Nucl Fusion.* 2006 **46** S799-S808

[3] Fattorini L, Lang P T, *et al.*, *Plasma Phys. Control Fusion.* 2008 **50** 125001

[4] Domier C W, Luhmann N C Jr., Roh Y, et al, *Rev. Sci. Instrum.* 2004 **75** 3868

[5] Tokuzawa T, Kawahata K, and Tanaka K, 2006 *Nucl. Fusion* **46** S670

[6] Luhmann N C Jr., Bindslev H, Park H, et al. 2008 *Fusion Sci. Thechn.* **53** 335

[7] van Gorkom J C, Schuller F C, Donne A J H, et al. 2001 *ECA.* **25A** 1401

[8] Mase A, Yokota Y, *et al.*, 2006 *Rev. Sci. Instrum.* **77** 10E916

[9] Yokota Y, Mase A, Kogi Y, Tokuzawa T, and Kawahata K. 2008 *Rev. Sci. Instrum.* **79** 056106

[10] Tsuda K, Nagayama Y, Yamamoto T, and Hasegawa H. 2005 *Annual Report of NIFS.* ISSN0917-1185 383

[11] Bruskin L G, Mase A, Yamamoto A, Kogi Y. 2001 *Plasma Phys. Control Fusion.* **43** 1333

Journal of Physics: Conference Series **227** (2010) 012035 doi:10.1088/1742-6596/227/1/012035

Argon ion velocity distributions in a helicon discharge measured by laser induced fluorescence

D Luggenhölscher[1], Y Celik[1], Y K Pu[2] and U Czarnetzki[1]

[1]Institute for Plasma and Atomic Physics, Ruhr University, 44780 Bochum, Germany
[2]Department of Engineering Physics, Tsinghua University, Beijing, China

E-mail: dirk.luggenhoelscher@ep5.rub.de

Abstract. A Helicon discharge in argon at low gas pressure of 0.1 Pa is generated with a flat coil antenna operated with 13.56 MHz. The power coupled into the discharge is around 1 kW which leads to an electron density of 10^{12} cm^{-3}. The velocity and temperature of the argon ions is measured by laser induced fluorescence (LIF) spatially resolved in radial and axial direction. Ambipolar diffusion accelerates the ions out of the heating region of the discharge with velocities up to 800 m/s. Due to homogenisation of the kinetic energy, the ion temperature increases from 0.15 eV in the centre to 0.4 eV at the edge of the plasma volume.

1. Introduction

Helicon discharges are driven by right-hand circularly polarised Whistler waves in bounded systems. An overview of physics and applications may be found in [1, 2]. The wave propagates along a static magnetic field of ten to 100 mT. Much higher plasma densities can be achieved with helicon discharges compared to other kinds of radio-frequency (RF) discharges operated with similar power densities. The neutral gas pressure is fairly low which makes helicons interesting for many kinds of applications. Although there are a lot of research activities on helicon discharges, the power coupling mechanism is still under discussion [2, 3, 4]. Here measurements of the argon ion velocity are performed in order to obtain the velocity profiles radially and axially resolved. Similar measurements have been performed before [5, 6] but not yet in this kind of flat coil helicon discharge.

2. Experiment

2.1. Helicon discharge

The plasma generating wave is launched by the induced electric field of a flat coil antenna [7, 8]. Details of the setup are described in [9], Langmuir-probe measurements of the plasma density and B-dot measurements of the wave field will be described in [10]. A scheme of the setup is shown in figure 1. The chamber is 50 cm high with a diameter between 35 and 50 cm. The antenna has a diameter of 20 cm and is separated from the discharge volume by a quartz jar. A solenoid is located 13 cm below the antenna outside the vessel and can provide a magnetic field of up to 30 mT. The antenna is powered by a 2 kW RF power supply at 13.56 MHz. With no current through the solenoid, the discharge is operated as a standard inductive discharge. With a magnetic field, a helicon discharge requires that the longitudinal wave vector k_l fulfills the condition $k_l L = m\pi$, with $L = 0.53$ m being the length of the field lines between the antenna and the bottom and m being an integer.

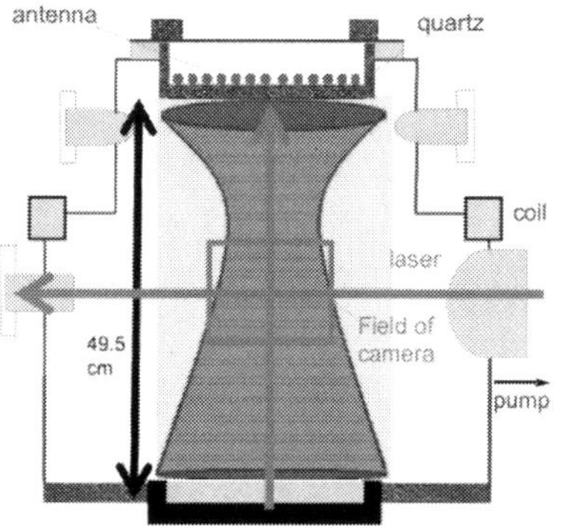

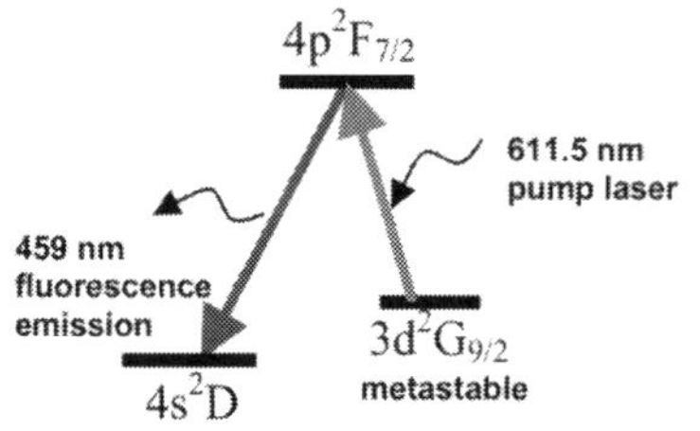

Figure 1. Helicon discharge vessel. The orientation of the laser beams and the field of view of the CCD camera is indicated.

Figure 2. Simplified term scheme used for excitation and detection of the argon ions.

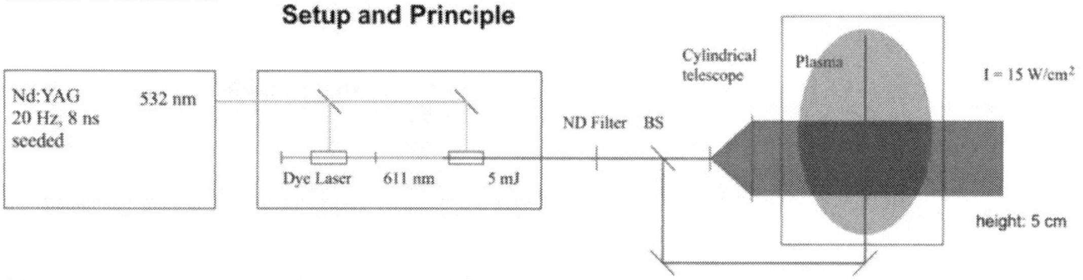

Figure 3. Laser system and laser induced fluorescence schematic.

With the assumption, that the electron density scales linear with the power P and the electron cyclotron frequency ω_c scales linearly with the magnetic field and therefore the current through the coil I, the following relation holds: $k_l\sqrt{k_l^2+k_r^2} \propto \omega_p^2/\omega_c \propto P/I$ [9], with k_r being the radial wave number. This means that only discrete ratios of power to magnetic field lead to a helicon mode of the discharge instead of a magnetised inductive discharge. Typical operation conditions are DC currents of 52 A providing a magnetic field of 9 mT in the plane of the coil and an RF power of 1 – 2 kW. The plasma density is a few 10^{12} cm^{-3}, measured with a Langmuir-probe [10]. In addition, B-dot probe measurements have been performed to determine the electromagnetic field of the helicon wave. An analytical model which contains Landau and collisional damping confirms these measurements indicating that these are the dominating coupling processes of the wave to the plasma [10]. The helicon is operated in the azimuthally homogeneous m = 0 mode.

2.2. Laser induced fluorescence setup

The spectroscopic scheme shown in figure 2 is used to excite metastable argon ions with photons of 661.5 nm and observe the fluorescence at 459 nm. The laser system used is a 20 Hz frequency doubled Nd:YAG laser with 8 ns pulse length, pumping a narrow band dye laser (figure 3). The bandwidth of the dye laser radiation is 1.0 pm and the pulse energy is around 5 mJ. The laser beam is split into two parts, one beam is send from the bottom to the top of the discharge chamber in the radial centre allowing for axially resolved measurements, the other one is spread to a sheet of 5 cm height by a cylindrical telescope and then passing radially through the plasma. In order to avoid saturation the intensity is reduced to 15 W/cm^2. Perpendicular to both beams an intensified CCD camera observes the fluorescence (figure 1) providing spatial resolution. The strong background emission is suppressed by an interference filter together with gating the camera synchronously to the laser pulses.

3. Experimental results

3.1. Radial measurements

LIF spectra are obtained by accumulating the fluorescence of typically 100 laser pulses at each

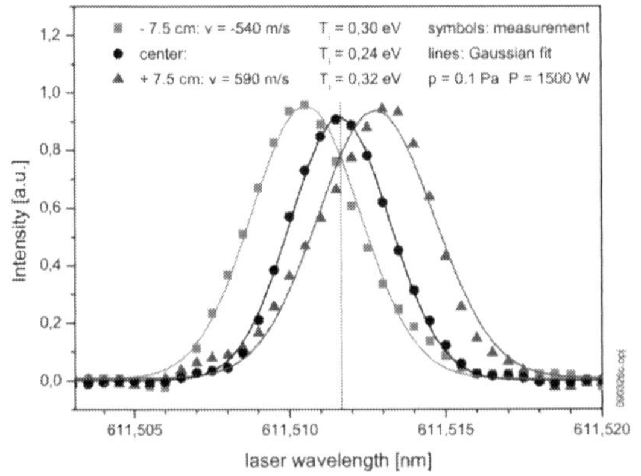

Figure 4. Fluorescence spectra at different radial positions. $P = 1500$ W, $p = 0.1$ Pa argon and 52 A coil current.

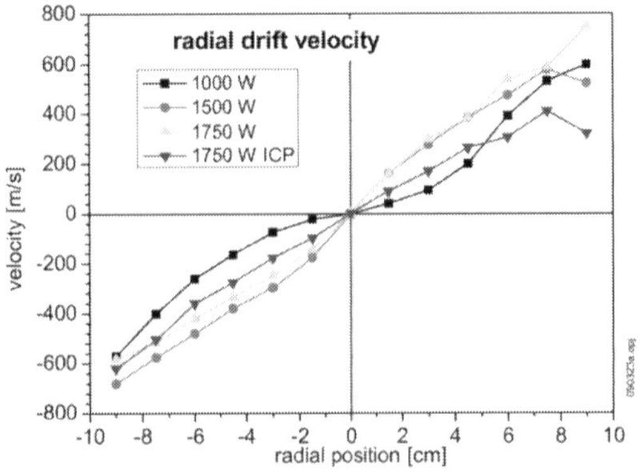

Figure 5. Radial velocity profiles. $p = 0.1$ Pa argon, 52 A coil current. Negative positions correspond to the left side of the axis in figure 1.

wavelength. From the series of pictures, spectra at different spatial positions are generated (figure 4). Gaussian profiles are fitted to the measurement to obtain the temperature from the Doppler width:

$$k_B T_i = \frac{mc^2}{8 \ln 2} \frac{\Delta\lambda_{FWHM}}{\lambda_0}$$

and the drift velocity v from the shift of the distribution:

$$v = c \frac{\Delta\lambda}{\lambda_0}.$$

Here m is the ion mass, $\Delta\lambda_{FWHM}$ the full width at half maximum of the spectrum, λ_0 the unchanged wavelength of the transition and $\Delta\lambda$ the shift of the central wavelength due to a drift. A clear shift of the ions is observable. Due to radial symmetry, the spectra in the centre are assumed to be unshifted, marking the spectrally unchanged transition λ_0. With this assumption, outward directed drifts with velocities of 500 – 800 m/s are observed. This drift is assumed to be caused by ambipolar diffusion since the discharge is operated in the $m = 0$ mode so the plasma generation is mainly concentrated on the axis given by the static magnetic field. Due to thermalisation the ion temperature increases from 0.24 eV in the centre to 0.3 eV 7.5 cm outside the radial centre.

Figure 5 shows radial velocity profiles taken in the axial centre at different RF powers for the helicon and in case of 1750 W for the inductive discharge. From 1000 W to 1500 W a strong increase in the measured velocities is observed while the profile at 1750 W stays more or less the same. This indicates that the condition of standing waves might not be perfectly fulfilled so it is no longer a pure helicon. The profile belonging to the ICP is clearly below the helicon ones, caused by the different power coupling. While the helicon is excited along the whole discharge length the ICP is generated only close to the antenna and after a few cm the electromagnetic field is strongly damped.

3.2. Axial measurements

With the laser passing along the axis from the bottom to the top, also a change of the line profiles is observed. In figure 6 three spectra are shown, all taken in the radial centre but at different axial positions: in the axial centre (0 cm), 6.8 cm and 13.5 cm below this centre. Again a shift and broadening is observed, the corresponding values are given in the legend. Assuming that the plasma is radially homogeneous in the centre the axial drift can be described in a one dimensional way by ambipolar diffusion which leads to a velocity u of the ions out of the axial centre given by [11]:

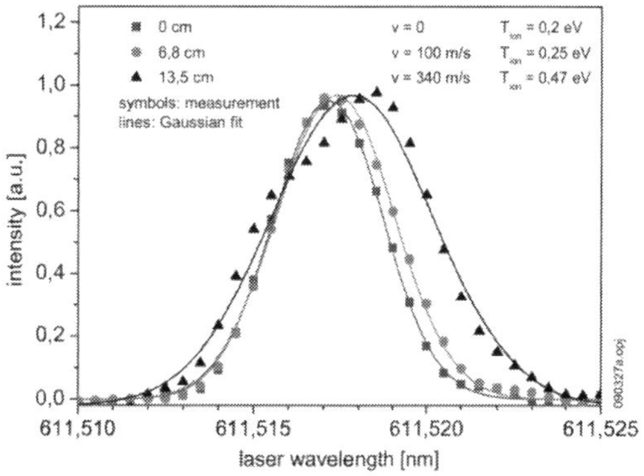

Figure 6. Fluorescence spectra at different axial positions relative to the axial centre P = 1500 W, p = 0.1 Pa argon, 52 A coil current.

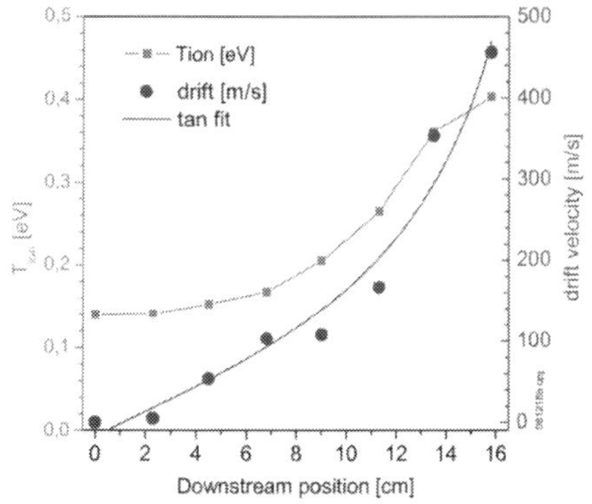

Figure 7. Axial profile of the ion temperature and velocity, same plasma conditions as in figure 6.

$$u = u_0 \tan(kz) \qquad \text{with} \qquad u_0 = \sqrt{D_a \nu_{iz}} \qquad \text{and} \qquad k = \frac{\nu_{iz}}{D_a},$$

with D_a the ambipolar diffusion coefficient and ν_{iz} the ionisation frequency of the electrons.

In figure 7 the ion temperature and velocities are shown with increasing distance to the centre. A tangent function is fitted to the measured velocities as described before and from the fit u_0 and k are obtained. From these values the ambipolar diffusion coefficient and the electron ionisation frequency are determined to be: $D_a = 21$ m^2/s and $\nu_{iz} = 1300$ s^{-1} respectively.

Conclusions

Ion drifts and temperatures in a helicon m = 0 mode discharge have been measured. A strong radial acceleration and weaker axial acceleration is observed. Parallel to the gain in drift velocity also a substantial increase in temperature occurs. The axial drift profiles agree well with a one-dimensional ambipolar diffusion model.

Acknowledgement

One of the authors (Y K P) would like to acknowledge the partial financial support from National Natural Science Foundation of China under Project 10775087.

References

[1] Chen F 1991 *Plasma Phys. Controlled Fusion* **33** 339
[2] Scime E, Keesee A and Boswell R 2008 *Physics of Plasmas* **15** 058301
[3] Clarenbach B, Krämer M and Lorenz B 2007 *J. Phys. D: Appl. Phys.* **40** 5117
[4] Kline J, Scime E, Boivin R, Keesee A and Sun X 2002 *Phys. Rev Lett.* **88** 195002
[5] Nakano T, Giapis K, Gottscho R, Lee T and Sadeghi N 1993 *J. Vac. Sci. Technol. B* **11** 2046
[6] Biliou C, Sun X, Choueiri E, Doss F, Scime E, Heard J, Spektor R and Ventura D 2005 *Plasma Sources Sci. Technol.* **14** 766
[7] Stevens J, Sowa M and Cecchi J 1995 *J. Vac. Sci. Technol. A* **13** 2476
[8] Shinohara S, Takechi S and Kawai Y 1996 *Jpn. J. Appl. Phys.* **35** 4503
[9] Crintea D, Isenberg C, Luggenhölscher D and Czarnetzki U 2008 *IEEE Trans. Plas. Sci.* **36** 1406
[10] Celik Y et al, to be published
[11] Lieberman M and Lichtenberg A 2005 *Principles of Plasma Discharges and Material Processing* (New Jersey: Wiley-Interscience)

Optimized in-situ window cleaning system by laser blow-off through optical fiber

A Alfier[1], R Pasqualotto[1], L Giudicotti[1,2], V Cervaro[1] and L Franchin[1]

[1] Consorzio RFX, Associazione EURATOM-ENEA sulla Fusione, Corso Stati Uniti 4, 35127 Padova, Italy

[2] Padova University, Electrical Engineering Department, Via Gradenigo 6/A, 35131, Padova, Italy

E-mail: alberto.alfier@igi.cnr.it

Abstract. An optimized in-situ window cleaning system by laser blow-off through optical fiber has been developed on the basis of a feasibility study previously presented. The beam generated from a Q-switched Nd:YAG laser (up to 330mJ output energy, pulse duration 5ns FWHM with 10Hz repetition rate) is launched into a high damage threshold optical fiber (Ø=1mm) through an f=80mm lens kept in a sealed box at 1mbar pressure. The fiber output is focused on the coated surface of a vacuum window previously exposed to the plasma of the RFX-mod experiment. We investigate the energy density threshold necessary to ablate the impurity deposition substrate: above threshold a single laser pulse recovers ~95% of the window transmission before its exposure to the plasma, while below it the efficiency of the cleaning process is too poor. The system so conceived can clean completely the largest window on RFX-mod (10^4mm^2 surface) in about 20minutes. We also present first results obtained firing the laser directly on a bundle of small core diameter fibers, showing performance similar to those attainable with commercial products.

1. Introduction

In present fusion devices impurity deposition processes, due to plasma wall interaction, affect vacuum windows used by optical diagnostics: for a Thomson scattering (TS) diagnostic, the window optical transmission degradation causes a distortion of the spectral transmission curve, which corresponds to a systematic underestimation of electron density n_e and to a misevaluation of electron temperature T_e. In the specific case of the main TS diagnostic installed on the modified Reversed Field eXperiment (RFX-mod) this impurity deposition mainly affects the signal level, gradually increasing the minimum density operational limit of the diagnostic[1]. To cope with this problem, an in-situ window cleaning system by laser blow-off, carrying out the laser radiation through an optical fiber, has been developed on the basis of a feasibility study previously presented [2].

2. The laser–fiber system

The Brilliant Easy laser from Quantel is used. This is a TEM_{00} Q-switched Nd:YAG laser, with 1064nm wavelength, 10Hz repetition rate, less than 50microrad pointing stability and 0.5mrad beam divergence. The energy of the 6mm diameter beam can be varied from 330mJ with 5ns pulse duration FWHM to 25mJ with 12ns FWHM, acting on the time delay between the Q-switch and the flashlamp

(Δt_{FLQS}). In Figure 1 we report the available combination of pulse duration and energy with respect to the Δt_{FLQS}. The pulse duration was measured with a 1ns response time photodiode read by an oscilloscope; the energy was measured with an optical energy meter and a thermopile detector, with 1.2kHz sampling rate. This energy range was scanned to determine the power threshold level in the cleaning process of the impurity layer.

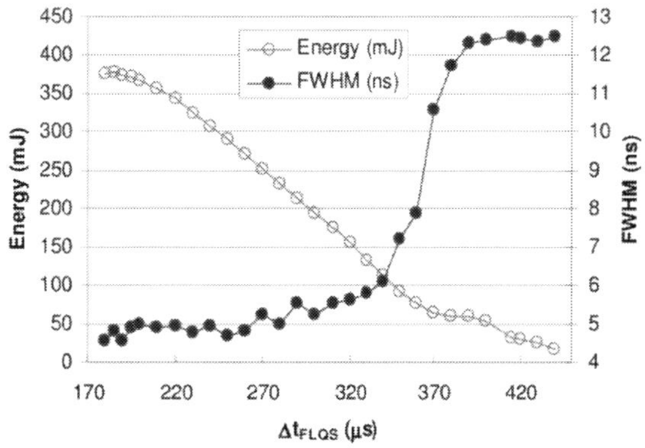

Figure 1. The output energy and the pulse duration FWHM of the Quantel Brilliant Eazy laser with respect to the time delay between the Q-switch and the flashlamp.

The laser beam is then coupled to the Ø=1mm optical fiber through a lens with focal length f=80mm, which focuses the beam ~10 mm before the fiber entrance face (not on the entrance face to avoid damages on the fiber face). The focal length matches the 0.36 numerical aperture (NA) of the fiber and the Ø=6mm beam. In case air breakdown occurs at the focus with lower NA fiber, the focal length should be chosen such as the NA is matched and the dimension of the breakdown spark is minimized: the longer the focal length, the longer the spark tube ($l \sim f^2$) and the higher the risk of damaging the fiber surface [3]; the lower the focal length, the shorter the distance of the fiber surface to the beam waist. In our case, on the basis of calculation reported in [3], the expected spark cylinder has $L=2.14\cdot\theta/d\cdot f^2=0.2$mm length and $D=2\cdot\lambda/3.14\cdot f/d$ = 10µm diameter (θ is the beam divergence and d the beam diameter on the lens). In order to avoid air spark in front of the fibre, a sealed box has been built (400x200x200 mm^3): the laser is launched into the fiber at 1mbar pressure. The laser beam enters the chamber though a BK7 vacuum window; an optical fiber feedthrough guarantees the vacuum seal allowing using the fiber to exit the vacuum chamber. Collimation between the laser beam and the fiber axis is done in atmospheric pressure, with the box open, running the laser at low energy (25mJ with 12ns FWHM). At the fiber output, two f=50mm lenses are used to first collimate the beam and then focus it onto the dirty surface of the window; two HR mirrors are used to deflect it to the desired direction. The beam hits the impurity substrate on the fused silica collection window from rear (i.e., passing through the window). The energy density on the window surface is varied changing the distance between the last lens and the window (i.e., varying the spot size) bewaring the laser is not focused inside the window in order to prevent any damage. This layout is compatible with the use on RFX-mod of an in-situ window cleaning remotely-controlled system: the collimated beam will be moved across the window surface and focused on the dirty surface from outside the RFX-mod vessel.

A damage energy threshold test was first done on a standard Ø=1mm fused silica fiber (silica core / silica cladding): as the energy output exceeds 45mJ with 8ns FWHM (5.6 MW/mm^2), the probability of melting the fiber is quite high (in the first 20 laser pulses): the damage is in the first 15cm from the fiber entrance face, since the damage occurs at the first reflections on the inner core/cladding interface where the beam energy is mainly absorbed. Below this threshold the laser causes no damaging. In order to further reduce the damage probability in case of spurious events (the laser is operating in the most unstable region of the curve shown in Figure 1), a Ø=0.94mm fiber with a high energy damage threshold (10 MW/mm^2) is used, manufactured by SEDI industries: this ensures almost a two-fold safety factor. Below this threshold, output energy from the fiber is found to be 50%-70% of that measured at its entrance (depending on the alignment accuracy and position of the fiber with respect to the launching optics), so that 80-100mJ in entrance are enough to obtain the required 40mJ.

3. Results

The system has been tested on two fused silica windows used in the main TS system [4] on RFX-mod; they have been removed in July 2009, after four months of exposure to various types of plasma discharges (RFP, helical RFP and Tokamak configurations) and wall conditioning treatment (one boronization and 20 - 60min daily He-glow discharges). The major oblong window was found to be almost completely darkened while the circular one only partially (see dashed black lines in Figure 2): the transmission was respectively 5%-30% and 60%-80% the transmission of the clean window.
The laser launching system was configured so that the fiber output energy was ~40mJ in 8ns (5 MW). The spot size on the dirty surface of the window was Ø=2.5mm, corresponding to an average energy density of ~8mJ/mm^2 (1 MW/mm^2). The window was moved with a step motion of 1mm perpendicularly to the laser beam: several regions of 20x5mm^2 area have been cleaned, each with different laser pulses per step, and their transmission function has then been measured.

With this set-up, a single pulse provides a recovery of 90%-97% of the initial transmission (see Figure 2 a-b) and increasing the number of pulses improves the results only of few %. Comparing the transmission recovered with a single pulse in the oblong and circular windows, the single pulse efficiency is more pronounced in case of thinner impurity substrate (circular windows), and so the need of further pulses. A relevant result is that in case of a too low energy density, the first pulse is not capable of recovery more than 60-70% of the initial values, and even further pulses per position cannot significantly improve the result. This indicates of the existence of an energy density threshold above which the cleaning process should be performed. This is more evident progressively reducing the laser energy without varying the spot size: with half energy density (~4mJ/mm^2 in 8ns), a single laser pulse produces a circular spot on the oblong window which shows a cleaner central region (r=0-0.5mm and ~90% of the original transmission) surrounded by a partially cleaned anular region (r=0.5-1.5mm and ~60% of the original transmission): the energy density in the central region is still above the threshold, while outside is below, and even with several laser pulses the transmission in this outer region does not exceed ~80% of the original transmission. Without a beam profiler we cannot estimate the shape of the beam at the fiber output: in the worst case of a Gaussian beam we can assume a FWHM of about 1.5mm at 40mJ (the entire cleaned spot), and the threshold energy value is estimated to be between 5 and 10 mJ/mm^2 in 8ns (again, 0.6-1.2 MW/mm^2). This also shows that the beam shape is an additional aspect that must be considered; an homogeneous beam is preferable because it allows to fine tune the beam energy above the impurity substrate ablation threshold value without changing the area of the cleaned surface: the tail and the top of a Gaussian shaped beam would be below the ablation threshold or above the damaging threshold respectively.

The system so conceived is more performing than those previously proposed in [2]: a 2.5mm spot size, with 10Hz repetition rate and the high energy output from the fiber, make it more attractive for its application to in situ window cleaning. The time required to clean the entire oblong major window of RFX with a single pulse / step and 1mm steps is only about 15 minutes, while it is 75minutes with 5 pulses per position (plus time required for motions, estimated to be 5 minutes): this is much less than 2 days of work required to replace the window (assuming an already vented vessel) or the 14 hours previously estimated with the less optimized laser-fiber system in [2].

Similar results can be obtained also launching the laser beam directly into the fiber with 330mJ energy pulse (e.g., running the laser at full energy), without a focusing lens, since the 1mm fiber partially collects the energy of the Ø=6mm beam, providing about 40mJ. As a general statement, it would be preferable to exploit all the laser output power, with the constraint of staying below the damage threshold of optical fibers. On the basis of this idea, a different strategy suitable for cleaning larger optics consists in firing the laser straight onto a bundle of fibers covering the entire beam diameter: in order to estimate the energy lost in the fibers interspaces we have tested a Ø=7mm fiber bundle (38mm^2), made of Ø=40μm fibers. Running the laser at full energy, 160mJ output energy is obtained from the bundle because of the loss in the interspaces (30%), reflections at both ends (10%), and damages into the fiber (10%): on a 25mm^2 spot area, a single pulse is sufficient to recover 90% of the nominal window transmission, as expected from the energy density value on the window surface (7mJ/mm^2 energy density in 5ns, or 1.3 MW/mm^2). A further advantage of this fiber bundle is that the output shape is square and not circular, so that the voids between two adjacent cleaned regions are negligible: a longer step can be adopted, so that the square shape can increase the cleaning speed by about 50%. With this technique, the time required to clean a surface is estimated to be more dependent on the beam motion time, i.e. 5minutes. The extracted power is comparable with that provided by commercial products, such as the Laserblast from Quantel: the basic model works at 50Hz, with 160mJ output at 25ns, corresponding to 6.4MW/mm^2. The main limits for our system is the repetition rate of the laser and the output energy which could be simply overcome with a new laser and a more transmissive fiber bundle respectively (200mJ at 100Hz in 10ns FWHM could be obtained). The most powerful Laserblast model provides 330mJ at 120Hz, which seems much further from our results; nevertheless the extracted energy density of 5-50 mJ/mm^2 is in the same range we obtain.

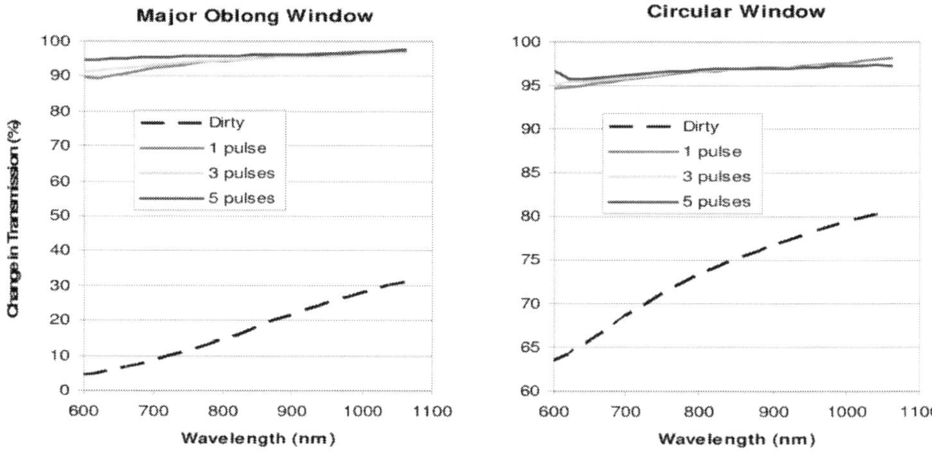

Figure 2. The transmission change for the major oblong and circular collection windows of the RFX-mod Main TS system after plasma operations and cleaning procedure with laser 1, 3, 5 pulses per step.

4. Future plans

A remotely controlled motion system will be designed, produced and tested on the basis of the Quantel laser actually available at Consorzio RFX focusing its beam onto the single high damage threshold fiber. Concurrently, on one side further tests will be done on a small diameter fiber bundle with higher transmission to verify the potential of this configuration and to make it more attractive for large optical surfaces used in large fusion devices such as JET and ITER. On the other side, first tests will be performed to verify the application of the laser blow-off cleaning technique also on metallic surfaces: damage tests will be first accomplished on clean surfaces, and then cleaning procedure will be applied to coated surfaces.

This work was supported by the European Communities under the contract of Association between EURATOM/ENEA. The views and opinions expressed herein do not necessarily reflect those of the European Commission.

References

[1] Neill D H 1990 *Rev. Sci. Instrum.* **61** 1263.
Alfier A and Pasqualotto R 2007 *Rev. Sci. Instrum.* **78** 013505.
Bonomo F et al 2009 *Nucl. Fusion* **49** 045011.
[2] Alfier A at al. 2008 *Rev. Sci. Instr.* **79** 10F338.
[3] Phuoc T X and White F P 1999 *Combustion and Flame* **119** 203.
[4] Alfier A and Pasqualotto R 2007 *Rev. Sci. Instrum.* **78** 013505.

Journal of Physics: Conference Series **227** (2010) 012037 doi:10.1088/1742-6596/227/1/012037

Velocity distribution of ions in UV laser induced plasma plumes

A Lorusso[14], V Nassisi[1], L Velardi[2] and M V Sicilano[3]

[1]Department of Physics of Lecce, Laboratorio di Elettronica Applicata e Strumentazione (LEAS), University of Salento, INFN of Lecce, Via per Arnesano, 73100 Lecce – Italy
[2]Department of Physics, University of Bari, Via Amendola, 70126 Bari – Italy
[3]Department of Material Science, University of Salento, Via per Monteroni, 73100 Lecce-Italy

E-mail: antonella.lorusso@le.infn.it

Abstract. Plasmas induced by UV laser ablation have been studied and analysed using time-of-flight measurements. A KrF laser beam of 23 ns FWHM time duration was focused into Si and Nb targets with a moderate laser fluence of 2 J/cm^2. A suitable theoretical expressions were derived for fitting the recorded ion current under the assumption of a "shifted" Maxwell-Boltzmann velocity distribution. The deconvolution of time resolved ion current signals, taking into account a multi-mode velocity distribution has revealed that the contribution of the different charge states of ions in the plasma induces electric field which accelerate the expanding plasma along the target normal.

1. Introduction

Interest in laser-produced particle beams is being stimulated by a series of applications during the last few years. In particular, the investigation of the characteristic of the ion component of the plasma plays a key role in understanding the processes on the basis of the laser-target interaction [1-3]. Furthermore, the ion energy distribution study is helpful to understand if the laser induced plasma is characterized by the presence of an ambipolar electric field known in literature as double layer (DL) [4, 5]. The DL is, in fact, formed as result of the spatial charge separation in the expanding plasma breaking the quasineutrality of the plasma. In the high power laser irradiance the presence of DL divides the expanding plasma into regions with thermal and accelerated ions resulting with a multipeak ion current signals [6, 7].

Nevertheless, the evidence of the ion energy distribution of the laser induced plasma at moderate laser fluence found also a possible explanation in the presence of an ambipolar electric field which is responsible of the ion acceleration. The presence of the DL in the laser induced plasma influences the kinetics and the angular spread of ions which are fundamental parameters for pulsed laser deposition of thin solid films as well as for the development of laser ion sources.

This work presents a study of the time resolved ion current signals by the deconvolution method taking into account the contribution of the different charge states of ions. The study is useful to understand if the kinetic energy distribution of ions is influenced by a possible presence of DL.

2. Experiment

The measurements of the ion currents are performed in a vacuum chamber of 10^{-5} mbar by Faraday cups (FC) placed in front to the target as shown in the Fig. 1a.

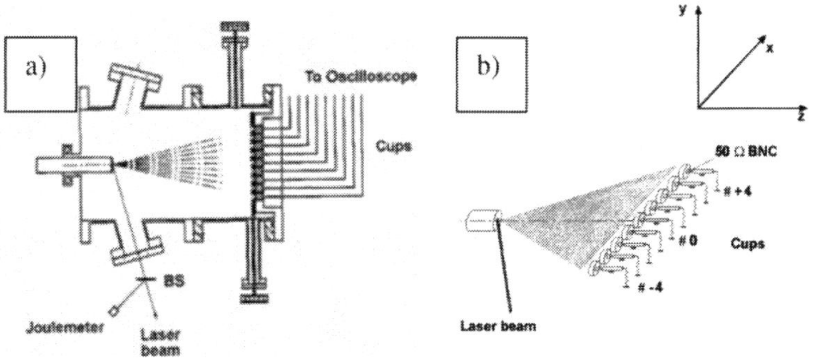

Figure 1. Sketches of the experimental apparatus: a) vacuum chamber; b) FC array placed in front to the target.

The operating circuit of FC is well described elsewhere [8]. The laser plasma is induced by using a KrF excimer laser (Lambda Physics, Compex) of 248 nm laser wavelength and FWHM pulse duration of 23 ns. The laser beam is focused by a 150 mm focal length lens with a resulting spot area of about 1 mm^2 and a laser fluence of 2 J/cm^2.

An array of nine FC is placed in front to the target at a distance of 90 mm in order to diagnostic the plasma characteristics. On the point of view of the target, the cups, placed on the left side, are labeled as +4, +3, +2, +1; the cup along to the target normal is labeled as 0, while the cups placed on the right side are -4, -3, -2, -1.

The cups are 9 mm in diameter and 11.5 mm distant from each other (see Fig. 1b). All cups are inserted into the grounded flange, insulated among them and coaxially connected to a BNC of 50 Ω which is able to derive the ion current signal to the oscilloscope.

3. Results and discussions

Typical time of flight ion signals recorded by the 9 cups are shown in the Fig. 2a and 2b for the Si and Nb targets, respectively.

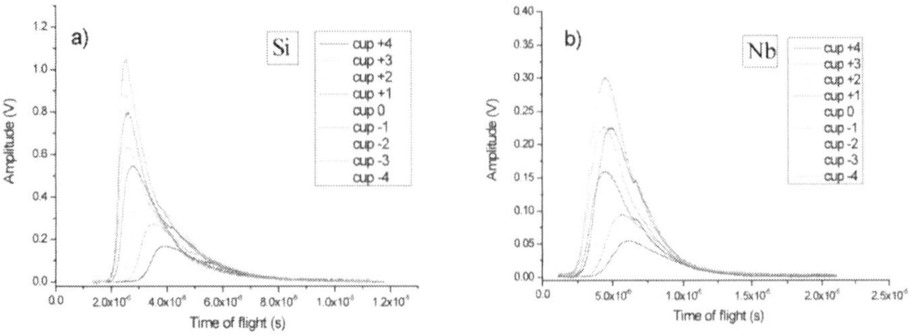

Figure 2. Time of flight ion current signals recorded by the different cups.

It is noteworthy to observe that the highest ion current signal is recorded by -1 cup and that the vast majority of ions are ejected within a small solid angle respect to the target normal while their abundance decreases along the oblique directions, e.g. +4 and +3 cups.

Furthermore, the cups in proximity of the target normal detect ion signals characterized by anticipated time of flight. Such a kind of ion acceleration could be justified in terms of ambipolar electric field formation. The formation of the DL is, in fact, associated with the generation of hot electrons ascribed to the absorption of the incident laser radiation due to inverse bremsstrahlung [9] and to three-body recombination processes [10]. The electrons escape mainly along the target normal inducing the formation of electric field with the ions of different charge states.

In order to confirm the above hyphotesis, let deconvolute the ion current signals taking into account a multi-mode velocity distribution along the contribution of the different charge states of ions carried out by the plasma.

Due to the collisions among the ions in the initial phase of the plasma expansion, the Knudsen-layer formation is responsible of the velocity distribution of ions which is well described by a shifted Maxwell–Boltzmann distribution. As it has been demonstrated [2, 8], it is possible to derive the equation for the ion current based on the shifted Maxwell–Boltzmann distribution (MBD):

$$I(t) \propto t^{-5} \exp[-\beta^2 (L - ut)^2 / t^2] \tag{1}$$

where β^2 stands for $m/2kT$, with m the atomic mass of the ion species, k the Boltzmann constant, T the Knudsen-layer temperature, L the target-cups distance and u the centre of mass velocity of ions.

The program Peakfit 4.12 version is utilized to perform the deconvolution analysis sharing the temperature parameter among the different charge states keeping the other parameters as free variables, e.g. the centre of mass velocity of the charge states, u_q.

An example of the deconvolution is reported in Fig. 3 concerning the Si ion signal recorded by -1and +4 cups. The dash lines of different colors represent the contribution of the different charge states of ions while the red continues line represents the fit of the experimental data (dot points) obtained by the sum of the different charge state contribution. It is noteworthy the presence of Si^{+4} for - 1 cup signal as well as a much more abundance of the difference charge states.

Cup +4

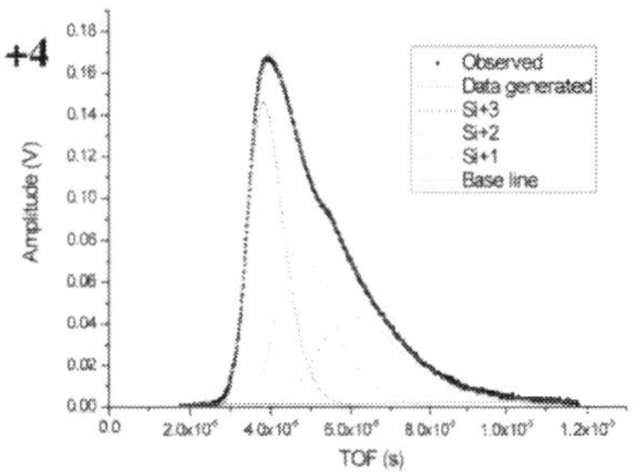

Cup -1

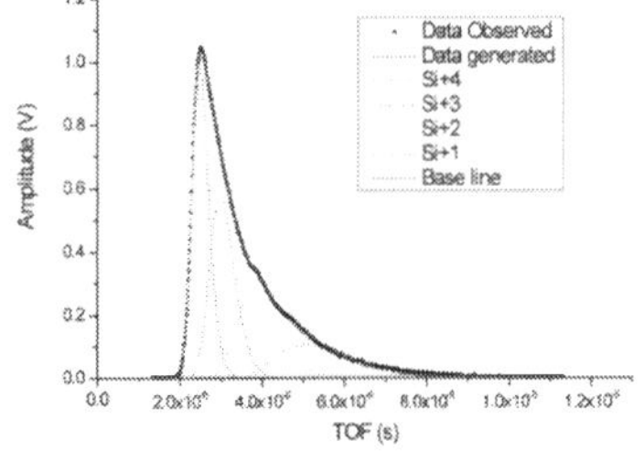

Si^{+q} Charge state	Collected charge (nC)	Temperature (eV)	Centre of mass velocity (m/s)
3	1.1	2.3	$21.7x10^3$
2	1.3	2.3	$16.8x10^3$
1	0.2	2.3	$12.1x10^3$

Si^{+q} Charge state	Collected charge (nC)	Temperature (eV)	Centre of mass velocity (m/s)
4	2.6	2.8	$35x10^3$
3	3	2.8	$28.4x10^3$
2	3.4	2.8	$21.6x10^3$
1	5	2.8	$15.2x10^3$

Figure 3. Deconvolution of time of flight signals for laser induced Si plasma. The temperature and the centre of mass velocity values are obtained by fitting the experimental data by Eq. (1)

Beside, in Fig. 4 the values of u_q are reported on the detector angle position for the different charge states. The enhancement of u_q is present in proximity of the target normal with the evidence of an abrupt change of slope for Si^{+4} and Nb^{+4}. Such result can be justified by the possible presence of DL in the expanded laser-induced plasma which much more influences the expansion of ions with higher charge state.

Si

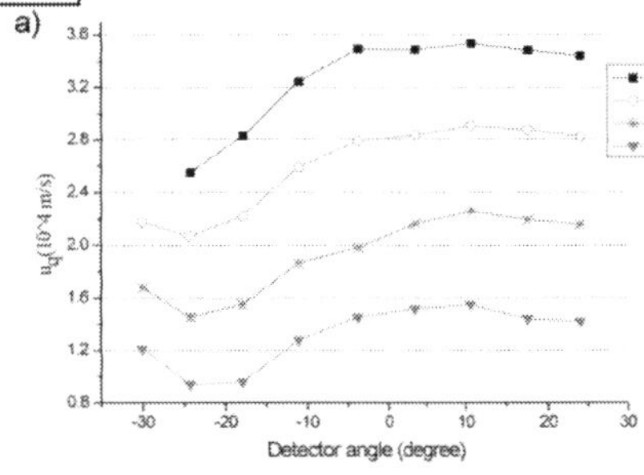

Nb

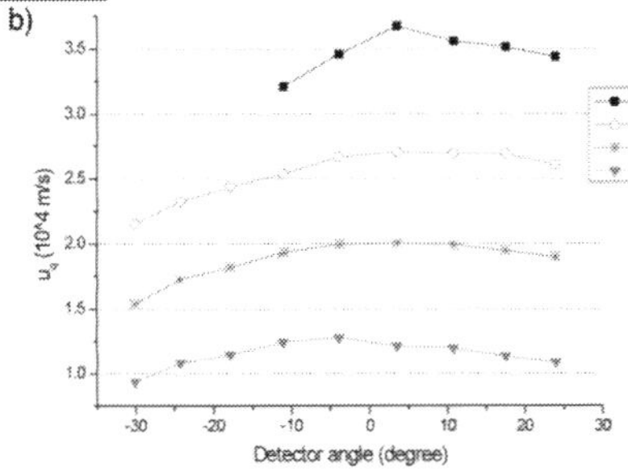

Figure 4. Dependence of the centre of mass velocity u_q of the different charge states of ions on the detector angle position.

4. Conclusions

The time of flight spectrometry with cups placed at different angular position respect to the target normal is very useful to study the characteristics of the UV laser induced Si and Nb plasmas. In particular, the time resolved ion current signals reveal a decreasing of the time of flight when the signals are recorded in the proximity of the target normal.

The justification of the experimetal results has to be found in the presence of an ambipolar electric field which accelerated the most energetic particles mainly along the target normal. The deconvolution of ion current signals is useful to give confirmation of our idea. The increasing of the centre of mass velocity on the different detector angle is ascribed, indeed, to the presence of an ambipolar electric field in the laser plasma.

References

[1] Amoruso S, Bruzzese R, Spinelli N and Velotta R 1999 *J. Phys. B: At. Mol. Opt. Phys.* **32 (14)** 131
[2] Kelly R and Dreyfus R W 1988 *Surface Sci.* **198** 263
[3] Doria D, Lorusso A, Belloni F, Nassisi V, Torrisi L and Gammino S 2004 *Laser Part. Beams* **22** 461
[4] Hairapetian G and Stenzel R L 1991*Phys. Fluids B* **3** 899
[5] Bulgakova N M, Bulgakov A V and Bobrenok O F 2000 *Phys. Rev. E* **62 (4)** 5624
[6] Metev S, Ozegowski M, Sepold G and Burmester S 1996 *Appl. Surf. Sci.* **96** 122
[7] Fukushima K, Kanke Y, Badaye M and Morishita T 1995 *J. Appl. Phys.* **77** 5406
[8] Doria D, Lorusso A, Belloni F and Nassisi V 2004 *Rev. Sci. Instrum.* **75** 387
[9] Pearlman J S, Thomson J J and C. E. Max 1977 *Phys. Rev. Lett.* **38** 1397
[10] Roudskoy V 1996 *Laser Part. Beams* **14** 369

Determination of electric field strength and kinetic temperature in the cathode fall region of a hollow cathode discharge

MI de la Rosa[1], C Pérez[1], K Grützmacher[1], AB Gonzalo[2] and JA del Val[2]
[1]Universidad de Valladolid, Facultad de Ciencias, 47071 Valladolid, Spain
[2]Universidad de Salamanca, Escuela Politécnica Superior, 05003 Ávila, Spain

E-mail: delarosa@opt.uva.es

Abstract. In this work, we demonstrate the high potential of two-photon excitation of the 1S - 2S transition of atomic hydrogen followed by optogalvanic detection, for measuring under identical experimental conditions, the kinetic temperature and the electric field strength in the cathode sheath region of a hollow cathode discharge. The first obtained results for both parameters are discussed in this paper.

1. Introduction

Experimental and theoretical investigations carried out for many years have provided a lot of information about low-pressure gases discharges. However, there is still a lack of knowledge concerning discharge dynamics; so that, new data are needed to complete the modelling of such complex phenomena, see e.g. [1-4], and the references therein.

For years our group has been involved in developing new experimental techniques for plasma diagnostic. In this work, we demonstrate the high potential of two-photon excitation of the 1S - 2S transition of atomic hydrogen followed by optogalvanic detection for measuring, under identical experimental conditions, the kinetic temperature (T_{kin}) and the electric field strength (E-field) in the cathode sheath region of a hollow cathode discharge (HCD).

In first measurements we applied the usual laser beam combination for this type of measurements: counter propagating beams for the Doppler-free measurements of the (E-field) and only one linearly polarized beam for measuring the Doppler broadened profile. The de-convolution of Doppler-free from the Doppler-broadened profiles results in the pure Doppler profile, hence the kinetic temperature. These first measurements of T_{kin} in the cathode fall revealed some curious results, i.e. temperature lower than room temperature. Analyzing the experimental conditions we came to the conclusion, that the Doppler-free profile corresponds to a local measurement given by the overlap volume of the two counter propagating beams, while the Doppler-broadened profile corresponds to an integrated measurement including the areas without E-field and cooler boundary layers of the plasma as well.

In order to overcome those limitations we remind, that the 1S - 2S transition is quite exceptional, because the conservation of the angular momentum is restricted to only $\Delta L = 0$, i.e. the two-photon excitation is only possible by two photons of opposite angular momentum (opposite circularly polarized laser beams). Using three appropriate circularly polarized laser beams, it is possible therefore to conduct four types of local measurements in exactly the same overlap volume which allow us to measure: a) homogeneous broadening, i.e. the Stark splitting without any Doppler broadened background, b) usual absorption spectra composed by the convolution of homogeneous and inhomogeneous broadening, c) the composition of a) and b), and d) is a zero signal control measurement.

2. Experimental arrangement and measurements

The experimental set-up is shown in figure 1. We used our advanced UV laser spectrometer, specially developed for plasma diagnostic. It provides tuneable 243 nm radiation needed for two-photon excitation of the 1S - 2S hydrogen transition: up to 10 mJ pulse energy in 2.5 ns with nearly Fourier-transform limited bandwidth of 300 MHz. Measurements are performed in a low density HCD plasma with a diameter of 15 mm and 50 mm length. For mapping the E-field and T_{kin} distributions, the HCD is mounted on a vertical translator stage. High spatial resolution is needed because of the very small sheath thickness of the cathode fall region (about 2.5 mm, depending on the discharge parameters). The spatial resolution is given by the overlap volume of the laser beams. Taken into account divergence of the beams and their crossing angle, the overlap volume is about 30 mm long with a central diameter of 200 μm. A comprehensive explanation of the whole experimental arrangement, figure 1, can be found elsewhere [6]. Our experimental laser spectroscopic arrangement, see figure 1, is based on three nearly perfectly circularly polarized laser beams of 243 nm (beams 1, 2 and 3).

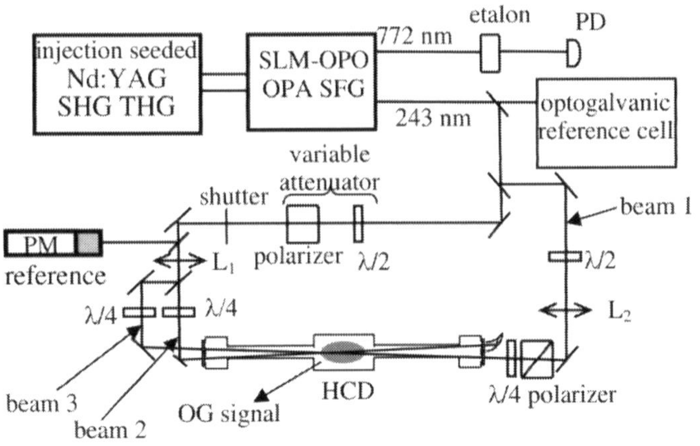

Figure 1. Experimental arrangement

Four different combinations of these three beams allow various types of measurements, see Table 1, which provide to measure the Doppler-free Stark splitting a), the inhomogeneous Doppler broadening b), the mixture of both c) as control for independent (linear) superposition of homogeneous and inhomogeneous broadening. Quite remarkable is the zero signal probe measurement d), because only two counter propagating beams with perfect opposite circular polarization should not give any signal at all, because the two photon absorption of the 1S - 2S transition is restricted to an angular momentum transfer of $\Delta L = 0$. In order to achieve nearly perfect circular polarization for all the beams; it is necessary to minimize any distortion of the linear and circular polarization states caused by other optical devices involved in the experiment. In addition, the temporal delay introduced between beams 2 and 3 must be as small as possible (in our case: less than 4 % of the laser pulse duration) in order to maintain the temporal overlap.

Figure 2 shows, as example, the four different spectra obtained by the four different laser beam configurations explained above, acquired for the following fixed discharge conditions of 400 Pa and 100 mA at a distance of 150 μm from the cathode surface.

Table 1. Characteristics and combination of the three laser beams which provide the four types of measurements and the corresponding spectra a) to d) shown in Figure 2

Figure 2	*Beam, propagation*	*Polarization, angular momentum transfer*	*Measurement*	*Broadening*
a	1 and 2, counter	opposite, $\Delta L = 0$	Doppler free	homogeneous
b	2 and 3, collinear	opposite, $\Delta L = 0$	Doppler broadened	inhomogeneous
c	1, 2 and 3, mixed	mixed, $\Delta L = 0$	mixed	mixed
d	1 and 3, counter	identical, $\Delta L = 2$	zero signal	-

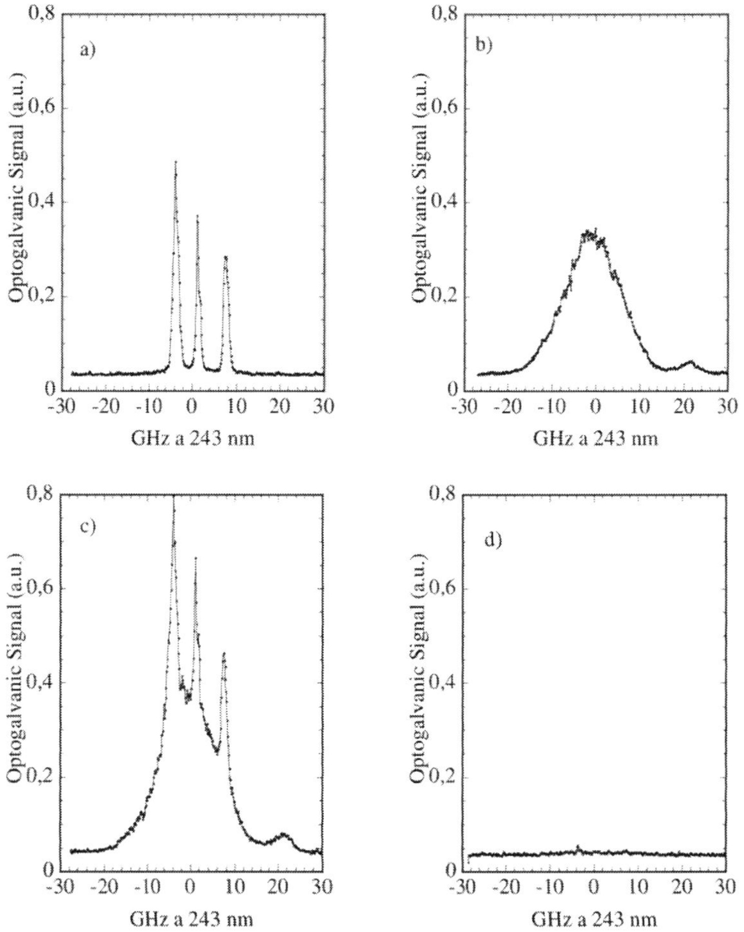

Figure 2. Four different spectra without any data evaluation, all obtained from the same measurement volume and fixed discharge conditions: 400 Pa, 100 mA, at 150 μm from the cathode surface

These spectra exhibit also very clearly the high polarization quality of the beams. Furthermore, the spectra c) is obviously sum of the spectra a) and b), hence the signals do not exhibit any saturation. Please note, that all the signals are only created where the laser beams overlap, this is the consequence of using the circularly polarized beams combined with the angular momentum transfer restricted to $\Delta L = 0$. For mapping the kinetic temperature (T_{kin}) and the electric field strength (E-field) we have measured in one measurement campaign the set of the four spectra at various distances from the cathode surface, starting at a minimum distance of about 150 μm in 8 steps of 250 μm.

3. Results

The Doppler-free profiles, figure 2 a), allow a reliable determination of the local E-field via the frequency shift of the red (blue) shifted $2P^{1/2}$ ($2P^{3/2}$) Stark components and the variation of the E-field in the cathode fall region [5-7].

The inhomogeneous broadening, i.e. the Doppler broadening, can be extracted by de-convolution [8] of the spectra a) from spectra b). This was done by calculating the inverse Fourier transform (FT) of the ratio of the FT of the Doppler broadened profile b) to the FT of the Doppler-free spectra a). As expected, the spectral profiles obtained after de-convolution exhibit very well Gaussian line shapes and provide therefore a T_{kin}, figure 3. However, the variation of T_{kin} with distance from the cathode surface is still a surprise and fairly the same for both discharge pressures investigated: 400 Pa and 900 Pa. Only for an E-field of about 1 kV/cm or lower, T_{kin} confirms the common understanding of low density discharges and reveals for both discharge pressures nearly the same reasonable value of about 375 K, i.e. somewhat larger than room temperature. However, with decreasing distance from the cathode surface (from 1.8 mm down to 0.15 mm for 400 Pa, and from 1.3 mm down to 0.15 mm for 900 Pa) and with the corresponding increasing field strength (up to 2.5 kV/cm for 400 Pa and 2.8 kV/cm for 900 Pa) T_{kin} decays more or less parabolic down to 140 K. We are convinced, that this reveals strong deviations from a homogeneous velocity distribution in the cathode fall region of the HCD. We have to remind, that the inhomogeneous broadening and the related T_{kin} which we can measure is only the projection of all atomic velocities present in the overlap volume onto the laser beam direction, while the projection of the atomic velocity distribution perpendicular to the laser beams, e.g. in the radial direction parallel to the local electric field is absolutely inaccessible. Our first interpretation of the "kinetic temperatures" determined in this study in the cathode fall region is quite simple therefore. The strong radial acceleration of the hydrogen ions towards the cathode and the large charge-transfer cross-section of hydrogen may be the reason for the pronounced discrepancy with respect to a homogeneous atomic velocity distribution. A more detailed analysis is demanded.

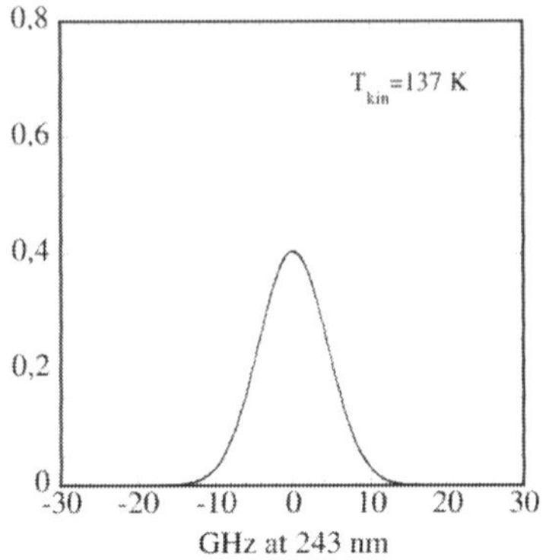

Figure 3. Gaussian profile obtained by de-convolution of the Doppler free profile from the Doppler-broadened spectra shown in figure 2.
In order to avoid noise of the profile due to the de-convolution procedure, the noise of the measured spectra was eliminated first by usual fit procedures, The convolution of the Doppler-free spectra figure 2a) with the Gaussian profile obtained by de-convolution agrees perfectly with the Doppler broadened spectra figure 2b).

References

[1] Sigeneger F, Donkó Z, Loffhagen D 2007 *Eur. Phys. J. Appl. Phys.* **38**, 161
[2] Martínez J P, Amare J C 1998 *J. Phys. D: Appl. Phys.* **31** 312
[3] Lister G G 1992 *J. Phys. D: Appl. Phys.* **25** 1649
[4] Dexter A C, Farrell T, Lees M I 1989 *J. Phys. D: Appl. Phys.* **22** 413
[5] de la Rosa MI, Pérez C, Grützmacher K, Gonzalo AB, Steiger A 2006 *Plasma Sources Sci. Technol.* **15**, 105
[6] de la Rosa MI, Pérez C, Grützmacher K, Fuentes LM 2009 *Plasma Sources Sci. Technol.* **18**, 015012
[7] Pérez C, de la Rosa MI, Grützmacher K 2009, see contribution (Poster) at this Conference
[8] Demtröder W, 1996 *Laser Spectroscopy, Basic Concepts and Instrumentation*, Springer Berlin, Second Enlarged Edition

Journal of Physics: Conference Series **227** (2010) 012039 doi:10.1088/1742-6596/227/1/012039

Characterization of hollow cathode fall field strength measured by Doppler-free two-photon optogalvanic spectroscopy via Stark splitting of the 2S level of Hydrogen and Deuterium

C Pérez, MI de la Rosa and K Grützmacher
Universidad de Valladolid, Facultad de Ciencias, 47071 Valladolid (Spain)

E-mail: concha@opt.uva.es

Abstract. Doppler-free two-photon optogalvanic spectroscopy has been applied to measure the strong electric field strength and the cathode fall characteristics of hollow cathode discharges operated in hydrogen and deuterium via the Stark splitting of the 2S level of atomic hydrogen isotopes. In this paper we show similarities and differences in the tendencies of the cathode fall characteristics of hydrogen and deuterium in a wide range of identical discharge parameters.

1. Introduction

The electric field strength (E-field) is one of the most relevant discharge parameters directly connected with charged particle fluxes, their energy distribution functions and charge densities. Spatial distribution of the E-field and its dependence on the plasma parameters are of basic importance for plasma discharge modelling and surface treatment technologies. Spectroscopic techniques for determining the E-field are based on the Stark effect of atomic or molecular spectra, i.e. splitting, shifting and intensity variations of spectral line components depend on the local E-field. Comparisons of the measured spectra with theoretical or calibrated values give the E-field. Laser spectroscopic techniques can provide high spectral, spatial and temporal resolution. They do not perturb the investigated plasma if nanosecond pulse duration and moderate irradiances are applied. Most of these techniques are based on the spectroscopic measurement of the Stark splitting, shifting or mixing of Rydberg states in different kinds of atoms and molecules using optogalvanic and/or fluorescence detection, see e.g. the overviews in references [1, 2]. Atomic hydrogen isotopes and helium are very good candidates for E-field measurements, because the Stark splitting of all their atomic levels is very well known, and atomic hydrogen also participates in many plasmas of industrial interest. In fact, different techniques for atomic hydrogen and helium have been published most of them included in [1, 2] and [3-10] respectively, and the references therein.

In this work we propose Doppler-free two-photon optogalvanic spectroscopy to measure the strong electric field strength (0.4 kV/cm to 4 kV/cm) in the cathode fall of hollow cathode discharges (HCD) via the Stark splitting of the 2S level of atomic hydrogen and its isotopes. This requires advanced pulsed tunable single longitudinal UV-laser spectrometers, which can provide spectral quality close to Fourier transform limit, i.e. about 300 MHz bandwidth at 243 nm with some nanosecond pulse duration. Pulse energies of about 1 mJ are sufficient for two-photon optogalvanic spectroscopy; hence the irradiance in the excitation region of the laser beams (diameter of about 200 μm) does not exceed

1 GW/cm^2. These conditions lead, in a first step, to the two-photon excitation; the absorption of a third laser photon results in subsequent photo ionization. Hence, the charged particles created in the overlap volume of the laser beams are accelerated by the E-field and the corresponding variation of the discharge impedance can be measured (OG signal). This technique provides a reliable determination of the local E-field via the frequency shift of the ($2P^{1/2}$) and ($2P^{3/2}$) Stark components, while the integral of the E-field versus distance from the cathode surface gives the cathode fall voltage drop. We are going to compare the cathode fall behavior of a HCD operated with hydrogen and deuterium for various discharge conditions: currents from 50 to 200 mA and pressures from 270 to 1350 Pa.

2. Experimental set-up, measurements and E-field determination

The experimental set-up is shown in figure 1. Measurements are performed in the low density plasma generated in a home-made HCD of cylindrical symmetry. The cathode is placed in between two hollow anodes to allow end-on excitation. The discharge operates at a constant gas flow (corresponding to the discharge pressure) entering at one anode side and leaving at the opposite. This device provides stationary stable non-LTE plasma, with low density and low kinetic temperature of atoms and molecules. Experimental conditions in this study result in a very quite discharge regime without any tendency to anode or cathode sparking.

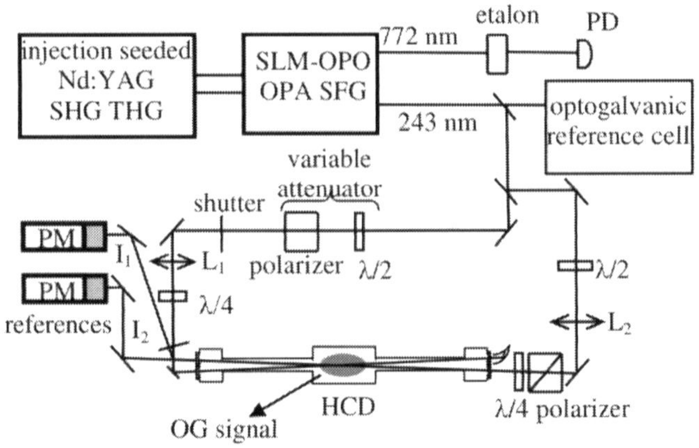

Figure 1. Schematic view of the experimental set-up

A comprehensive explanation of the whole experimental arrangement is given elsewhere [11, 12]. The laser system consists of a 10 Hz injection-seeded Q-switched Nd:YAG laser (Continuum, Powerlite) and a modified OPO-OPA system (Continuum, Mirage 500). This laser offers tunable radiation in the visible and near-IR range (426 nm to 2.12 μm). The required 243 nm radiation is obtained by sum-frequency generation (SFG) of 772 nm and the third harmonic of the Nd:YAG laser in a BBO crystal. This concept provides up to 10 mJ pulse energy in 2.5 ns and 300 MHz bandwidth [13]. The frequency detuning of the laser, i.e. the OPO signal-wave (772 nm), is controlled by a monitor etalon with free spectral range (FSR) of 7 GHz and a photodiode (PD). From this, we can derive that the regularity of the frequency detuning is typically about 1 % of the FSR of the etalon during a complete scan. An accurate frequency reference (ν_0) of the 1S – 2S transition for zero E-field is provided by the optogalvanic reference cell (in figure 1), which contains atomic deuterium or hydrogen, generated by thermal dissociation at a pressure of about 500 Pa.

A moderate amount of the laser energy is introduced in the experiment (figure 1). The 243 nm laser radiation is divided in two counter propagating circularly polarized laser beams providing Doppler-free two photon excitation for the 1S - 2S transition, according to the selection rules for this transition $\Delta L = 0$. Each beam is focused with a 1 m lens into the plasma. In order to achieve the best spatial resolution, both beams cross in a horizontal plane parallel to the HCD axis, and the measurements are taken in the upper central part of the cathode fall region. The spatial resolution in the experiment is given by the overlap of the two laser beams. The overlap volume is about 30 mm long and has a focus diameter of 200 μm. Vertical displacements of the HCD permit performing measurements at various distances from the cathode surface. Taking into account the beam divergence (4:1000) and the focus diameter, the closest proximity to the cathode wall we can reach without irradiating the cathode surface is 150 μm.

A detailed description of the Stark effect in the 1S - 2S transition of atomic hydrogen isotopes can be found in [11] and theoretical calculations in [14]. The measured spectra of the 1S - 2S transition show clearly the theoretical evolution of the three underlying Stark components as a function of the electric field. The red shifted component ($2P^{1/2}$) starts growing and shifting in the wing of the central component ($2S^{1/2}$), while the blue one ($2P^{3/2}$) starts growing and shifting more slowly far off the central component which decreases and undergoes a very weak blue shift.

The comparison of the calculated Stark shifts with the experimental ones gives the corresponding E-field. In order to obtain the frequency shift of the three Stark-components with respect to the zero E-field frequency ν_0, the measured spectra were fitted to lorentzian profiles. In the case of deuterium we fitted to three lorentzians. For the Stark splitting of atomic hydrogen one also has to account for the hyperfine structure underlying each Stark component, this means six lorentzians for the fit.

3. Results

After having determined the local E-field for all the spectra measured at various distances from the cathode surface, we can plot the E-field variation in the cathode fall region for the corresponding discharge parameters. In figures 2 and 3 we show the E-field vs. radial distance from the cathode surface for a fixed current of 100 mA, and several pressures: 270, 400, 600, 900 and 1350 Pa in deuterium and hydrogen. For 1350 Pa pressure a linear decay of the E-field is observed and with decreasing pressure the maximum field strength decreases, the dependence is more parabolic and the cathode sheath thickness is expanding. Only for the lowest pressure of 270 Pa the tendency of the sheath thickness behaves differently and contracts again. As it can be seen in the figures similar behaviour is observed for deuterium and hydrogen, although the E-fields at the cathode surface are larger for deuterium.

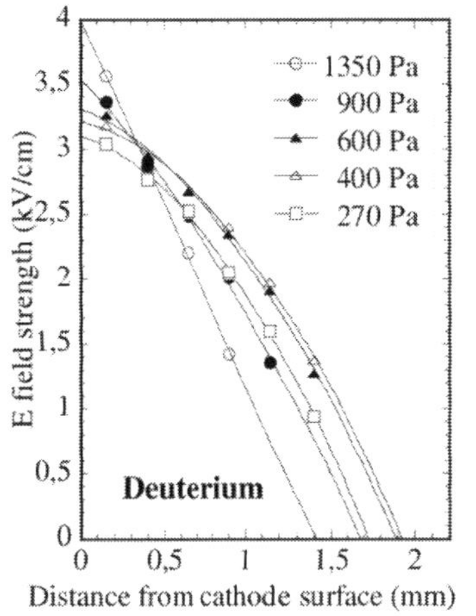

Figure 2. E-field vs. radial distance from the cathode surface for a fixed current of 100 mA and several pressures, in deuterium

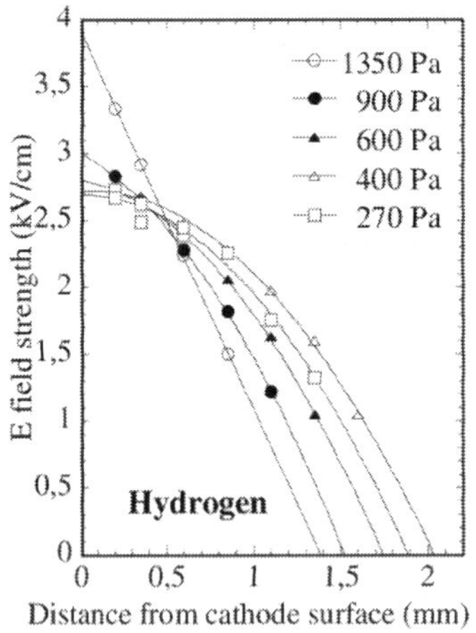

Figure 3. E-field vs. radial distance from the cathode surface for a fixed current of 100 mA and several pressures, in hydrogen

Figures 4 and 5 show a complement to the former figures, i.e. for both isotopes the measured variation of the E-field in the cathode fall region versus the radial distance from the cathode surface, for a fixed pressure of 900 Pa, and several currents: 50 (or 60), 100, 150, and 200 mA. These figures confirm, once more, a similar behaviour observed in deuterium and hydrogen; this means as the current increases the maximum E-field increases while the cathode fall region becomes more compressed. Similar results are found for a pressure of 400 Pa and the same discharge currents. Absolute values for the E-field are bigger in deuterium in all cases. The E-field is always larger in deuterium although the thickness of the cathode fall region is quite similar compared to hydrogen.

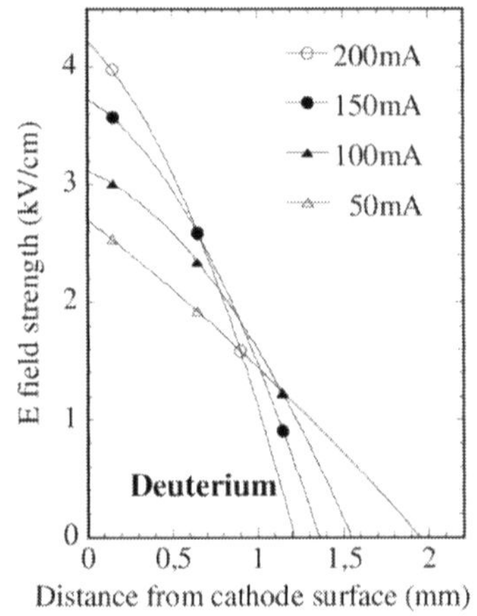

Figure 4. E-field vs. radial distance from the cathode surface, for a pressure of 900 Pa and several currents, in deuterium

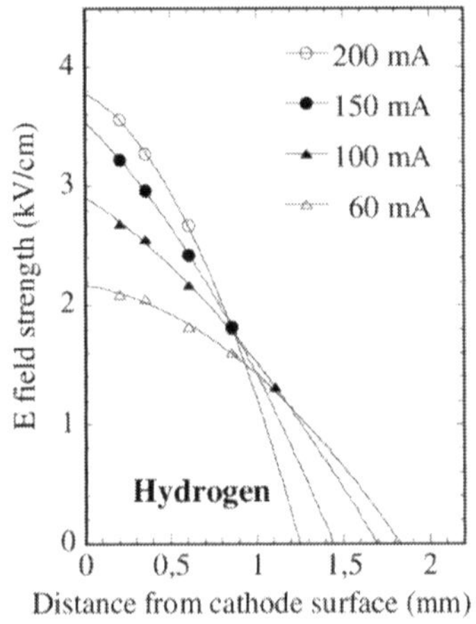

Figure 5. E-field vs. radial distance from the cathode surface, for a pressure of 900 Pa and several currents, in hydrogen

References

[1] Gravilenko VP 2006 *Instruments and Experimental Techniques*, **49** 149
[2] Muraoka K, Maeda M 2001 *Laser-Aided Diagnostics of Plasmas and Gases*, (Institute of Physics Series in Plasma Physics IOP)
[3] Barbeau C, Jolly J 1991 *Appl. Phys. Lett.* **58** 237
[4] Booth JP, Fadlallah M, Derouard J, Sadeghi N 1994 *Appl. Phys. Lett.* **65** 819
[5] Booth JP, Derouard J, Fadlallah M, Cabaret L, Pinard L 1996 *Optics Communications* **132** 363
[6] Czarnetzki U, Luggenhölscher D, Döbele HF 1998 *Phys. Rev. Lett.* **81** 4592
[7] Czarnetzki U, Luggenhölscher D, Döbele HF 1999 *Plasma Sources Sci. Technol.* **8** 230
[8] Czarnetzki U, Luggenhölscher D, Döbele HF 2001 *Appl. Phys. A* **72** 509
[9] Cherkasova EK, Gavrilenko VP, Zhuzhunashvili AI 2006 J. *Phys.: Appl. Phys.* **39** 477
[10] Gemisic Adamov M, Steiger A, Grützmacher K, Seidel J 2007 *Phys. Rev. A* **75** 013409
[11] de la Rosa MI, Pérez C, Grützmacher K, Gonzalo AB, Steiger A 2006 *Plasma Sources Sci. Technol.* **15** 105
[12] de la Rosa MI, Pérez C, Grützmacher K, Fuentes LM 2009 *Plasma Sources Sci. Technol.* **18** 015012
[13] Grützmacher K, de la Rosa MI, Gonzalo AB, Steiger M, Steiger A 2003 *Appl. Phys. B* **76** 775
[14] Seidel J 2000 private communication (unpublished)

Journal of Physics: Conference Series **227** (2010) 012040 doi:10.1088/1742-6596/227/1/012040

Electric field measurements in near-atmospheric pressure nitrogen and air based on a four-wave mixing scheme

Sarah Müller[1], Tsuyohito Ito[2], Kazunobu Kobayashi[3], Dirk Luggenhölscher[1], Uwe Czarnetzki[1], Satoshi Hamaguchi[3]

[1]Institute of Experimental Physics, Ruhr-Universität Bochum, Bochum, Germany
[2]Frontier Research Base for Young Researchers, Graduate School of Engineering, Osaka University, Osaka, Japan
[3]Center for Atomic and Molecular Technologies, Graduate School of Engineering, Osaka University, Osaka, Japan

E-mail: `sarah.mueller@ep5.rub.de`

Abstract. Electric fields are measured for the first time in molecular nitrogen at atmospheric pressures. Measurements are performed in either pure nitrogen or air. The laser spectroscopic technique applied here is based on a CARS-like four-wave mixing scheme originally developed for measurements in molecular hydrogen by Ochkin and Tskhai in 1995. The technique is ideal for investigation of microdischarges at atmospheric pressures. The frequencies of two focussed laser beams in the visible are tuned to match the energy difference between the two lowest vibrational levels in nitrogen. The presence of a static electric field then leads to the emission of coherent IR radiation at this difference frequency. The signal intensity scales with the square of the static electric field strength. Parallel to this process also anti-Stokes radiation by the standard CARS process is generated. Normalization of the IR signal by the CARS signal provides a population independent measurement quantity. Experimental results at various pressures and electric field strengths are presented.

1. Introduction

An important parameter for understanding the physics of discharges is the electric field strength. Knowledge of the spatial and temporal distribution of the field strength can provide information about the distribution of charged particles, current density, and dissipated power. Particularly microdischarges in the atmospheric pressure regime have attracted much attention in recent years. However, the small size, extreme plasma parameters and often fast change of discharge conditions complicates the investigation of such systems. Well-known diagnostic methods applied in low-pressure discharges are often intrusive or not feasible due to high collision rates.[4] Here, we introduce a unique laserspectroscopic technique which is based on a CARS-like four-wave mixing-scheme. Originally, it was developed in hydrogen environment by Ochkin and Tskhai [1, 2, 3]. In this article, we show that this novel method is also applicable with nitrogen as the probe gas.

2. Theoretical basis and experimental setup

The electric field measurement is based on recording the intensity of infrared (IR) radiation generated through anti-Stokes Raman scattering on the gas molecules in the investigated medium. The excitation scheme used for the field measurement is presented in figure 1(a).

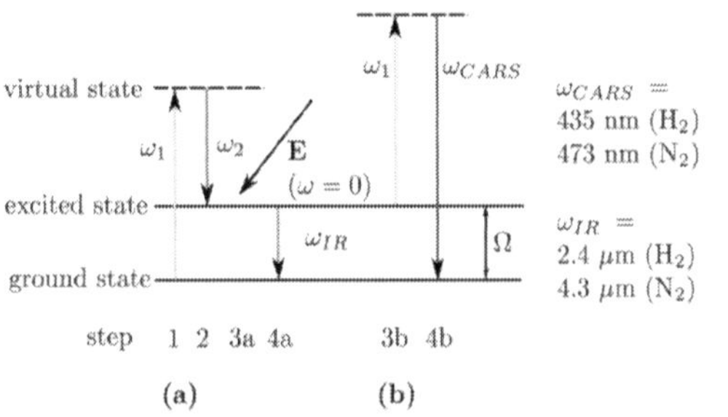

Figure 1. Excitation scheme for laser electric field measurements. Part (a) describes the generation of field dependent radiation. Part (b) belongs to generation of field-independent anti-Stokes radiation.

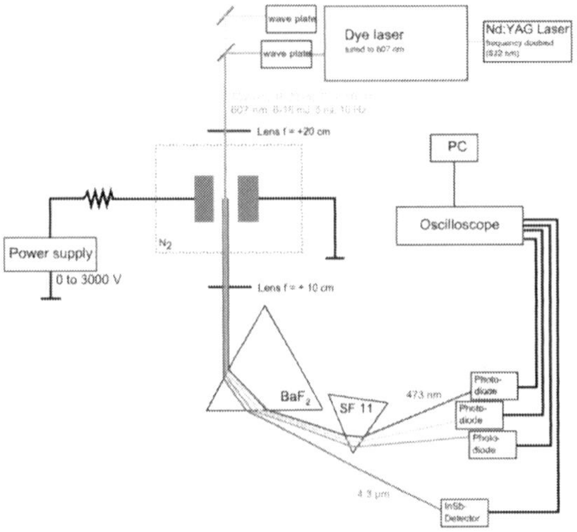

Figure 2. Schematic drawing of the experimental setup for laser electric field measurements.

The process can be described as four-wave mixing based on the third order nonlinear susceptibility that leads to generation of coherent radiation with field dependent intensity: The first two waves are provided by two lasers (intensities I_1 and I_2 respectively), which are collinearly phase-matched and superimposed. The third wave involved in this excitation process is effectively represented by the quasi-static electric field. The beam mixing leads to generation of a fourth wave, whose intensity is proportional to the squared electric field strength. In addition to the field dependent signal, the anti-Stokes radiation (473 nm) is detected. Both signals depend on the laser intensities and the population difference between the $\nu = 0$ and $\nu = 1$ states in the molecular medium:

$$I_{IR} \propto (N_{ground} - N_{excited})^2 I_1 I_2 |E|^2 \quad (1)$$

$$I_{CARS} \propto (N_{ground} - N_{excited})^2 I_1^2 I_2 \quad (2)$$

By normalizing the static field dependent IR signal to the blue anti-Stokes signal, the dependence on the poplation distribution and the total gas density cancels out. Therefore, the electric field should linearly depend on the square root of the measured signal ratio:

$$|E| = C\sqrt{I_1 \frac{I_{IR}}{I_{CARS}}} \quad (3)$$

The laser intensity I_1 remains in the equation and can be recorded in parallel. In the experiment, this beam is provided by an solid state laser which has a rather stable output (±3 %) so that I_1 is effectively a constant. In case of fields inhomogeneous along the propagation direction of the laser, the static field has to be replaced by the line average field. In any case, only the focal region contributes due to non-linearity of the process.

Figure 2 shows a schematic diagram of the experimental setup used for the laserspectroscopic measurement of electric fields. Two laser beams are collinearly superimposed and focused into a gas-containing test chamber. A system of prisms is used to disperse the two incident beams and the generated radiation in order to record their intensities seperately. The laser system employed in the experiments consists of two pulsed lasers with 3-5 ns pulse duration and a repetition rate of 10 Hz. A frequency-doubled Nd:YAG-laser (Continuum Surelite III-10) at 532 nm is used for the four-wave mixing as well as a pumping light source for a dye laser. The field-dependent signal radiation originating from this transition is at 4.29 μm, the field-independent anti-Stokes signal occurs at 473 nm.
After focusing both lasers collinearly into the test chamber, both generated beams as well as the incident laser beams are separated by a BaF_2 prism. This material is feasible for application in the infrared regime. However, its dispersion is not sufficient to separate all involved lines for detection. Therefore, an additional prism made of SF 11 is employed to divide the visible beams. For measurement of the IR signal intensity, an InSb detector is employed. The visible beams are detected by photodiodes. Each detector is equipped with a filter which only allows transition of the appropriate radiation. The signals from the detectors are read and stored by an oscilloscope. Every data point is an average over 100 single shots.
The electrode system for generation of a uniform electric field consists of two parallel stainless steel electrodes with 17 mm diameter and flat surface. The gap length between the two electrodes is 3.15 mm. One electrode is supplied with a dc voltage generated by a power supply, the other one is grounded. The applied electric field ranges from 0 up to about 1 kV/mm. The electrodes are placed within a gas container filled with pure nitrogen or a mixture of argon and nitrogen at 1 bar up to 2.5 bar gas pressure. The entrance window is made of standard borosilicate, whereas a polished sapphire disc is used as exit window.

3. Results

Figure 3 shows measurements with different pressure conditions. The linear dependency of signal intensity on the squared electric field is confirmed by the measurements: the measured data points appear nearly perfectly on a straight line. At lower field strength, the data shows a stronger variation from the best-fit line. In this field range, lower signal intensity leads to higher statistical variation.

After collecting data series at 1.0 bar and 1.2 bar, the optical alignment is slightly changed in order to compensate mechanical alterations of the beam path due to the room climate. After realigning, the calibration constant is different. Therefore, the full lines cannot be directly compared with the dashed lines. Moreover, this finding indicates the necessity to perform a calibration measurement prior to every field measurement to determine the actual calibration constant. Furthermore, the best-fit lines seem to vary with the pressure. However, the signal intensity should be pressure indepenent after normalization to the blue CARS signal. Pressure dependent swelling of the sapphire window may cause a refraction of the signal beam and accordingly variation of the imaging on the detector spot (diameter 0.3 mm). In a reference measurement at fixed applied field of 1 kVmm^{-1}, the nitrogen contents in a nitrogen/argon mixture was varied from 1 bar up to 2 bar keeping the total pressure constant at $p = 2$ bar. In this case, the signal varied only within a few percent. By contrast, the signal reduced by 10 percent when increasing the nitogen partial pressure and total pressure from 1 bar to 2 bar.
The statistical error can be estimated from figure 3. This gives a minimum detectable field of 100 V/mm. It should be noted that the signal scales like $(N \cdot l)^2$ and other sensitivities will result at different focal or discharge lengths l or gas densities N ($l \sim 7$ mm in this case).
In Figure 3(b) two series of measurement in open air are shown. The measured data clearly confirms again the proportional relation (3). With this measurement we can successfully show the applicability of our technique even in open environments.

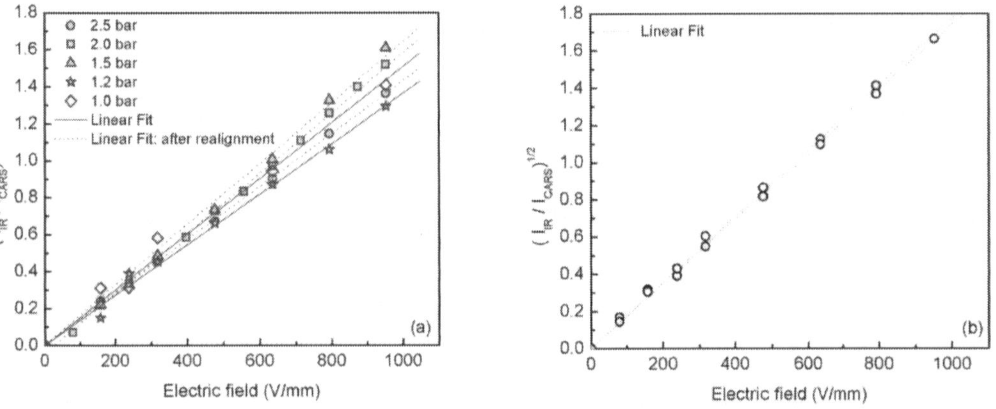

Figure 3. Figure (a) shows two series of electric field measurement at different pressure conditions. Optical alignment for obtaining the data with dashed line is slightly different from the one corresponding. Two series of calibration measurements in open air environment in figure (b). In this case, the optical alignment has not changed after running the first series.

4. Summary

The feasibility of electric field measurements based on field-induced coherent Raman scattering is demonstrated for the first time in a nitrogen containing gas at atmospheric or higher pressure, including open air. The technique is especially useful for the determination of temporal and spatial profiles of the electric field in air-based microdischarges, where nitrogen is abundant. In our current experimental setup, the minimum detectable field strength is about 100 V/mm, which is sufficiently small compared with the average field present in typical microdischarges. No further knowledge of other gas/plasma parameters such as the nitrogen density is required.

5. Acknowledgements

Part of the research was supported by the DFG in the frame of Research Group 1123. Part of research was performed at the Frontier Research Base for Global Young Researchers, Osaka University, with funding through the program Promotion of Environmental Improvement to Enhance Young Researchers Independence, the special coordination funds for promoting science and technology, Japan ministry of education, culture, sports, science and technology (MEXT). This research has been also funded in part by Grants-in-Aid for Scientific Research from Japan Society of Promotion of Science (JSPS)/MEXT.

References

[1] Evsin O A, Kupryanova E B, Ochkin V N, Savinov S Y, Tskhai S N, 1995 *Quantum Electronics* **25** 278 (1995)
[2] Zheltikov A M, Korotev N I, Maumov A N, Ochkin V N, Savinov S Y, Tskhai S N, *Quantum Electronics* **29** 73 (1999)
[3] Ochkin V N, Tskhai S N *Physics-Uspekhi* **46** 1214 (2002)
[4] Czarnetzki U, Sasaki K *J. Plasma Fusion Res.* **83** 215 (2007)
[5] Ito T, Kobayashi K, Mueller S, Luggenhölscher D, Czarnetzki U, Hamaguchi S *J. Phys. D* **42** 092003 (2009)

Journal of Physics: Conference Series 227 (2010) 012041 doi:10.1088/1742-6596/227/1/012041

Advances of Thomson scattering diagnostic on the TEXTOR tokamak

M.Yu.Kantor[1,2,3] **and the TEXTOR team**

[1]Ioffe Institute, RAS, Saint Petersburg 194021, Russia
[2]FOM-Institute for Plasma Physics Rijnhuizen*, Association EURATOM-FOM, P.O. Box 1207, 3430 BE Nieuwegein, The Netherlands
[3]Institute for Energy Research – Plasma Physics*, Forschungszentrum Jülich GmbH, Association EURATOM-FZJ, D-52425 Jülich, Germany
*Partners in the Trilateral Euregio Cluster

E-mail: m.kantor@fz-juelich.de, m.kantor@mail.ioffe.ru

Abstract. A Thomson scattering (TS) diagnostic based on the laser multi-pass intra-cavity plasma probing has been developed and successfully implemented in the TEXTOR tokamak [1, 2]. The laser system operates in a burst mode delivering up to 45 laser pulses 50 J of probing energy each at 5 kHz frequency. The diagnostic has provided the measurements of fast evolution of electron temperature and density at accuracy of 2% and 1% correspondingly at a spatial resolution <1 cm along the whole plasma diameter during a 9 ms time interval. The developed TS diagnostic combines unique features: large amount of collected photons, a high repetition rate of the measurements, a high spatial resolution along the whole plasma diameter and a high spectral resolution with hundred spectral points. Some of new advances of the TS diagnostics and its possible applications on the TEXTOR tokamak are briefly discussed in the report.

1. Introduction

Thomson scattering diagnostic on the TEXTOR tokamak has achieved unique capabilities in measurements of fast evolution of electron temperature (T_e) and density (n_e) profiles due to laser intracavity multipass probing [1, 2]. In this report some new developments of the TS diagnostic are proposed and briefly discussed. The realization of these plans will allow getting new essential physical data in plasma experiments.

2. Electron distribution and plasma current density

TS diagnostics deal with light spectra scattered by free electrons in plasma from a laser beam, but usually only two parameters are extracted from the TS spectra: the width and amplitude corresponding to local T_e and n_e. In principle, TS diagnostic could provide other parameters like the shift of the spectrum, distortions of the spectral wings and etc which relate to important physical quantities, but collected scattered photon fluxes are too low for these measurements.

The new TS system in TEXTOR can open a possibility to study TS spectra and electron distribution, but not only T_e and n_e quantities. Figure 1 shows TS spectra measured in TEXTOR with the double- (blue) and 12-pass (red) probing systems from a $8*12*2$ mm^3 plasma volume at the line electron density ~$2*10^{19}$ m^{-2} in a single laser pulse. The 12-pass system allows a significant reduction of the statistical errors and expansion of the spectral range of the measurements from 2.5 keV to 6 keV in terms of electron energies. When averaging the scattered signals over all pulses in the laser train results the measurement accuracy of scattered spectra achieves 1-2% in the plasma centre. It is illustrated by black points with error bars in the figure. These capabilities are quite essential for study of ECR heating and current drive when EC waves can disturb the distribution of high energy electrons.

Another possibility relates to the measurements of plasma current density in TEXTOR. The line of sight of the TS collection optics is turned to the toroidal axis by ~81° in the horizontal plane. Therefore the electron current drift velocity $v_D=j/en$ has a small projection on the differential wave vectors $\Delta k_{1,2} = k_s - k_{i1,2}$ resulting in a distortion of the TS spectra [3]. Here k_s is the wave vector of radiation scattered to the TS collection optics and $k_{i1,2}$ are wave vectors of the direct and return laser beams

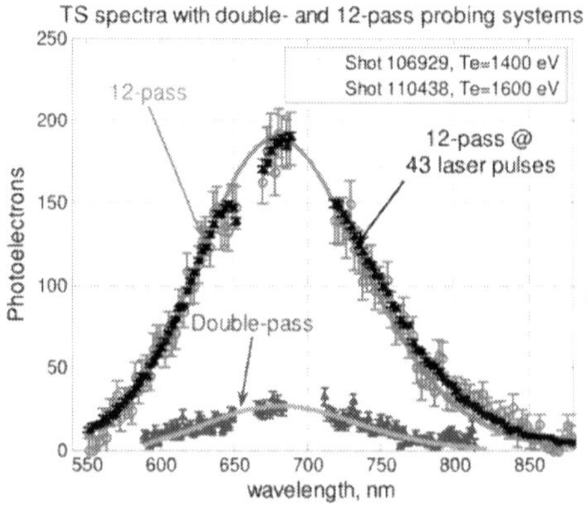

Figure 1. TS spectra measured with double- and 12-pass laser probing systems

The relative change of the scattered spectra at the thermal velocity v_{Te} is $\sim 2 * v_D / v_{Te} * Cos\eta$ [3], where η is the angle between the plasma current and the differential wave vector. For the TEXTOR geometry ($Cos\eta$ ~0.16) the expected changes of scattered spectra deviations range from ~0.4% to 1% in a 300 kA plasma discharge.

The accuracy of the measurements with the 12-pass system in individual spectral points (Figure 1) ranges from ~6% in the spectral core to ~15% in the spectral wings. Enlargement of the observation volume and averaging the scattered signals over 40 pulses of a laser train allow the reduction of the accuracy by a factor of 10 and direct measurements of electron current density in a single plasma discharge. Note, such measurements have been proofed in the RTP tokamak by collecting data from 20-40 plasma discharges [4].

This study, certainly, requires a perfect calibration of the TS system which should provide the instrumental errors less than the achieved statistical errors. A development of special fitting procedures is also required.

3. TS measurement at a low plasma density

A study of plasma-wall interaction on the TEXTOR tokamak [5] utilizes low temperature plasmas at the electron densities 10^{16}-10^{17} m^{-3}. The achieved laser probing energy is already high enough to deliver a significant number of scattered photons from this plasma, while the TS spectrometer [6] was not designed for low temperature measurements. As the result, most of the collected photons are lost in a spectral blocking region around the laser wavelength formed by a notch filter and a gap in a two part mirror [6].

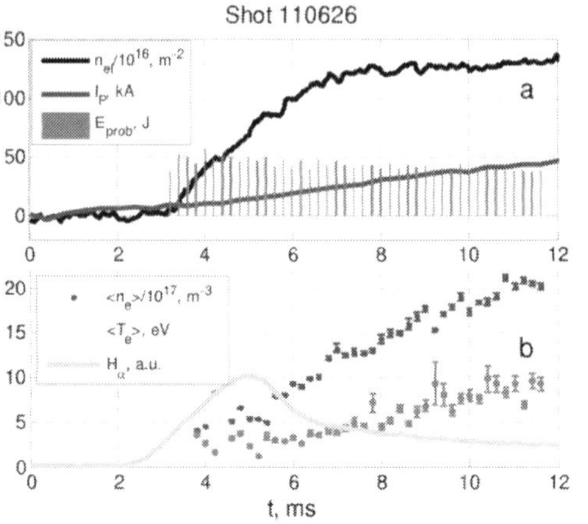

Figure 2.
a. Line electron density (HCN) and plasma current along with laser probing energy.
b. Averaged T_e and n_e (TS) and D_α emission.

Nevertheless, the TS system allows an assessment of the diagnostic capabilities for these studies from TS measurements at the very start of TEXTOR plasma discharge when the electron density develops from the zero level to ~10^{18} m^{-3}. Laser pulse probing energy, developing line integrated electron density and plasma current are shown in figure 2. Developing electron temperature and density measured by TS diagnostic and averaged along their spatial profiles as well as D_α line emission are shown in section b) of the figure.

Figure 3 presents two contour plots of measured TS spectra at 3.2 and 4 ms. The TS images are transformed so that the ruby wavelength is straight (red line). White curves are the borders of the blocking region.

While T_e<3 eV the collected spectra are completely blocked in the spectral gap. The measurements become possible after 4th ms as soon as the blue wing of a TS spectrum comes out from the gap at Te>3 eV. The wing at Te=3 eV contains only ~10% of the total collected photons, but the density measurement accuracy is rather good ~15% which corresponds to a hundred registered photoelectrons in the spectral wing.

Some measured profiles are shown in figure 4. One can see that hot plasma starts from the central region (blue circles) expanding to the edges during first few ms.

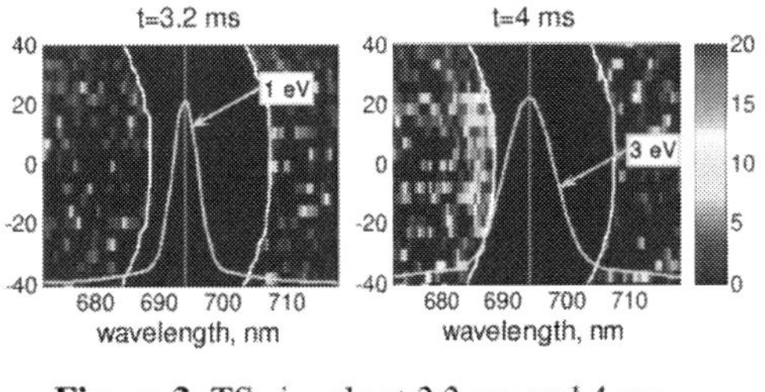

Figure 3. TS signals at 3.2 ms and 4 ms (n_e~3*10^{17} cm^{-3}).
Red line is the ruby laser wavelength;
White curves are the blocking region borders

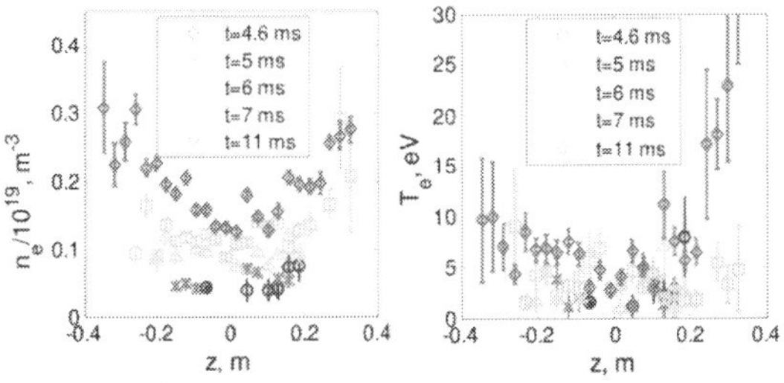

Figure 4. Electron density and temperature profiles measured by TS at the start of plasma discharge.

In a steady state, the measurement accuracy can be improved by 4-6 times by summing up scattered photons along the whole laser train enabling TS measurements at ne~10^{16} m^{-3} if Te>3 eV. TS measurements of lower T_e require:

1. a higher dispersion grating;
2. a more narrow central blocking region well matched to the laser wavelength shape in the intermediate image plane of the spectrometer. This could be achieved by shifting the scattered spectra from the entrance slit and dumping the laser wavelength in a curved mask;
3. a more narrow band notch filter, e.g. highly doped ruby [7, 8];
4. development of a fitting procedure for low T_e measurements.

4. Local measurements of deuterium atom density

Local measurements of hydrogen or deuterium atom density in plasma with Laser Induced Ionization (LII) are based on laser ionization of excited atoms and quenching their line emission [9]. One of the advantages of the LII over widely applied Laser Induced Fluorescence [10] technique is a possibility to use a ruby laser for the measurements.

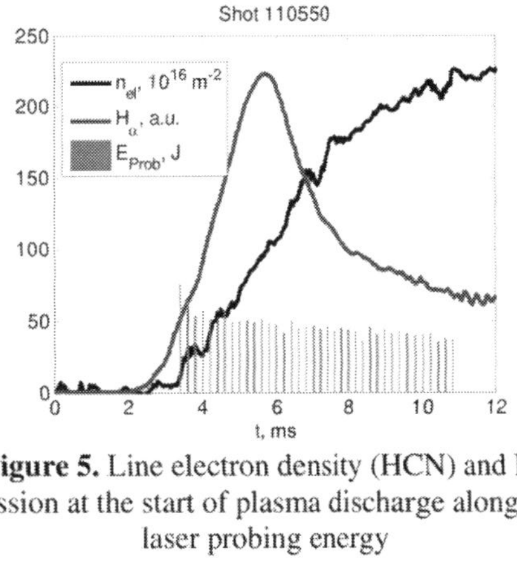

Figure 5. Line electron density (HCN) and D_α emission at the start of plasma discharge along with laser probing energy

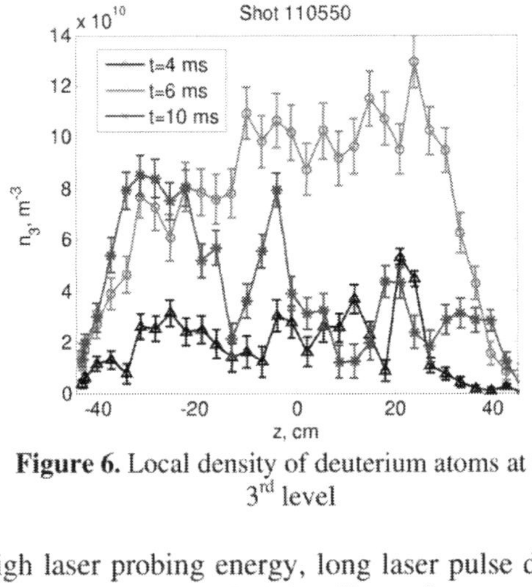

Figure 6. Local density of deuterium atoms at the 3rd level

The multipass intracavity probing provides a high laser probing energy, long laser pulse duration and an extension of the radiated plasma volume along the line of sight. These factors are very important for the LII diagnostic, increase considerably its sensitivity and made possible the LII measurements at the plasma start in TEXTOR.

The evolution of line electron density and D_α emission at the start are shown in figure 5. A 100 MW/cm^2 laser probing intensity (50 J during 1.5 μs through ~0.3 cm^2 laser cross section in plasma) is quite enough to ionize nearly all hydrogen and deuterium atoms from the 3rd excited level on the laser beam path in plasma [9]. The ionization results in partial quenching D-alpha emission collected by the TS diagnostic from plasma. The quenching signals were determined as the difference between the signals taken with and without laser pulses. The local density of the excited atoms on the laser beam path can be found directly from the measured signals [9, 11]:

$$n_3 = \frac{\Delta N}{V_{Las}\Omega\eta\, A_{32}\tau}$$

Here ΔN is the quenching signals in photoelectrons, V_{Las}~1.5 cm^3 is the volume on the laser beam path from which the signals are collected, Ω~0.005 sr is the solid angle of light collection, η~0.03 is the ratio of the registered photoelectrons to the photons collected from plasma, A_{32}= 0.44*10^8 1/s is the probability of spontaneous radiation decay from the 3th to 2nd atomic level, τ~1.5 μs is the duration of laser pulses.

The local densities of the excited atoms measured in different times at the spatial resolution ~3 cm are shown in figure 6. The statistical errors are in the range from 10% to 30% and the diagnostic sensitivity is ~10^{10} m^{-3}. The density of the ground state atoms can be calculated from a collisional-radiative model (see e.g. [12]) using local T_e and n_e data. The sensitivity of the LII diagnostics in respect to the ground state atom density is expected to be ~10^{13} m^{-3} in TEXTOR. These estimations can be even improved by several times at the steady state by summing up the quenching signals from several laser pulses.

The excited atom density is significantly reduced in a hot plasma core and the quenching signal is much worse distinguished against intense D_α emission from the plasma edge. A ~10-fold enhanced sensitivity of the LII diagnostic for the plasma core measurements can be achieved by a ~2-3-fold enlargement of the laser beam diameter in plasma with a special multipass system [11]. In this case the sensitivity of the LII diagnostic in TEXTOR is estimated to be ~10^9 for excited and 10^{12} for ground state deuterium atoms in a cubic meter. In a steady state conditions the sensitivity can be additionally improved by about the order of magnitude after averaging quenching signals along a laser pulse train.

5. Conclusion

TS diagnostics equipped with a multipass intracavity laser probing system provides not only perfect TS measurements, but opens new applications of the diagnostic for studies of tokamak plasmas from measurements of the electron energy distribution, plasma current density, electron density and temperature fluctuations, TS measurements at very low plasma density and local measurements of neutral hydrogen atom density.

Acknowledgments: This work was supported by NWO-RFBR Centre-of-Excellence grant 047.018.002. This work, supported by the European Communities under the Contract of Association between EURATOM-FOM, was carried out within the framework of the European Fusion Programme. The views and opinions expressed herein do not necessarily reflect those of the European Commission.

References

[1] Kantor M.Yu. et al. 2009 *Plasma Phys. Control. Fusion* **51**, 055002
[2] Kantor M.Yu. et al. 2009 *Proc. 36th EPS Conf. on Plasma Physics, Sofia, Bulgaria* P1.184
[3] Segre S.E. 1978 *Plasma Physics* **20**, 287
[4] Karelse F.A. et al. 2001 *Plasma Phys. Control. Fusion* **43,** 443
[5] Samm U. 1999 *Plasma Phys. Control. Fusion* **41,** B57
[6] Barth C.J. et al. 2001 *Review Scientific Instruments* **72**, 1138
[7] Govers C. et al. 1988 *Applied optics* **27**, 3625
[8] Thomas C.E. et al. 1986 *Review Scientific Instruments* **57**, 1819
[9] Gladushchak V.I. et al. 1995 *Nuclear Fusion* **35**, 1385
[10] Razdobarin G. et al. 1979 *Nuclear Fusion* **19,** 1439
[11] Kantor M.Yu. et al. 2007 *13th International Symposium on Laser-Aided Plasma Diagnostics. Takayama, Japan. NIFS-PROC-68*, p. 104
[12] Johnson L.C. and Hinnov E. 1973 *J. Quan. Spectr. Rad. Trans.* **13,** 333

Progress in development of the KSTAR Thomson scattering system

Jongha Lee and Seung Tae Oh

National Fusion Research Institute, 113 Gwahangno, Yuseong-gu, Daejeon, 305-333, Korea

E-mail: jhlee@nfri.re.kr

Abstract. The design of Thomson scattering system for the Korean Superconducting Tokamak Advanced Research (KSTAR) device is described. The system includes a laser beam guiding system, laser input port, laser beam dump, collection lens, shutter, and cassette system. The laser guiding system has a collimating lens set for reducing the laser power loss. The laser beam dump will be attached inside of the vacuum vessel and hence it has been designed compactly. The preliminary design of collection lens, shutter system, and the cassette were done by PPPL and the engineering design is being executed by NFRI. The collection lens system has two sets of lenses, one set is designed for the core and the other set is designed for the edge. Two sets of pneumatic shutter systems are designed for independent remote control. Most of the KSTAR Thomson scattering system design is already finished and we are planning to install the whole system by March, 2010. We will measure the plasma parameters (T_e, n_e) in KSTAR during the third campaign.

1. Introduction

The Korea Superconducting Tokamak Advanced Research (KSTAR) tokamak [1] achieved the first plasma on the last day of June 2008 at the National Fusion Research Institute, Daejeon, Korea. Thomson scattering has become an important diagnostic for measuring electron temperature and density profiles in most tokamaks. [2][3][4] Therefore we plan to install the KSTAR Thomson scattering system before the KSTAR 3rd campaign in 2010. The goal of the Thomson scattering diagnostic on KSTAR is to provide reliable electron temperature T_e, and density n_e, profiles in both core and edge regions during long-pulse tokamak plasma discharges. To meet this goal, we are designing each part of the KSTAR Thomson scattering system, such as laser beam guiding system, laser inlet system, and laser beam dump system inside the KSTAR vacuum vessel and collection optics systems with a cassette. For the laser system, we will use a commercial Nd:YAG laser (1064 nm, 10 Hz, ~1.5 J, Coherent Co.) for the 3rd campaign. For the KSTAR 4th campaign in 2011 and the next 5 years, a high-power high-performance (> 3 J) high repetition rate (> 50 Hz) Nd:YAG laser system will be used which is being developed by JAEA, Japan. [5] Next year in 2010, the design requirements for parameters to be measured are as follows; time resolution: 100 msec (10 Hz), spatial resolution: 15 mm in the core and 25 mm at the edge, temperature range: 10 eV < Te < 1.5 keV in the core and edge, density range: 3×10^{18} m^{-3} < ne < 2×10^{20} m^{-3}. [6]

In this paper, we describe the engineering design of the laser beam dump, laser shutter systems etc., and the 1st calibration data for the polychromators.

Journal of Physics: Conference Series **227** (2010) 012042 doi:10.1088/1742-6596/227/1/012042

2. Thomson scattering apparatus on the KSTAR tokamak

The KSTAR Thomson scattering system is configured with 90° scattering optics, including a single tangential beam, as illustrated in Figure 1. The laser beam, transported for a distance of about 40 m and focused by a 5.5 m focal length lens, is directed into the vacuum vessel through the mid-plane port of Bay L (the *Lm*-port). The focal lens and steering mirror are mounted on an optical bench which is located in front of the *Lm*-Port. Finally the laser beam reaches the beam dump that is located at the *Bm*-port wall in the vacuum pumping duct. The collection optics with cassette system is located in the *Nm*-port and shares that space with the ECH launcher system (Figure 2). Most of the other parts of the Thomson system such as laser, polychromators, and A/D system will be located in the diagnostics room which is well shielded from radiation.

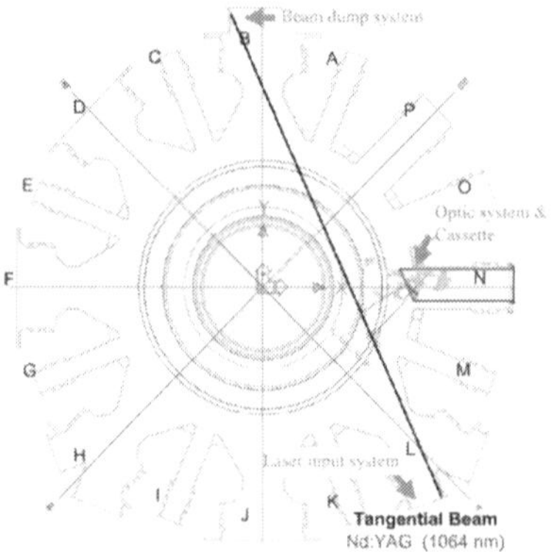

Figure 1. Top view of the KSTAR Thomson scattering system

Figure 2. Photograph of the KSTAR Thomson cassette port inside the vacuum vessel. (The ECH launcher system is shown on the right side of the photo)

2.1. Lens design and cassette

The preliminary design of the Thomson collection lenses was done by PPPL [6], but recently we have changed the design using lens design software. PPPL's design consisted of four lenses for the core and edge collection optics systems and each optics module had the same design. However, we have designed the core and edge parts of the lenses differently because each part of the KSTAR Thomson collection lenses has different viewing angles. Finally we have obtained the optimum focusing spot size for core and edge lens systems using quartz glass. In our design, the spot size of the core part is around 20 micrometer and the edge part is 500 micrometer. The cassette and shutter were also designed by PPPL. We will finish this cassette system early in 2010 and will test the collection optics positioning and alignment.

2.2. Laser input port and beam dump design

The laser input port of the KSTAR Thomson system is located on the mid-plane *Lm*- port. The Thomson laser input system shares the available space with the resistive bolometer system. The laser input port contains five parts of the system; an electro-pneumatic operation gate valve, ceramic break, bellows, baffles, and vacuum window.

The laser beam dump design is very simple as shown at Figure 3. This wedge type beam dump is made with SUS316L outside and thick carbon sheet inside. Of course, this beam dump has a baffle and straight knife edges around the edge of a doughnut hole that are located at the front of this wedge dump system. This simple beam dump will be attached at the end of *Bm*- port because of the limited number of KSTAR tokamak vacuum windows.

2.3. Polychromator and calibration

The KSTAR polychromator was developed by the Japan - Korea nuclear fusion joint cooperation program. This program completed the polychromator design and 25 polychromators with power supplies were made by NIFS, Japan. These polychromators have 5 different wavelength channels (filter combinations in nm): [890, 990], [990, 1030], [1030, 1050], [1050, 1060], [1062, 1066]) (Figure 4), and avalanche photo diodes (APD). Recently we performed calibrations, including detector position, wavelength, light intensity, pulse width, and voltage-out calibration for the 25 polychromators. The detector position calibration used tungsten light and we rotated the x- and y- axis to find the maximum voltage from the APD signals. The next calibrations were the wavelength and light intensity calibrations. In these calibrations, we use a calibrated detector (Melles Griot silicon detector) and a monochromator (CVI Co.). Table 1 shows an averaged central wavelength and Full width at half maximum (FWHM) of wavelength with standard deviation for each polychromator channel from this calibration. Figure 4 is an example of filter combination spectrum data for the polychromator #1 from the calibration of light intensity. For the pulse width calibration, we made a diode laser pulse light source that had a pulse width of ~19 nsec. Table 2 shows the average pulse width with standard deviation for each polychromator channel measured using an oscilloscope. Finally, we measured the output voltage from the APD. From these five different calibrations, we confirmed the performance of the polychromators. From this result, we will measure the KSTAR electron temperature next campaign, 2010, using these polychromators that measuring range will be 10 eV < Te < 1.5 keV. (Figure 6)

Table 1. Filter wavelength calibration data for the polychromators.

	Channel 1		Channel 2		Channel 3		Channel 4		Channel 5	
	λc^a	δ^b	λc^a	δ^b	λc^a	δ^b	λc^a	δ^b	λc^a	δ^b
Avg.[c]	1063.6	11.24	1054.5	5.32	1043.4	10.8	1010.5	32.8	944.4	103.2
σ^d	0.8	0.66	0.51	0.48	0.49	0.48	0.59	0.44	1.50	1.37

[a] Center of wavelength (nm), [b] Full width at half maximum(FWHM) of wavelength (nm)
[c] Average wavelength (nm), [d] Standard deviation

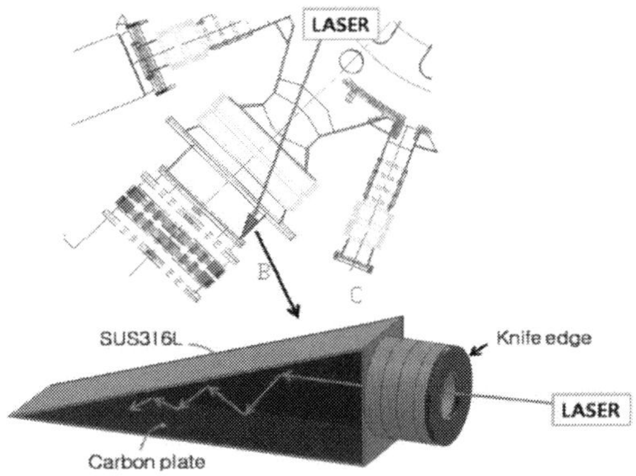

Figure 3. Beam dump design for the KSTAR Thomson scattering system.

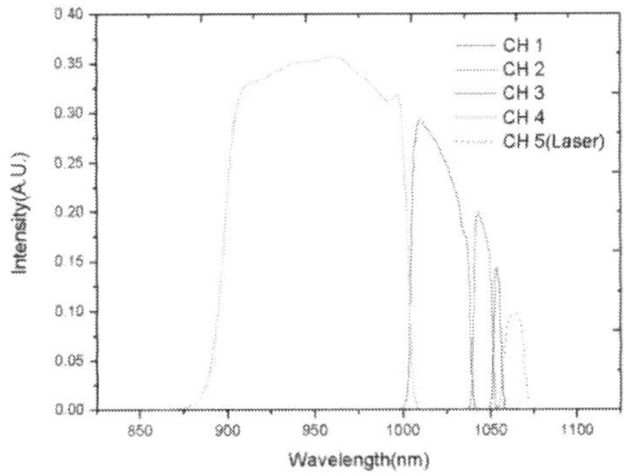

Figure 4. Polychromator filter combination (example: polychromator #1).

Table 2. Laser pulsewidth(FWHM) calibration data for the polychromators.

	Channel 1	Channel 2	Channel 3	Channel 4	Channel 5
Avg. [a]	19.68	19.74	19.24	19.16	19.04
σ [b]	0.69	1.35	1.23	1.25	1.4

[a] Average pulsewidth at FWHM(nsec), [b] Standard deviation

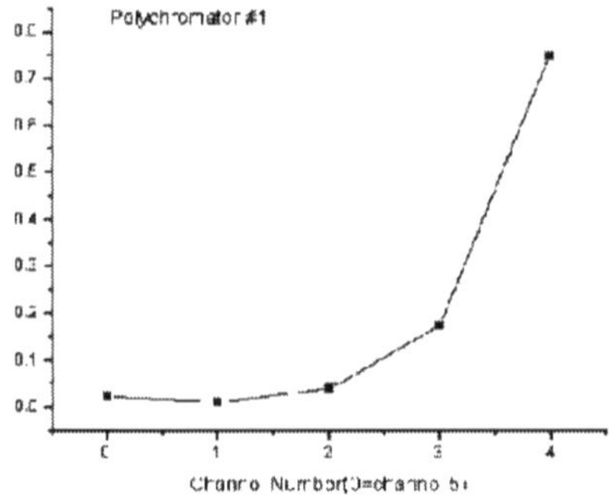

Figure 5. Normalization data which is result of light intensity calibration for the polychromator #1 .

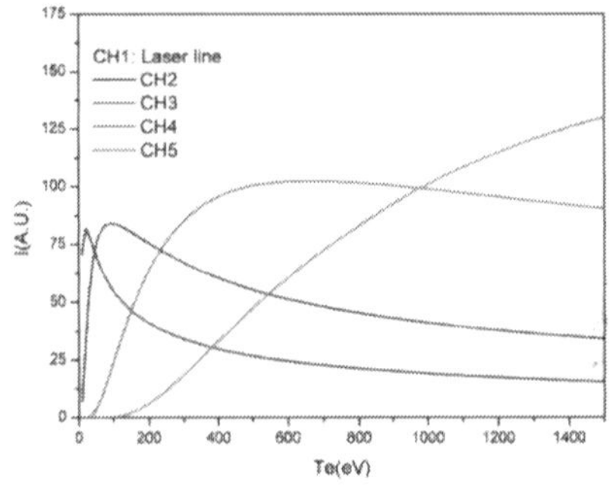

Figure 6. Calibration curve for the electron temperature measurement.

3. Installation schedule for the KSTAR Thomson scattering system

To measure electron density and electron temperature profiles on KSTAR using Thomson scattering diagnostics, most of system design has been completed now and the system must be fabricated this year. However the cassette system is actually quite complicated to design for the KSTAR Thomson diagnostics because of the long narrow port size and, in addition, sharing port space with the ECH launcher system. For this reason, the Thomson cassette will be finished next February. The optical fiber and collection lens are designed and these are now undergoing a final check.

The 1st calibration of KSTAR polychromator was successfully finished, and moreover at the end of this year we will perform a 2nd calibration. The 3rd calibration will be performed just before the KSTAR 3rd campaign. These calibrations will give reliable results for the KSTAR plasma electron temperature and density.

The KSTAR Thomson scattering diagnostic system should be installed next year to measure KSTAR 3rd plasma, 10 spatial points for 2010, 25 spatial points for 2011, and 40 spatial points for 2012.

References

[1] Lee GS et al. 2000 *Nucl. Fusion* **40** 575
[2] Peacock N J et al. 1969 *Nature (London)* **224** 488
[3] Murmann H et al. 1992 *Rev. Sci. Instrum.* **63** 4941
[4] Carlstrom T N et al. 1992 *Rev. Sci. Instrum.* **63** 4901
[5] Hatae Takaki et al. 2004 *J. Plasma Fusion Res.* **80** 870
[6] Lee HG et al. 2004 *Rev. Sci. Instrum.* **75** 3903

Journal of Physics: Conference Series **227** (2010) 012043 doi:10.1088/1742-6596/227/1/012043

Thomson scattering diagnostics for ITER divertor

EE Mukhin[1], AS Kukushkin[2], S Yu Tolstyakov[1], MM Kochergin[1], VV Semenov[1], GS Kurskiev[1], AG Razdobarin[1], KA Podushnikova[1] and P Andrew[2]

[1]Ioffe Physical Technical Institute, 194021, St.Petersburg, Russia

[2]ITER Organization, Cadarache, 13067 Saint-Paul-Lez-Durance, France

E-mail: e.mukhin@mail.ioffe.ru

Abstract. The ITER design has highlighted the fundamental need to monitor the machine operation in more detail. The mission of the Thomson scattering diagnostics in the ITER divertor research/operation is discussed with due attention paid to challenges and capabilities of the existing diagnostic design.

1. Introduction

Detailed measurements of electron parameters in ITER divertor will be an important part of the experimental programme. The measurements will be used in studies of the divertor ability to adequately control plasma position during disturbances with adequate screening of impurities released as a result of intense plasma-surface interaction. The Thomson scattering (TS) system for the ITER divertor [1] is well suited for this purpose. The diagnostic will provide the measurements in the outer divertor leg with the provisional target requirements according to Table 1.

Parameter	Range or coverage	Spatial resolution	Time resolution
T_e	1 – 100* eV	5cm along leg	15 ms
n_e	10^{19} - 10^{21} m^{-3}		(60 Hz)

* the upper limit of measured electron temperature is discussed in [2]

Table 1. Provisional target requirements for the TS diagnostics in the outer divertor leg

The difficulty for Thomson scattering in the divertor is associated generally with limited access to the plasma and with survivability of optical components that have to be placed close to this region. Here we review briefly the measurement requirements and the role of the divertor TS with special attention to challenges and capabilities of the approach, omitting the technological problems of the in-vessel diagnostics design as well as dedicated R&D on survivability of in-vessel diagnostic components.

2. Potential Role of Thomson Scattering System in ITER Divertor

Measurements of electron parameters in the divertor pursue two basic aims: monitoring inter-mode transitions (e.g., attached/detached conditions of the divertor plasma) and measuring the divertor impurity concentrations and flows. The requirements to the range and accuracy of the electron temperature and density measurements in the ITER divertor are based on the results of modeling of operational conditions and follow from the main objectives of the measurements. Simulated with the

SOLPS4.3 code [3], 2D distributions of electron temperature and density (see Figure 1) show the range of plasma parameters in the divertor region to be expected in different operational modes.

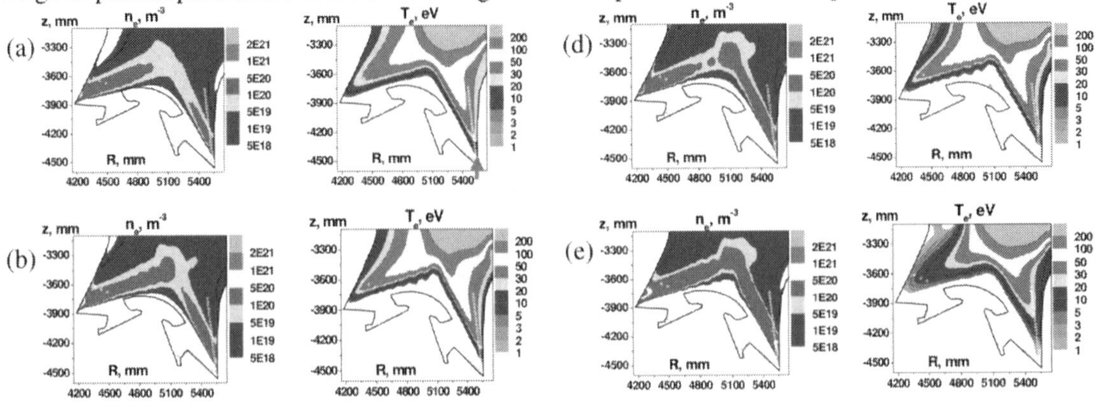

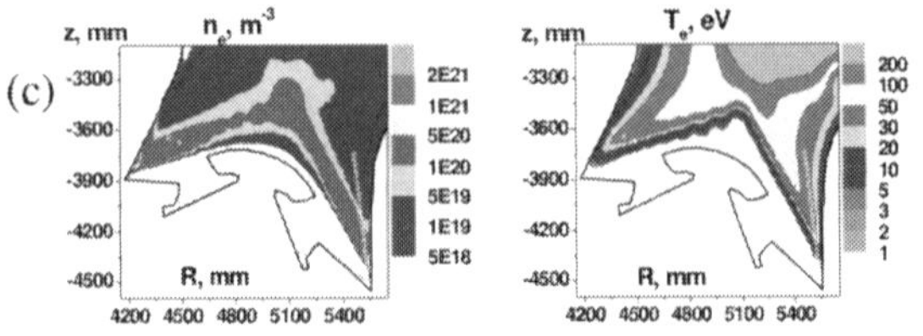

Figure 1. 2D distribution of electron temperature (left) and density (right) for discharges with loads on divertor plates: (a) 15 MW/m^2 (run 1537), (b) 10 MW/m^2 (1511), (c) 8 MW/m^2 (1514), (d) 6.5 MW/m^2 (1540), (e) 0.5 MW/m^2 (1538). Dots along the probing chord (arrow) mark the provisional spatial channels of the diagnostics.

The operational range of the divertor corresponds to long time operation with a power loading on the divertor plates of 6.5 to 10 MW/m^2 (see Figure 1-b,c,d). The lowest load (Figure 1-e) corresponds to excessively high plasma density in the divertor region and then, deterioration of the main plasma confinement and potential discharge disruption can be expected. The highest load is beyond the design limits, and the operation mode shown in Figure 1-a can be allowed for several seconds only. Figure 2 shows the distribution of electron parameters along the given probing chord in the outer divertor leg. As seen from the SOLPS simulations, the expected range of the divertor plasma parameters varies strongly along the probing chord, especially near the ionization front, and is rather restricted in each spatial point under the inter-mode transitions. The maximum power flux on the divertor targets is to be controlled within 20% around 8 MW/m^2. The corresponding variation of the plasma parameters distribution along the probing chord can be used to diagnose a shift of the ionization front position. As shown in Figure 3, the most pronounced effect is expected in the downstream region of the outer leg. One can see that the parameter variations corresponding to a 10% change of the power loading of the target can be detected.

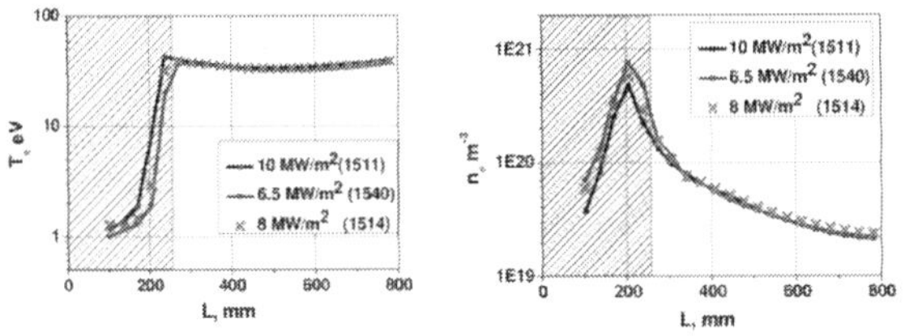

Figure 2. T_e and n_e distribution along the probing chord of the TS system in the outer leg of the ITER divertor for three modes within the operational window. The central one is shown with red crosses, and the two others marked by blue and black lines are the limiting cases appropriate for the long time operation.

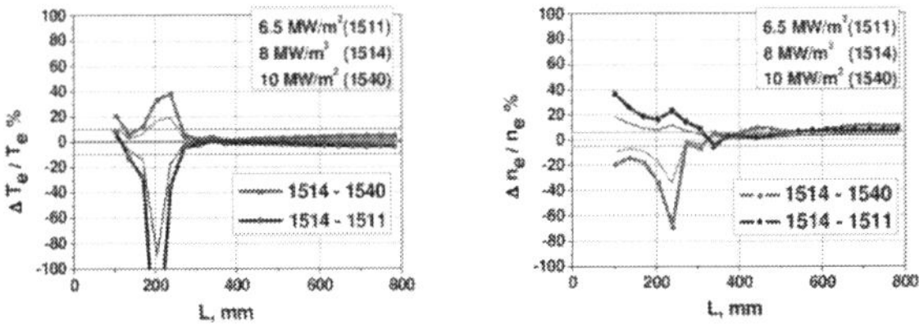

Figure 3. Variation of T_e and n_e along the probing chord. The bold blue line corresponds to deviation of the maximum power flux density on divertor plates from 8 to 10 MW/m^2 and the black one from 8 to 6.5 MW/m^2. The thin lines correspond to a 10% variation of the power loading. The red lines mark the expected limits of the measurement accuracy.

Journal of Physics: Conference Series **227** (2010) 012043 doi:10.1088/1742-6596/227/1/012043

The information from the upper part of the leg cannot be used for the detection of the simulated transition (although it can help with interpretation of spectroscopic data). However, the position of ionization front characterized by strong gradients of T_e and n_e can move due to the expected delay of the plasma position control response to transients [4]. In addition, the planned divertor exchange strategy implies that operational experience gained on CFC targets will be required for tungsten. In the new divertor design the extended tungsten baffle will allow magnetic configurations in which the separatrix strike points are placed on the upper, tungsten part of the target. Therefore the uncertainty of the ionization front position and, hence, the position of the strong gradients of the electron temperature and density necessitates a uniform distribution of the spatial diagnostic channels along the probing chord.

3. The Challenge for the Thomson Scattering System in the ITER Divertor

Plasma characteristics vary strongly along the probing chord, where actually two areas can be identified (see Figure 2), each demanding a specific instrumental approach. The upper part of the probing chord has rather normal plasma parameters for an edge plasma in fusion experiments: $n_e \sim 10^{19} - 10^{20}$ m^{-3} and $T_e \sim 20 - 50$ eV and the lower part is characterized by strong gradients of the electron parameters $n_e \sim 10^{19} - 10^{21}$ m^{-3} and $T_e \sim 1 - 50$ eV. The extra-low T_e and resulting requirement to measure extra-narrow Thomson spectra will be aggravated by extremely high stray light expected just in the area due to possible dustiness and proximity to the walls. We suggest using different spectral equipment for these two parts of the probing chord. One of them, based on the filter spectrometers, is a conventional and reliable approach for modern Thomson diagnostics, and it is logical to use it for analysis of the upper part of the probing chord. A grating spectrometer with extremely high rejection of stray light and fine resolution is a good solution for the TS spectrum measurements in the bottom of the outer leg. Figure 4 shows the error calculations as a function of temperature for the expected experimental conditions with n_e of 10^{19} m^{-3}. The working range of a grating spectrometer is from 1 eV to 100 eV, if an accuracy of better than 10% for T_e and 5% for n_e measurements is required. The low temperature limit of the filter spectrometer is several eV. The uncertainty of the ionization front position requires a flexible strategy for the spectral device distribution to keep the diagnostic operational during the temporal displacement (see above) of the ionization front. One of the solutions is to interleave the more reliable filter spectrometers with the more expensive but more powerful grating spectrometers in the upper part of the probing chord.

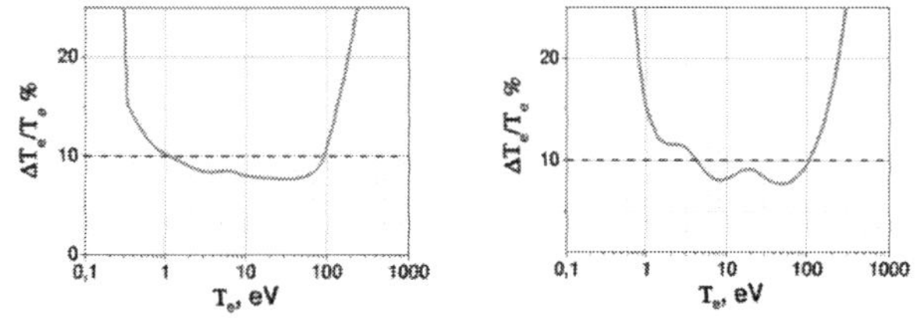

Figure 4. Predicted error values vs electron temperature for the grating (left) and filter (right) spectrometers. Calculations are done for the laser energy of 1.5J, $n_e = 10^{19} m^{-3}$ (underestimated) and background light 100 photons/nm (overestimated).

One of the most challenging requirements is the measurement of extremely low T_e (~ 1eV range) near the ionization front. In the cool and dense plasma the laser wavelength approaches the Debye length and the deviation of the shape of the scattered spectra from Gaussian becomes very pronounced. The Salpeter parameter (defined as the inverse product of the change in the scattered wave vector times the plasma Debye length) shows the degree of collective behaviour in the scattering: higher values correspond to the collective scattering and lower to the individual particle scattering. Usually classical analysis of scattered spectra with assumption of Gaussian shape of the scattered spectra is considered valid when the Salpeter parameter does not exceed 0.25, whereas the expected maximum of the Salpeter parameter in the proposed TS geometry can reach 0.43 at the highest n_e considered for the ITER divertor operation, Fig.5. The calculated spectra for the classic case (electron density ~10^{19} m^{-3}) and for the maximal Salpeter parameter 0.43 (which can be expected under deterioration of plasma confinement) are shown in the Figure.6 along with transmission of the first spectral channels of grating spectrometer. Obviously, for the high density case, a signal of the first channel needs to be excluded from consideration. It is quite acceptable, because the signal intensity is proportional to the electron density and therefore the signals in the second and third channels will be sufficiently intensive.

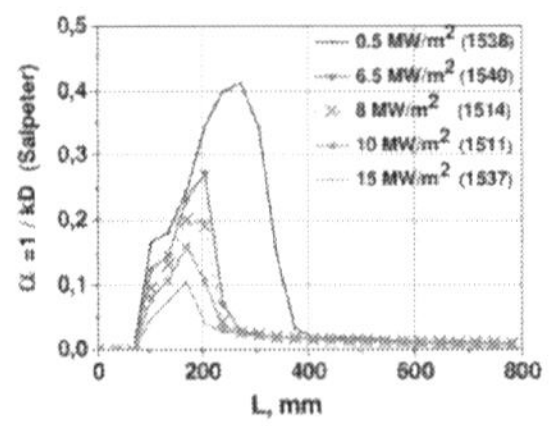

Figure 5. Value of the Salpeter parameters along probing chord for operational modes with different power loading of the divertor plates.

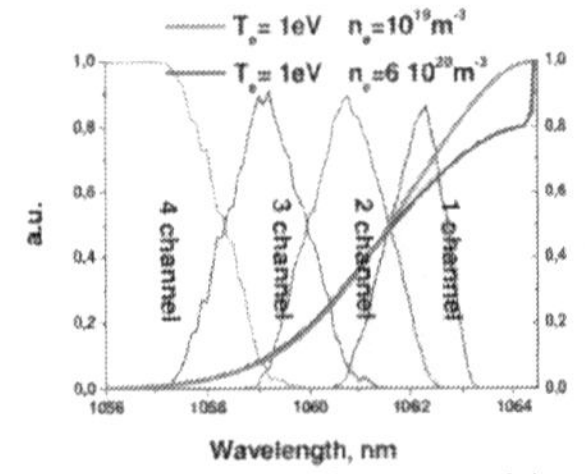

Figure 6. Thomson spectra shown with red (n_e~10^{19} m^{-3} α~0.043) and blue (n_e~6 10^{20} m^{-3} α~0.43) curves. Thin lines show spectral transmission of the grating spectrometer spectral channels. The first spectral channel is colored yellow.

4. Conclusions

Measurements along the divertor leg can provide useful information to support the proper divertor operation. In particular, they can provide the data on the location of the ionization front for the interpretation of spectroscopic data and validation of models of divertor performance. Due to strong plasma non-uniformity near the ionization front and uncertainty of the ionization front position, a uniform spatial resolution is needed along the whole probing chord. Two different kinds of spectral instrumentation are suggested for assessment of the top and bottom parts of the probing chord to cope with the extreme conditions in the bottom part (measurements of extra-narrow Thomson spectra aggravated by expected intensive stray light). One of the most challenging requirements is diagnosing the cool, dense plasma near the ionization front, where the collective behavior in the scattering can be important. For the most problematic case of high density and extremely low temperature, it is shown that the expected distortion of Gauss-like Thomson spectra is not insurmountable and can be avoided by excluding the signal of the first spectral channel from consideration.

Acknowledgments

This report, supported by Rosatom (contract -N.4z.41.03.08.258), was prepared as an account of work for the ITER Organization. The Members of the Organization are the People's Republic of China, the European Atomic Energy Community, the Republic of India, Japan, the Republic of Korea, the Russian Federation and the United States of America. The views and opinions expressed herein do not necessarily reflect those of the Members or any agency thereof. Dissemination of the information in this paper is governed by the applicable terms of the ITER Joint Implementation Agreement.

References

[1] Razdobarin GT, Mukhin EE 1998 Status of Thomson Scattering Diagnostic Design for ITER X-Point and Divertor Plasmas// *Diagnostics for Experimental Thermonuclear Fusion Reactors 2* Plenum Press New York London 237-246
[2] Tolstyakov SYu, Mukhin EE, Kochergin MM et al. 2009 *Near-infrared tokamak plasma spectroscopy in support of divertor Thomson scattering diagnostic development for ITER//* LAPD-14
[3] Kukushkin AS, Pacher HD, Loarte A et al. 2009 *Nucl. Fusion* **49** 075008
[4] Pitts RA, Kukushkin AS, Loarte A et al. 2009 Status and physics basis of the ITER divertor// *12th International Workshop on Plasma Facing Materials and Components for Fusion Applications*

Calibration methods for ITER core LIDAR

R Pasqualotto[1], L Giudicotti[1,2], A Alfier[1], M J Walsh[3]

[1] Consorzio RFX, Associazione EURATOM-ENEA sulla Fusione, Corso Stati Uniti 4, 35127 Padova, Italy
[2] Padova University, Electrical Engineering Department, Via Gradenigo 6/A, 35131 Padova, Italy
[3] EURATOM/UKAEA Fusion Association, Culham Science Centre, Abingdon OX14 3DB, United Kingdom

E-mail: roberto.pasqualotto@igi.cnr.it

Abstract. The standard technique for the relative calibration in present Thomson scattering diagnostics, can be used in ITER only if the full collection path is included in the calibration: indeed, the in-vessel optics will be exposed to neutron and gamma irradiation and particle flux, that will likely cause a distortion of the spectral transmission curve. The standard scheme would ask to position a diffuser in front of the first collection mirror, e.g. on the protective shutter, and to illuminate it with a light source nearby. Alternatively, a back illumination scheme could be used, with a more efficient retroreflector array. A completely different approach is based on TS measurements from two lasers with different wavelengths: this self-calibrating method does not require any in-vessel tool and will provide both the electron temperature and the relative sensitivity of spectral channels.

1. Introduction

In a Thomson scattering (TS) diagnostic, calibration of relative sensitivity, necessary to calculate the electron temperature T_e, is usually performed with a white light source coupled to a diffuser, which simulates the radiating scattering volume and illuminates all or more frequently only a reduced set of the collection optics. The relative sensitivity of the i-th spectral channel is calculated as the ratio of the signal from the lamp detected by that channel and the expected signal calculated as the convolution of the spectral emissivity of the lamp, the spectral reflectivity of the diffuser, the spectral transmission curve of that channel and the spectral response of collection optics; these factors are measured in advance with a calibrated tunable monochromatic light and are supposed not to change with time. If a broadband filter polychromator is used, only average changes of the collection optics transmission curve over the spectral bands of the spectrometer channels are detected and these are interpreted as modifications of the relative sensitivities. This way it is not possible to discriminate changes within each channel spectral band.

Ideally the light source should be positioned in the vacuum vessel, to include the entire collection system, while usually it is positioned just outside the collection vacuum window, so that all the collection system but the window is included in the calibration. Unfortunately the window is the element more susceptible to changes in transmission because of plasma and neutral deposition on the internal face. This is not just a change in absolute transmission, which would only affect the absolute calibration and then the estimate of the electron density n_e, but also a change in the transmission spectral curve which affects the relative calibration and then the estimate of T_e.

For ITER TS diagnostics, high accuracy and reliability are required, but operating conditions are very different from existing tokamaks: high neutron dose environment, in-vacuum mirrors and extremely

long plasma discharges. T_e and n_e profiles along a diameter in the midplane will be measured by the core LIDAR TS system, with spatial coverage r/a < 0.9, T_e and n_e ranges respectively of 0.5-40 keV and $3{\cdot}10^{19}$-$3{\cdot}10^{20}$ m^{-3} [1,2]. In this system the vacuum window is far from the plasma and not susceptible to deposition, but there are still two sources of degradation of the spectral transmission: a) two in-vessel mirrors are susceptible to plasma deposition, especially the first mirror, directly viewing the plasma; b) the high radiation level changes the optical properties of not sufficiently shielded optics, producing absorption and photoluminescence in refractive components, like the vacuum window. We then expect some degradation of the spectral transmission of in-vessel optics, probably at a high rate, with the need of frequent recalibrations.
The standard scheme would require to position a diffuser in front of the first collection mirror and to illuminate it with a light source nearby. The diffuser could be mounted on the shutter protecting the first mirror from deposition. Alternative and possibly more feasible schemes are being investigated: the internal light source can be replaced by a back illumination scheme and the diffuser could be replaced by a more efficient retroreflector array.

2. Back-illumination scheme

The basic requirement for the calibration source, either internal or external, is to produce a continuous, smooth and stable spectrum in the entire detection range of the ITER core LIDAR (from 350 nm to 1100-1400 nm). Its intensity has also to be high enough to produce a detectable signal, once attenuated by the diffuser or retroreflectors mentioned above. So far, the best candidate light source is an halogen lamp, but its use in ITER is challenging and requires to first solve few issues. The glass envelope would be darkened by the high radiation and neutron flux, changing the lamp emissivity spectrum, while a bare thoriated tungsten filament would too rapidly evaporate, thus reducing the filament lifetime. However radiation darkening might be not that severe, as the glass is thin and heats up, tending to self anneal. Two other potential problems are operation in vacuum at high temperature and in a high magnetic field. In addition lamps with highest available color temperatures (> 3000K) are required, to still have meaningful intensity level at 350-500 nm. The diffuser used with an internal source or with a back illumination scheme could be the back of the first mirror shutter coated with plasma sprayed alumina and it should be protected from plasma deposition when closed.
If the source can not be positioned next to the diffuser, this could be illuminated from outside the bio-shield, with a back illumination scheme: the light is sent through the collection optics backwards, compared to the TS signal direction, then it illuminates the diffuser or a retroreflector array; the diffused or reflected light is then gathered through the collection optics again but in the forward direction this time and finally it is measured with the broadband LIDAR polychromator or a higher resolution monochromator. The diffuser, supposed Lambertian, diffuses uniformly over a 2π solid angle, while the retroreflectors send the light prevalently back along the incident direction, being much more efficient than the diffuser and thus requiring a less powerful light source.
A benefit of a flexible choice of light source is that the spectral transmission of the collection optics can be separately measured. This requires using a scanning wavelength source. It also assumes that the spectral reflectivity of the diffuser or retroreflectors is known and does not deteriorate.
The back illumination scheme has some other open issues. In principle the double pass through the collection optics should be easily taken into account, but this might complicate the measurement as the transmission should be separately measured with a higher resolution monochromator. Another possible issue is stray light: light is fed into the collection optics by inserting a beamsplitter, which gets light backwards into the collection optics from the source. This method can cause stray light into the detector, especially if a diffuser is used instead of a retroreflectors array, because of the different efficiency and because stray light can also be generated by reflections from collection optics and from the walls of the port, even though this can be minimized by imaging the porthole onto the detector.
The back-illumination scheme, using an array of micro retroreflectors in front of the first mirror, is being proposed for the calibration system of the divertor impurity influx monitor [3,4]. Micro retroreflectors are necessary there, to avoid excessive losses, since a fan of parallel rays would not be

reflected back parallel by a retroreflector if imaging optics are inserted in the optical path. If the retroreflectors are sufficiently small (0.2 mm pitch), the divergence of the reflected rays and than the losses are tolerable. The smaller the pupils of the optical system, the more relevant this problem is.

This scheme could be implemented also for the core LIDAR. A Zemax ray tracing model has been used to define the maximum tolerable size of the retroreflectors for the LIDAR case. A 30 mm diameter uniform source is positioned just before the spectrometer aperture, back-illuminating the collection optics. A retroreflector array is positioned in front of the first mirror and with the same diameter, retroreflectors pitch ranging from 2 to 200 mm. Fig.1 shows the incoming and reflected light through the vacuum window (in-window and out-window), and the power lost by the retroreflector array (scattered) and by the two toroidal mirrors between the retroreflectors and the vacuum window (lost). The result is that arrays with a pitch up to 20 mm can be used without significant deterioration of the performance. Using a Lambertian diffuser instead, the reflected power through the vacuum window is about 10^2 smaller.

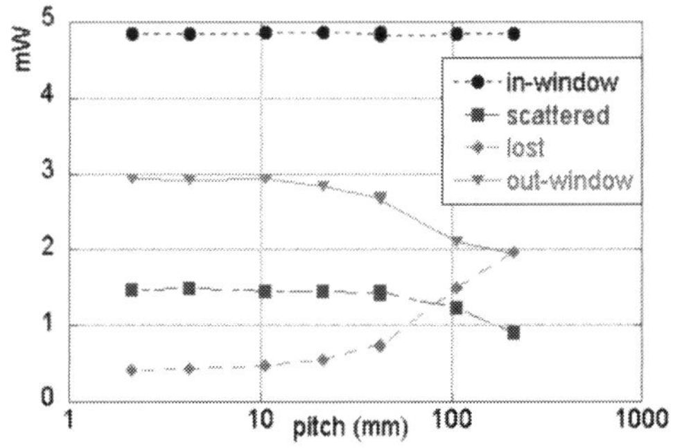

Figure 1. Performance of retroreflector array

3. Dual laser self calibration

The original Design Description Document of the ITER core LIDAR (DDD 5.5C.01) suggests to use TS measurements from two Q-switched lasers with different wavelength [5], as preferred solution for the relative calibration of the system. This method would not require any in-vessel tool, or to replicate the illumination characteristics of the TS signal and would provide both the electron temperature T_e and the relative sensitivities of spectral channels. T_e is calculated from the ratio of TS signals from the two lasers on the same spectral channel, without requiring knowledge of relative sensitivities. In addition the spectral transmission curve is obtained from comparison of measured and expected signals, once T_e has been derived. Restrictions on usable wavelength combinations apply: a) a spectral interval with significant scattered signals from both lasers, such as to have detectable signals from both lasers in at least two spectral channels, is required to get T_e; b) each wavelength of the full spectral range of interest must have a detectable signal from at least one laser to obtain the relative sensitivity of all channels. One weakness of the dual wavelength method is that it assumes small variations of transmission curve inside each channel spectral band. To resolve the transmission changes inside a spectral channel, a separate higher resolution spectrometer should be used.

The laser combination originally proposed was Ti:Sapphire (800 nm) as main laser and duplicated Nd:YAG (532 nm) as calibration laser. At least two channels have to cover the spectral range where we have detectable signals from both lasers in the same channel. If we assume that both signals are detectable if their ratio if < 20 or $> 1/20$, at $T_e = 5$ keV this spectral range is 450-750 nm, where likely two spectral channels are present. This spectral region is smaller for lower T_e and wider for higher T_e.

The present ITER requirement of measuring higher T_e (< 40 keV) than previously assumed requires a revision of the proposal, connected to a new evaluation of the optimum lasers combination.

Considering that Nd:YAG @ 1064 nm is presently being considered as a good candidate for TS measurements, possibly complemented by doubled Nd:YAG @ 532 nm for lower T_e [2], two combinations look promising for the dual laser calibration: (a) Nd:YAG @ 1064 nm + Nd:YAG @ 532 nm or (b) Nd:YAG @ 1064 nm + ruby @ 694 nm. For (a), measurable signal ratio is obtained at 10 keV in spectral range 450-800 nm, while for (b) at 5 keV in spectral range 620-950 nm. Below about 10 keV for the combination (a) and below 5 keV for (b) it is likely that the useful spectral range is too narrow to well fit two spectral channels in, thus being not sufficient to estimate T_e.

These roughly estimated threshold values did not consider the finite bandwidth of spectral channels. They have been more carefully derived with a full simulation of the TS signals in realistic conditions and with two realistic sets of six spectral channels, depending on the choice of NIR detector. The full numerical simulation confirms the results from the rough estimate (see fig.2), with operational condition of about $T_e > 8$ keV for (a) and $T_e > 3$ keV for (b). Combination (a) would have the advantage of possibly using already foreseen lasers, while (b) could be used earlier in the ITER lifetime.

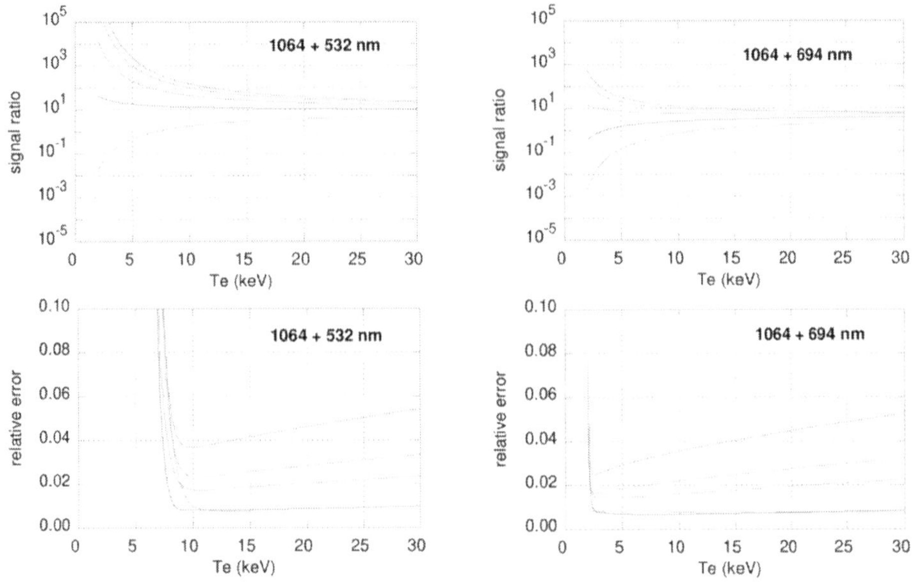

Figure 2. TS signals ratio (top) and relative error of calibration coefficients, derived from propagation of signal errors, (bottom) for two cases: Nd:YAG @ 1064nm + Nd:YAG @ 532nm (left) and Nd:YAG @ 1064nm + ruby @ 694nm (right)

Considering the unfamiliar and hostile ITER environment, a conservative strategy is proposed, operating both calibration methods: the back-illumination scheme being more reliable in the initial period of operation, the dual wavelength later on. In fact the dual wavelength method starts working above a threshold T_e, then probably some time after first day Hydrogen plasma, while the diffuser/retroreflector scheme might degrade with time loosing reliability. In addition, plasma Bremsstrahlung could play the role of internal light source, but this relies on repeated identical plasma discharges.

This work was supported by the European Communities under the contract of Association between EURATOM/ENEA. The views and opinions expressed herein do not necessarily reflect those of the European Commission.

4. References

[1] Walsh M J, Beurskens M, Carolan P G, Gilbert M, Loughlin M, Morris A W, Riccardo V, Xue Y, Huxford R B and Walker C I 2006 *Rev. Sci. Instrum.* **77**, 10E525

[2] Walsh M J et al. 2008 Performance Evaluation of ITER Thomson Scattering Systems *Proceedings 22nd IAEA Fusion Energy Conference (Geneva, Switzerland, 13-18 October 2008)* paper IT/P6-25

[3] Sugie T et al. 2008 "Spectroscopic Measurement System for ITER Divertor Plasma: Impurity Influx Monitor (divertor)" *Burning Plasma Diagnostics* ed G Gorini et al. (New York: AIP) **988** p 218-221.

[4] Ogawa H, Iwamae A, Sugie T, Kasai S, Kawano Y, Kajita S and Kusama Y 2008 Research and Development of Optical Diagnostics for ITER *Proceedings 22nd IAEA Fusion Energy Conference (Geneva, Switzerland, 13-18 October 2008)* paper IT/P6-23

[5] Smith O R P, Gowers C, Nielsen P and Salzmann H 1997 *Rev. Sci. Instrum.* **68**, 725

Journal of Physics: Conference Series **227** (2010) 012045 doi:10.1088/1742-6596/227/1/012045

Near-infrared plasma spectroscopy in support of divertor Thomson scattering diagnostics development for ITER

SYu Tolstyakov[1], EE Mukhin[1], MM Kochergin[1], GS Kurskiev[1], VV Semenov[1], A G Razdobarin[1], KA Podushnikova[1], AV Zabuga[1] and Globus-M[1] team, VS Lisitsa[2], MG Levashova[2], VA Soukhanovskii[3] and NSTX[4] team, M Beurskens[5], S Brezinsek[6], A Meigs[5] and JET[5] team P Andrew[7]

[1] Ioffe Physical Technical Institute, 194021, St.Petersburg, Russia
[2] Kurchatov Institute of Atomic Energy
[3] Lawrence Livermore National Laboratory, Livermore, CA, USA
[4] Princeton Plasma Physics Laboratory, Princeton, NJ, USA
[5] UKAEA Fusion Association, Culham Science Centre, Abingdon, Oxon, OX14 3DB, UK
[6] Institut für Plasmaphysik, EURATOM-Association, Jülich, D-52425, Germany
[7] ITER Organization, Cadarache, 13067 Saint-Paul-Lez-Durance, France
E-mail: Sergey.Tolstyakov@mail.ioffe.ru

Abstract. One of the main challenges of the implementation of divertor Thomson scattering system on ITER is weak laser scattering signal to be detected against intense background plasma radiation. The paper review briefly the line and continuum radiation data from present magnetic fusion devices in the spectral range of interest to TS diagnostics. The results will form the basis of design and development of the TS diagnostics for the ITER divertor.

1. Introduction

The operation of divertor Thomson scattering (TS) system on ITER will be under adverse conditions: the measurement will have to be made with the high radiation loads acted on optical elements which operates in a harsh environment and exposured to particles and dust. Additional challenges include a limited diagnostic access to the plasma and a weak laser scattering signal to be detected against intense background radiation. Here we review briefly the line and continuum radiation data from present magnetic fusion devices and calculation results of ITER divertor radiation in the spectral range of interest to TS diagnostics – the near infrared (NIR) region in the vicinity of 1000 nm. As it is seen from Figure 1 there are 21 viewing chords crossing the outer divertor leg and the divertor area under the dome.

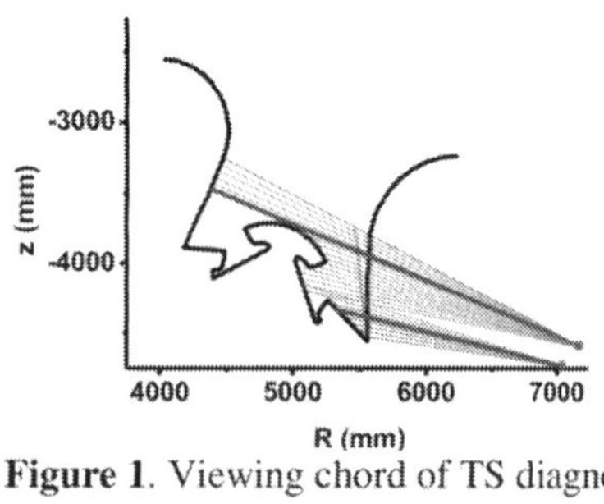

Figure 1. Viewing chord of TS diagnostics in the ITER divertor

A spectral survey of divertor plasma emission in the NIR range has been undertaken in the Globus-M Spherical Tokamak (St.Petersburg, RF), JET (Culham, UK) and NSTX (Princeton, USA) devices. The survey provided initial estimates of the impact of line and continuum emission on the ITER TS measurements accuracy. Based on the measurements it was possible to shape spectral characteristics of TS polychromator channels to minimize contributions from background plasma line radiation to the

TS signal. Based on estimates of the continuum and black body radiation intensities (from divertor plasma and divertor plates, respectively), the calculations of T_e accuracy were performed. The results will form the basis of design and development of the TS diagnostics for the ITER divertor.

2. Continuum background radiation within TS spectra

The background radiation in spectral channels of divertor TS in ITER is the continuous plasma radiation, thermal radiation of viewing surfaces, line radiation and stray light from laser radiation scattered on the dust and close walls. The specially developed grating spectrometer based on principle of subtractive dispersion [1] should manage the problem of the expected high intensive stray light. The continuous radiation of divertor plasma has two inputs: blackbody radiation of heated walls that is substantially target plates / dome surfaces and the strong bremsstrahlung radiation ($\sim n_e^2$) of high dense plasma $n_e \sim 10^{20}$-$10^{21}\,m^{-3}$ of ITER divertor. The background radiation is temporally constant within laser pulse scale and will be measured before laser pulse and subtracted. Nevertheless, accuracy of the measurements will drop as square root from sum of TS and background signal. Both types of background radiation are important for 11 spatial channels under the dome due to: (1) the upper viewing covers the heated surface attackable by plasma flows and (2) observation chords penetrates over the more extended plasma regions. Moreover, here the detected TS spectrum (and background) is wider in accordance with the higher temperatures under the dome.

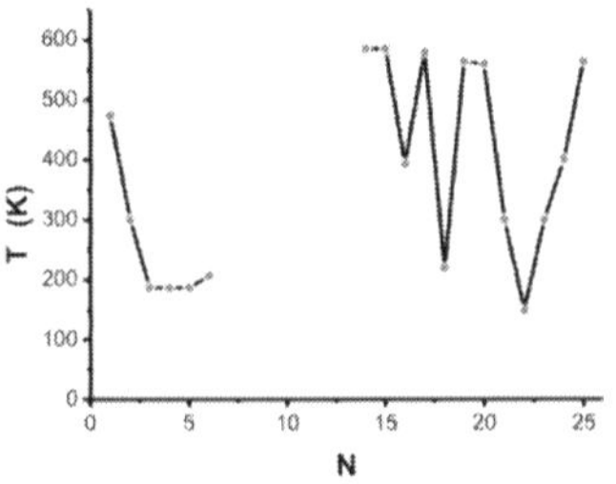

Figure 2. The temperature estimation along target plates. N corresponds to the squared points in Figure1.

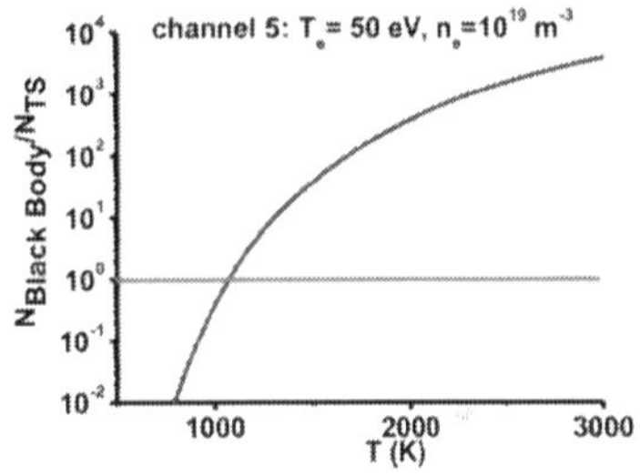

Figure 3. Ratio of heated surface radiation to TS signal for different temperatures of the surface

The most intensive black body radiation is expected for the upper chords viewing the dome surface, the estimated temperature here is 600 K (see Figure 2). As it is seen from Figure 3, the ratio of TS signal to surface radiation is ~ 2 for this temperature and is not crucial for noise consideration. The measurement limitations will arise at temperatures more than 1500 K. But care should be taken to the surface color temperature that can strongly differ from predicted bulk temperature. The matter is that carbon films on the divertor surface can have a rather high temperature [2]. Actually, JET experiments show the black body emission from the divertor with temperature reaching 2000 K [3]. It is worth to note that the detailed calculation of surface radiation intensity for ITER divertor plates is complicated now, but it is necessary to keep in mind the possible additional noise source.

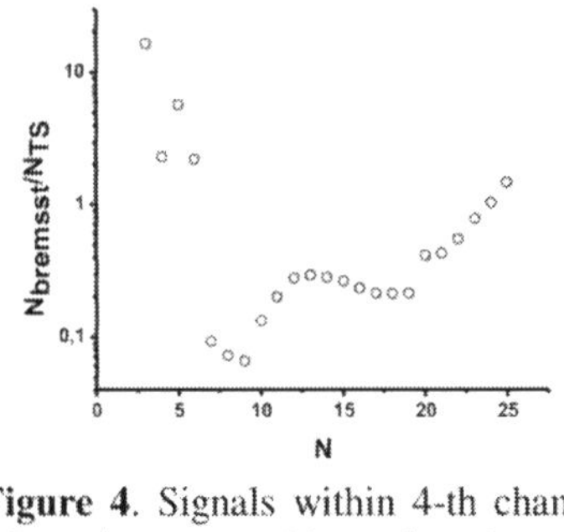

Figure 4. Signals within 4-th channel – TS and bremsstrahlung for observation chords.

The bremsstrahlung input estimation is carried out from calculated plasma parameters distribution along viewing chords. The distribution was extrapolated from 2D matrix of plasma parameters simulated with the SOLPS4.3 code [4] for the main operational mode with 8MW loading on divertor targets.

The bremsstrahlung input is mainly pronounced for six spatial channels on the top where the viewing chords directed under the dome and the path through divertor plasma is several times longer than for other chords (see Figure 2). As it is seen from comparison of TS signal and bremsstrahlung signal in spectral channels of grating spectrometer (see Figure 4), the estimated level of bremsstrahlung can strongly exceed TS signal for the upper spatial points of probing chord and can markedly limit the diagnostics sensitivity.

3. Line emission within TS spectra

The ranges of measured electron parameters in ITER are determined by the technical requirements to ITER diagnostics [7]. One of the disputed electron parameters in ITER divertor is maximal T_e which is declared in [7] as 200 eV. At the same time the upper limit of T_e guaranteed by technical specification of divertor TS (see [6]) is 100 eV. The complexity of the TS measurements in the range from 100 eV to 200 eV is the strong line radiation in the range from 1004 nm to 1012 nm that interfere with TS spectra. The only neatly study of the spectral gap can help to fit the diagnostic specifications to the technical requirements. To calculate the intensity of plasma radiation within TS spectral range it was used the divertor plasma parameters simulated with the code SOLPS4.3 [5]. The intensity of hydrogen line emission and bremsstrahlung was calculated by the code CRETIN [4] along two observation chords directed under the dome and in the outer leg (see Figure 1). Results of these calculations are presented in Figure 5. As is seen, the line intensity strongly exceeds (up to ~2 orders in the line peak) the continuous radiation and the reduction of TS signal/noise ratio can be prohibitive. The use of filters blocking the spectral lines will limit the polychromator spectral range and as a result T_e range. Therefore, the spectral shaping should be a compromise between a minimal line income and a maximal covering of TS spectra which means that the line broadening is one of the main critical points for TS spectrometer spectral channel shaping. The thorough simulation (see Figure 6) of the broadening of the P7 line integrated along two considered chords (see marked by blue chords in Figure 1) involves a number broadening mechanism - Zeeman splitting, ion static/dynamic and electron impact Stark broadening, Doppler broadening. It was shown, the basic mechanisms of the broadening are Zeeman splitting and ion static Stark broadening. As is seen from Figure 6 the resulting FWHM of the lines does not exceed 1 nm. But the acceptable radiation level for the most background lighted chords directed under the dome is one order more than bremsstrahlung level and corresponds to the line width of ~4 nm.

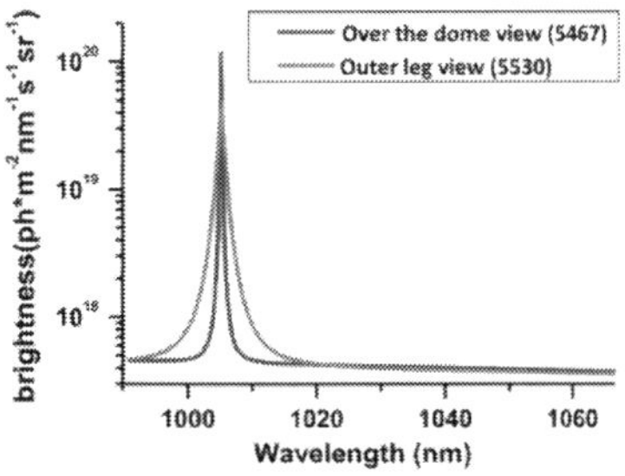

Figure 5. The H lines and bremsstrahlung emission (CRETIN code).

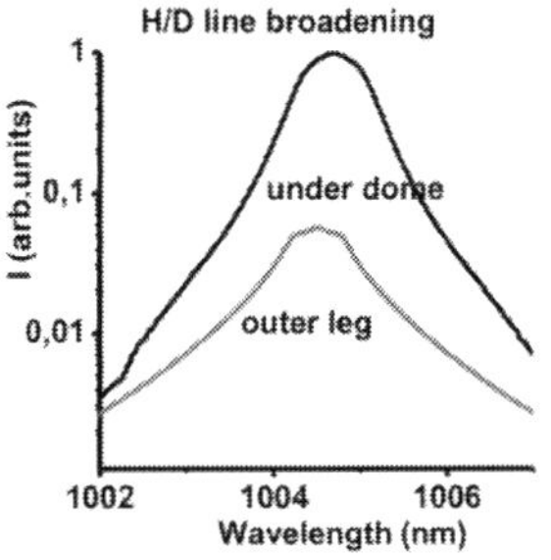

Figure 6. P7 line broadening.

In support to the calculations the experimental study of the NIR plasma emission was performed in tokamaks NSTX, JET and Globus-M. The measurements were made in the divertor region where the thermal loading on the divertor targets are similar one to the ITER (~several MW/m^2). The coincidence of absolute signals integrated along the viewing chords can be explained by equivalent properties of the SOL plasma that is conventionally the main source of radiation. In this case the results of the experiments can be extrapolated on ITER and can be considered as a starting point to estimate background condition in ITER divertor. As is seen from the spectra (Figure 7) the main obstacle lines are the lines of H/D from Pashen series and He-II superimposed

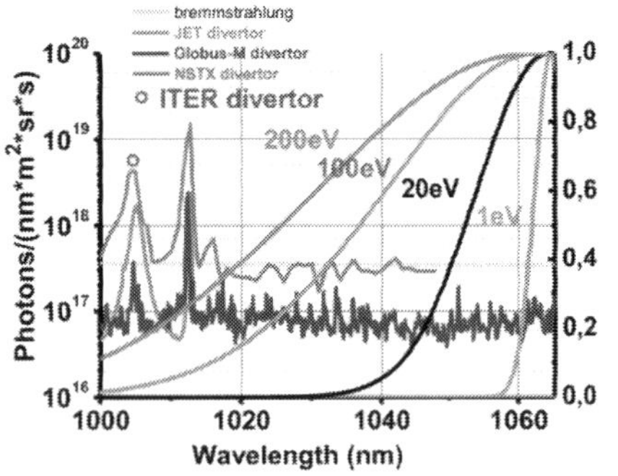

Figure 7. Data on spectroscopic measurements in the Globus-M and NSTX divertor and TS spectra.

with carbon line. The spectral intensities of the lines measured in the divertor of NSTX and Globus-M are nearly equal and two order more than expected intensity of bremsstrahlung in ITER divertor.

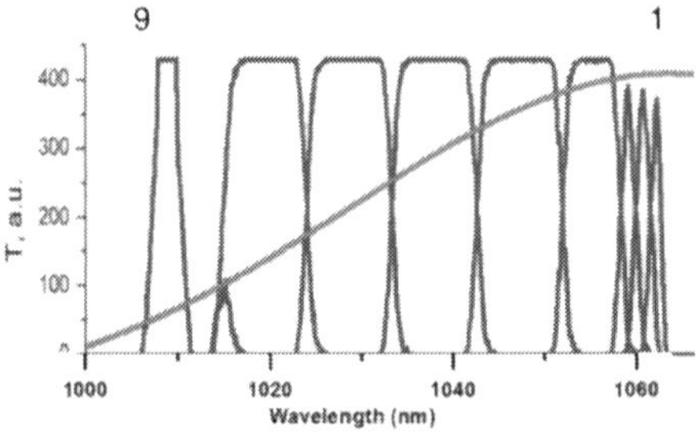

Figure 8. Spectral channel 1-9 of the designed grating polychromator with TS spectrum 200 eV

The performed spectroscopic calculations and measurements help to shape the spectral characteristics of TS polychromators. The spectral channel of designed grating polychromator are presented in Figure 8. The first three narrow channels are intended to measure the low temperatures $\geq$ 1eV. Next six channels serve temperature range up to 200 eV. The channel characteristics was optimized to provide the even T_e measurement accuracy within full range of TS temperature range [5]. At the same time according to previous consideration the He/C and broadened hydrogen lines became the limiting factor to measure temperatures ~200 eV. To manage this complications the characteristic of 9-th channel is corrected by filter (Figure 7) notching the line emission.

4. Conclusions

The spectroscopic measurements and calculations were used as the basis for TS polychromator design to provide minimal income of plasma radiation into the TS measurements accuracy. The density and temperature ranges of TS diagnostics following from [4] are limited within $n_e=10^{19}$-10^{21} m^{-3} and T_e=1-200 eV under accuracy $\leq$5 % and $\leq$10 % respectively. The predicted accuracies of n_e and T_e along the probing chord (21 points) for Principal operation mode 8 MW/m^2 (run #1514 [5]) are presented in Figure 9. The intensity of bremsstrahlung background light was taken multiplied by 10 that is conventional for TS accuracy simulation [DDD5.5]. As is seen from Figure 8 the needed accuracy of 5% for density and 10% for temperature does not exceeded in this simulation for the main ITER operation regime.

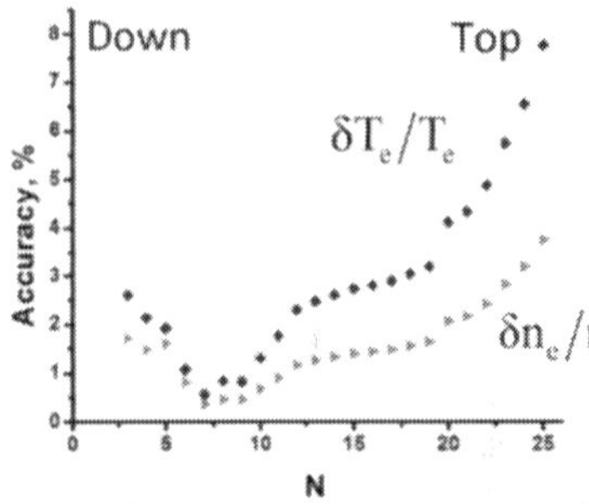

Figure 9. Measurement accuracy of n_e (triangle) and T_e (squares) distribution along probing chord.

Acknowledgments

The authors wish to thank personally Dr. A.B.Kukushkin (Kurchatov Institute of Atomic Energy) for the help in spectroscopic calculations and useful considerations. Dr. H.A.Scott (LLN Laboratory, Livermore, CA, USA) is acknowledged for the Cretin code support. This report, supported by Rosatom (contract -N.4z.41.03.08.258), was prepared as an account of work for the ITER Organization. The Members of the Organization are the People's Republic of China, the European Atomic Energy Community, the Republic of India, Japan, the Republic of Korea, the Russian Federation and the United States of America. The views and opinions expressed herein do not necessarily reflect those of the Members or any agency thereof. Dissemination of the information in this paper is governed by the applicable terms of the ITER Joint Implementation Agreement. Also this work was performed in part under the auspices of the U.S. Department of Energy under Contracts DE-AC52-07NA27344 and DE-AC02-76CH03073. This work was supported by NWO-RFBR Centre-of-Excellence grant 047.018.002. The NIR Spectroscopy Detectors OMA-V was provided by Ioffe Fusion Technology (S.Petersburg, Russia).

References

1. Mukhin E E et al 2004 Double-grating polychromator for laser-aided plasma diagnostics, Rev. Sci. Instrum. 75, 1261
2. Reichle R et al 19.12.2004 Preliminary final report for the EFDA task TWP2002 TW2-TPDS-DIASUP-231 Task 2.3Thermography * part II : CEA
3. Stamp M, 3rd Sept, 2003 Hot Inner Divertor at JET, TFE meeting, JET
4. Scott H A, 2001 J. Quant. Spectrosc. Radiat. Transfer 71 689
5. Kukushkin A S, et al 2009 *Nucl. Fusion* 49 075008
6. Mukhin E E et al 2009 Thomson scattering diagnostics for ITER divertor, this conference
7. ITER diagnostics technical requirements

Journal of Physics: Conference Series **227** (2010) 012046 doi:10.1088/1742-6596/227/1/012046

Electron density measurement in a pulsed-power plasma by FIR laser beam deflection and/or interferometry

P Kempkes, F Mackel and H Soltwisch

Institute for Experimental Physics V, Ruhr-University Bochum, Universitaetsstrasse 150, 44780 Bochum, Germany

E-Mail: henning.soltwisch@ruhr-uni-bochum.de

Abstract. A pulsed-power experiment has been designed to produce arc-shaped magnetic flux tubes similar to ascending solar flares. The tubes are filled with hydrogen plasma (electron temperature $\leq$ 10 eV, electron density $2...3\times10^{20}$ m^{-3}) and expand with a velocity of ~2.5 cm/µs, while keeping their cross section constant at a radius of about 1.5 cm. For measuring the spatial electron density distribution within the moving flux tube, a single cw laser beam can be used. The information taken from the laser beam, which traverses the vacuum vessel perpendicular to the plane of the plasma arch, can be either the phase shift or the beam deflection due to the density gradient. Assuming a parabolic distribution with a central electron density of 2×10^{20} m^{-3}, the maximum deflection angle occurring at an impact parameter of 0.7 amounts to $\gamma_{max}/\text{deg} \cong 10^{-5} \times (\lambda/\mu m)^2$. Hence, a FIR laser operating at λ = 433 µm would be deflected by γ_{max} = 1.9° only. Alternatively, a beam passing through the plasma centre would experience a plasma-induced phase shift of $\Delta\phi_{max}/\text{rad} \cong 10^{-2} \times (\lambda/\mu m)$, yielding 4.3 rad for a FIR laser (λ = 433 µm) and 0.1 rad for a CO_2 laser (λ = 10.6 µm). While the former is readily detectable in a standard interferometer, the latter requires a more advanced technique of measurement to achieve the necessary resolution. On the other hand, the short wavelength compared to FIR radiation allows for a very narrow beam and hence for a high spatial resolution. For these reasons a so-called coupled-cavity scheme for a CO_2 laser interferometer is presently under development.

1. The *FlareLab*-Experiment

FlareLab is a laboratory experiment that has been built to perform detailed studies into the behavior of plasma configurations with initially loop-shaped magnetic field lines and to analyze the observed dynamics of the magnetic flux tubes by means of extensive 3-dimensional MHD simulations [1]. Well-known examples of such configurations are arched Solar prominences, which often result in spectacular eruptions like Coronal mass ejections (CMEs).

The lay-out of the experiment follows a scheme which was proposed and successfully operated by Bellan and Hansen [2]. A sketch of our present setup is shown in figure 1. In order to produce a plasma, hydrogen gas is puffed into the space above the electrodes, and a discharge is ignited. The current causes the plasma column to pinch along the guiding B-field and to form an expanding loop structure (see figure 1 for a snapshot of a discharge taken 5 µs after ignition with an exposure time of 50 ns by means of a fast ICCD camera).

At present our set of diagnostics is still limited to standard electro-technical devices for measuring currents and voltages, and to an electrostatic triple probe (supported by elementary spectroscopy) for

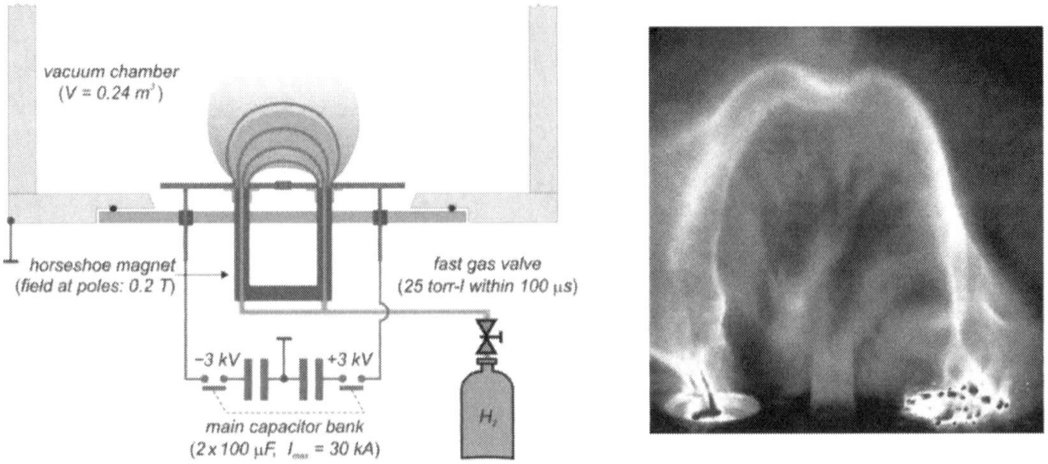

Figure 1. Sketch of the experimental setup and snapshot of a discharge (5 μs after ignition).

estimating electron densities and temperatures. In addition, a movable pick-up coil has been installed to obtain space- and time-resolved information about the magnetic field distribution near the flux tube. It turns out that the loop expands with a fairly constant speed of about 2.5 cm/μs while keeping its cross section at a diameter of ≤ 3 cm. Triple probe and spectroscopy provide electron temperatures and densities of the order of a few eV and a few times 10^{20} m^{-3}, respectively, with little variation throughout the loop expansion. On the other hand, MHD simulations of the dynamical behavior [1] have shown that the underlying model of the density evolution plays an important role for matching the experimental observations. Hence there is an urgent need to improve our diagnostic capabilities for a more precise determination of the electron density distribution.

2. Laser-aided electron density measurements

In our case, already a single cw laser beam (directed perpendicular to the plane of the flux loop) will be sufficient to resolve the spatial density profile. After ignition the plasma arch ascends from its foot points until it reaches the beam and travels across. As a result, the beam is scanning the whole plasma cross-section and provides all the information needed to apply standard Abel inversion techniques in order to convert the measured data into a spatial density distribution (assuming cylindrical symmetry and neglecting possible profile changes during the passage). The changes imposed on the beam are time-varying phase shifts and beam deflections due to the density gradient. While the former are routinely utilized in various interferometric schemes, the latter are less frequently exploited in special schlieren and shadowgraph techniques. We will briefly consider both options for our purpose.

2.1. Plasma-induced beam deflection

If the gradients of an axisymmetric refractive index distribution $\mu(r)$ are large enough to cause detectable deflections of a test beam, it is possible to calculate $\mu(r)$ from the knowledge of the deflection angle γ as a function of x (see figure 2). Using the approximate expression for the refractive index of a fully ionized plasma without magnetic field (i.e. $\mu = (1 - \omega_p^2/\omega^2)^{1/2} \approx 1 - \omega_p^2/(2\omega^2)$ where ω_p denotes the electron plasma frequency and $\omega = 2\pi c/\lambda$ is the probing wave frequency), it can be shown that the electron density distribution $n_e(r)$ derives from the deflection profile $\gamma(x)$ according to [3,4]:

$$n_e(r) \approx \frac{2\varepsilon_0 m_e}{\pi e^2} \cdot \omega^2 \int_r^a \gamma(x) \frac{dx}{(x^2 - r^2)^{1/2}} . \quad (1)$$

Conversely, for a parabolic density distribution $n_e(r) = n_{e0}\,(1 - r^2/a^2)$ the deflection profile is given by

$$\gamma(\xi) \approx 2\kappa\,\xi(1-\xi^2) \quad \text{where } \xi = \frac{x}{a} \text{ and } \kappa = \frac{\omega_{p0}^2}{\omega^2} = \frac{e^2}{4\pi^2 c^2 \varepsilon_0 m_e} n_{e0}\,\lambda^2 . \tag{2}$$

It has a maximum value of $\gamma_{max} = \gamma(\xi_1) = \kappa$ at a beam position of $\xi_1 = 2^{-1/2} = 0.707$ (cf. figure 2).

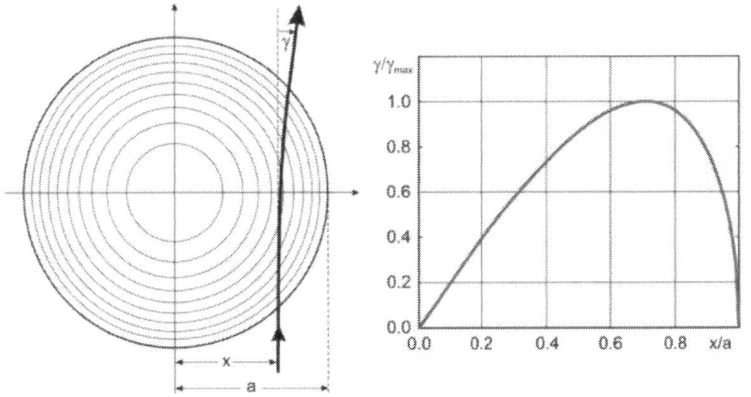

Figure 2. Beam deflection caused by an axisymmetric plasma with a parabolic distribution of the electron density.

Taking a characteristic central density of $n_{e0} = 2\times10^{20}$ m^{-3} and asking for a maximum deflection angle of $\gamma_{max} \geq 3°$ to achieve reasonable detectability, the probing wavelength must be $\lambda \geq 540$ µm. At such long wavelengths it would be difficult to restrict the laser beam to a ray which is much narrower than the plasma radius of about 1.5 cm. Moreover, suitable detector arrays providing high enough spatial and temporal resolution in the FIR regime would hardly be available. Therefore we have to conclude that beam deflection measurements are not a viable option for our experiment.

2.2. Plasma-induced phase shift

Interferometric phase-shift measurements are well established for determining line-averaged values of the electron density in a plasma. The relation between phase shift $\Delta\varphi$ and density n_e is given by [5]:

$$\Delta\varphi = r_e\,\lambda \int_L n_e\,dl \quad \text{where } r_e = e^2/(4\pi c^2 \varepsilon_0 m_e) = 2.82\times10^{-15} \text{ m} \tag{3}$$

(the integration extends over the path length L). For an axisymmetric distribution $n_e(r)$ as shown in figure 2 this expression can be transformed into an Abel integral equation according to

$$\Delta\varphi(x) = 2r_e\,\lambda \int_x^a n_e(r) \frac{r\,dr}{(r^2-x^2)^{1/2}} \quad \Leftrightarrow \quad n_e(r) = \frac{-1}{\pi r_e \lambda} \int_r^a \frac{d\,\Delta\varphi}{dx} \frac{dx}{(x^2-r^2)^{1/2}} . \tag{4}$$

Taking again a parabolic profile with a central density of $n_{e0} = 2^{20}\times10$ m^{-3} and a radius of $a = 1.5$ cm, a beam passing through the plasma centre at $x = 0$ would experience a wavelength-dependent phase shift of $(\Delta\phi_{max}/\text{rad}) = 1.13\times10^{-2} \times (\lambda/\mu\text{m})$, resulting in 0.12 rad for a CO_2 laser ($\lambda = 10.6$ µm). In terms of interference fringes this number corresponds to 1/50 of a fringe, which is close to the resolution limit of common interferometers (as, e.g. Mach-Zehnder or Michelson configurations). On the other hand, the comparatively short CO_2 wavelength allows for a fairly narrow probing beam, and the technical equipment (including detectors) is well developed and easy to operate.

3. Interferometer setup and first results

A technique, which is well suited for measuring small phase shifts, makes use of a coupled-cavity laser interferometer [6,7,8] as shown in figure 3. Here, the radiation of a grating-tuned CO_2 laser is reflected back into the cavity by means of an external mirror on the opposite side of the discharge chamber. The mirror is placed at a distance of two times its focal length and mounted on a 3-axes micrometer table. The mounting contains a piezo actuator for shifting the mirror within a range of

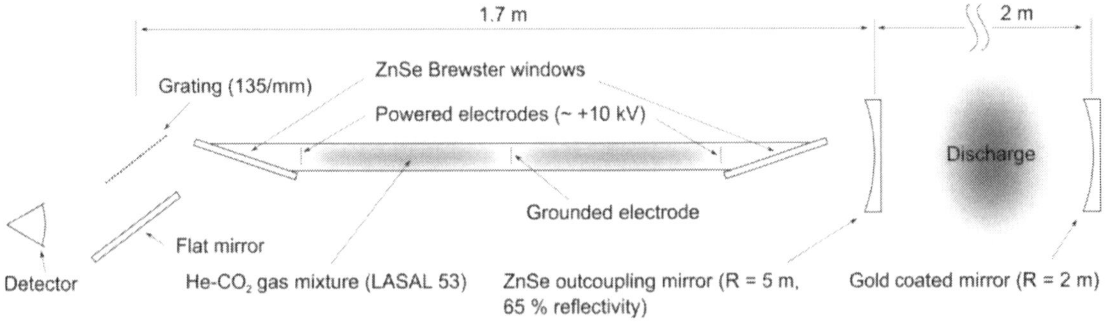

Figure 3. Coupled-cavity CO_2 laser interferometer.

appr. 20 μm. Depending on the position of the external mirror, the radiation reflected back into the laser is in phase or out of phase relative to the standing wave inside the laser resonator. This results in a change of the laser power as a function of the position of the external mirror (or the optical path length), which is monitored by collecting the residual light from the resonator at the grating by means of a photodiode detector.

In order to test the sensitivity of the device, the external mirror was moved by the piezo actuator, and the detector signal was recorded as a function of the mirror shift (figure 4). At the steepest slope a sensitivity of 3.7 V/μm is achieved over a range of roughly 1 μm. Since the detector noise is well below 100 mV, changes of the optical path length of less than 0.03 μm (corresponding to 0.3% of a wavelength) can be detected, which will be sufficient to determine the electron density distribution in our FlareLab experiment.

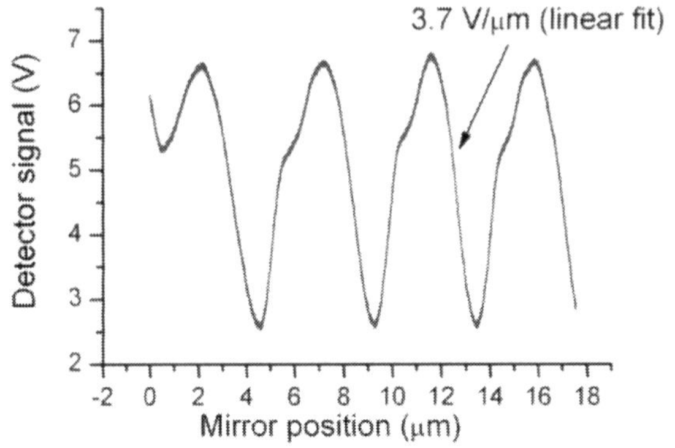

Figure 4. Variation of the laser output power as a function of the position of the external feedback mirror. The laser is operated at the 9R20 line (~9.3 μm).

References

[1] Arnold L, Dreher J, Grauer R, Soltwisch H and Stein H 2008 *Phys. Plasmas* **15** 042106
[2] Bellan P M and Hansen J F 1998 *Phys. Plasmas* **5** 1991
[3] Born M and Wolf E 1959 *Principles of Optics* (Oxford: Pergamon)
[4] Kogelschatz U and Schneider W R 1971 *Quantitative schlieren techniques applied to high current arc investigations* (Baden: Brown Boveri Research Report KLR-71-36)
[5] Soltwisch H 1991 Interferometry for measuring electron density distributions *Diagnostics for contemporary fusion experiments* ed P E Stott (Bologna: Editrice Compositori) pp 75-94
[6] King P G R and Steward G J 1963 *New Scientist* **17** 180
[7] Mitchell I H and Wheeler C B 1988 *J. Phys. E: Sci. Instrum.* **21** 696
[8] Pargmann C, Singh S V and Soltwisch H 2005 *Rev. Sci. Instr.* **76** 113506

Journal of Physics: Conference Series **227** (2010) 012047 doi:10.1088/1742-6596/227/1/012047

Developing in-situ Ellipsometry for Tokamak Discharges in KSTAR

Suk-Ho Hong and Woong-Chae Kim

National Fusion Research Institute,113 Gwahangno, Yusung-Gu, Daejeon, 305-333, Korea

E-mail: sukhhong@nfri.re.kr

Abstract. An in-situ ellipsometer based on four-detector polarimeter(FDP) is under development at KSTAR. In-situ ellipsometer for tokamak discharges will measure the characteristics of thin films deposited onto a quartz window near the edge region in real time. These characteristics contain local deposition/erosion rates as well as hydrogen to carbon ratio, which have to be measured in-situ, for more clear insight view of plasma-wall interaction in tokamak edge plasmas. This paper reports the status of in-situ ellipsometer development for tokamak discharges at KSTAR, to study plasma-wall interaction and fuel retention. Basic concept, design and construction of the ellipsometer are described.

1. Introduction

Ellipsometry is an optical technique for the characterization of an interface or a film between two media. It is based on exploiting the change of polarization state of a beam that is reflected from or transmitted through the media. An ellipsometric measurement does not destroy the sample surface or interface during the measurement: It does not disturb deposition/etching processes, hence it is suitable for in-situ measurements. Also, the change of film properties can be characterized during the process with ultra high sensitivity for the change of the surface condition (accuracy of a monolayer detection). Thus, ellipsometry is a standard diagnostics in low temperature processing plasmas to control the thickness of thin films fabricated.

Why we need in-situ ellipsometry for tokamak discharges? Plasma-wall interactions during the tokamak discharge lead to the erosion of plasma facing components (PFCs) [1]. Eroded materials are transported to relatively cold area of the vacuum vessel and re-deposited as thin layers on the surface. Erosion and re-deposition of PFC materials, especially carbon containing materials cause several issues on the operation as well as safety. As the thickness of layers grows up, the flaking of thick layers leads to the creation of carbon dusts in the vacuum vessel during the plasma operation [2]. Furthermore, since carbon is always deposited as hydrocarbon molecules, fuel retention in the layer is another important issue [3]. ITER has set safety limit of tritium in the vacuum vessel [4], thus fuel retention has to be controlled effectively when it is necessary: Exact erosion and re-deposition rate of PFC materials have to be studied quantitatively. Furthermore, output of an in-situ ellipsometry provides complex refractive indices of the deposited layers. In the case of carbon materials, complex refractive indices are directly connected to hydrogen to carbon ratio in the layer [5]. Thus, measuring complex refractive indices give direct information on the fuel retention.

We have designed and constructed a four-diode-ellipsometer for tokamak discharges to measure the re-deposited layer thickness and optical constants in-situ. Basic concept is based on the use of four-detector photopolarimeter (FDP) suggested by Azzam et al [6]. The FDP consists of four photo-diode detectors and a laser, that measure complete polarization state of laser light reflected by substrate-thin film system. Dittmar et al. have constructed an in-situ ellipsometer based on FDP for TEXTOR, but the measurements during the tokamak discharge have little success due to the thermal erosion of protective coatings on the mirror [7]. In order to control the temperature of the optical components, we have used cooling units, that are based on thermoelectric effect. We have finished the design and construction of the ellipsometer, and planned to be tested at a small vacuum vessel. After the test of the ellipsometer, it will be installed in KSTAR at a mid-port location to measure the growth of re-deposited a-C:H layers in-situ.

2. Principle: Four-Detector Polarimeter

In order to measure the optical characteristics of film-substrate system, it is needed to know complete polarization state of a light beam reflected by the surface. There are numbers of methods to measure the polarization state [8]. These devices measure the polarization state as ellipsometric angles Ψ and Δ, that are connected to Fresnel's coefficients

$$\tan \Psi e^{i\Delta} = \frac{r_p}{r_s} e^{i(\delta_p - \delta_s)}, \tag{1}$$

where r_p and r_s represent the reflection coefficients of parallel(p) and perpendicular(s) polarization state, and δ is the phase difference. Ellipsometric angles Ψ and Δ provide characteristics of sample under study. Note that, in tokamak environments we are dealing with high level of heat flux and magnetic fields. Thus, the use of conventional optical components will face problems such as Faraday effect, cooling problem, and experimental setup using stepper motors will not be available. To overcome these problem a static photopolarimeter, so-called "four-detector polarimeter (FDP)" can be used: The state of polarization is measured by splitting light by diffractive elements onto four photodetectors and measuring their response. The concept of FDP is described by Azzam in 1985, for the first time [6], and used for various polarization measurements, e.g., for Kerr effect measurement [9]. The incoming light is reflected under an oblique angle by the first three detectors and finally absorbed by fourth detector at normal incidence. At each detector a part of light will be absorbed and generate photocurrent I_k, where k=0,1,2,3. Azzam has showed a linear relation between photo-current I and stokes vector S,

$$I = AS \tag{2}$$

where A is a 4×4 matrix called *instrument matrix*, which describes the relation between I and S. Instrument matrix contains all characteristics of the response of the FDP to incident light. Once the matrix A is determined, complete polarization state of the light can be determined.

3. Design and Construction of in-situ Ellipsometer for Tokamak Discharges

As we have mentioned above, most of diagnostics for tokamak discharges have restricted access to the plasma due to the high heat flux and high magnetic fields: Areas in tokamaks at which physical quantities of plasma-wall interaction can be measured are relatively restricted. This leads to the limitation of size, design, geometry and installation of a diagnostic. An in-situ ellipsometry for tokamak discharge measures thin film formed by plasma-wall interaction at edge region. It can be installed at a vertical or at a mid-port. KSTAR has 24 vertical ports (12 at top and 12 at bottom) and 16 mid-ports. Because the size of mid-ports is much larger than vertical ones and they have more flexibilities, we have designed an in-situ ellipsometer which will be installed at a mid-port in KSTAR. A picture of designed and constructed prototype ellipsometer is shown in figure 1. The basic concept of the ellipsometer is to construct a simple FDP. Normally, conventional ellipsometers measure the polarization state of ambient-thin film-substrate system.

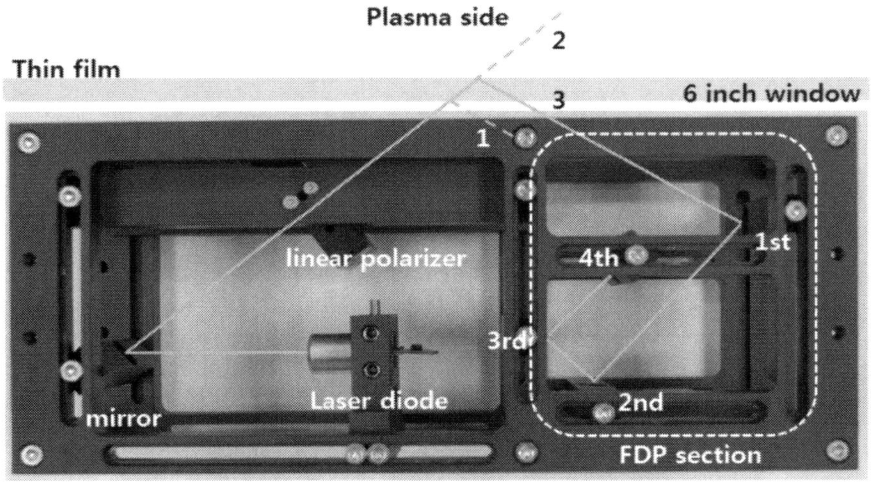

Figure 1. Picture of in-situ ellipsometer based on FDP constructed at KSTAR. Solid line represents the light beam from a green laser diode.

Figure 2. Picture of in-situ ellipsometer assembled into aluminum case attached to 6 inch window.

A laser beam is passing through the ambient and thin film and reflected by the substrate, then detected by a photodoide with an (rotating) analyzer. In order to measure a sample exposed to a tokamak edge plasma, the laser beam has to pass through and to be reflected back along a long path from the laser to the detector. Thus, it is impossible to construct an ellipsometer based on such a simple "reflection geometry" in a tokamak. An alternative configuration was suggested by Dittmar et al [7]: This is based on the backside reflection measurement. The laser is placed at the opposite side of the plasma and FDP measures the back-reflected laser beam. At KSTAR, we have adopted only the concept of Dittmat et al [7], and designed for KSTAR mid-port installation (see figure 1). The dimension of the ellipsometer part in figure 1 is 205 mm × 90 mm × 80 mm. A 35 mW green diode laser is horizontally mounted and laser beam is launched. The beam is reflected by a photodiode of the same type as detector (act as mirror). The beam passes through a linear polarizer to be sure of the polarization state of the incident beam. A part of the beam will be reflected at the surface of a 6 inch window (label

1 in figure 1) and the rest of the beam will pass through the window (label 2 in figure 1). At the backside of the window where the deposition occurs due to the plasma-wall interaction, a part of the beam is again reflected (label 3 in figure 1) and finally detected by FDP. The mirror and FDP are mounted on movable rods so that the incident angles and the horizontal/vertical position of each component can be adjusted in a fine scale. Light intensities at each detector will be recorded separately by a personal computer (5 channels). The acquisition speed will be in the kHz range. By analyzing the signal, the state of the polarization of the beam at the detector can be determined. Note that, lower power laser can be used for a conventional ellipsometer. Due to the concept of backside reflection, the light intensity reaching at the detectors is significantly reduced by a factor of 10^3 [7]. Preamplifiers can be employed in case of bad signal to noise ratio at detectors. In order to test the ellipsometer before it is installed in KSTAR, further test and design optimization are needed. This is a reason for the use of a 6 inch quartz window: The constructed ellipsometer can be mounted easily to any vacuum system which has 6 inch port for further test. The case of the ellipsometer is made of aluminum. The outer part of the case will be covered by graphite tiles to protect system from high heat flux and a electrothermal cooling system will be installed inside.

4. Summary and Outlook

We have designed and constructed a prototype of an in-situ ellipsometer for tokamak discharges. Assembly and alignment of optical components were accomplished and the detectors are ready to record signals. The next step of the development is calibration of the ellipsometer. Using laser light of known polarization state, the response of FDP can be calibrated. This procedure determines the instrument matrix A [10]. For the calibration, we have constructed a "calibration unit" which is exactly the same as ellipsometer unit shown in figure 3, but assembled as mirrored at the upper part of ellipsometer unit without 6 inch window, with an additional $\lambda/4$ plate after the polarizer. This calibration unit produces any "known" polarization state for the calibration. After the instrument matrix A is determined, the system will be ready to measure any polarization state incoming towards FDP. The system will be mounted to a small device where a-C:B:H thin films are deposited to optimize the boronization in KSTAR, and the growth of thin film will be measured and the ellipsometer will be intensively tested and optimized. In 2010, the ellipsometer will be installed in KSTAR (Faraday effect on window has to be considered also) and plasma-wall interaction at edge region will be studied: Information on local erosion and deposition, hydrogen to carbon ratio can be directly obtained during plasma shots.

Acknowlegment

This study was supported by the Korean Ministry of Education, Science and Technology under the KSTAR project.

Reference

[1] Federici G et al., 2001 Nuclear Fusion, **Vol. 41**, 1967
[2] Pegourie B et al., 18th International Conference on Plasma Surface Interactions (2008)
[3] "Generic Site Safety Report - Volume III - Radiological and Energy source terms - July 2004", G 84 RI 3 R0.2.
[4] "Generic Site Safety Report - Volume VIII - Ultimate safety margins - July 2004", G 84 RI 7 R0.2.
[5] Schwarz-Selinger T et al, J. Appl. Phys., **Vol. 86**, No. 7, 1 October 1999
[6] Azzam R M A, Optics Letters, **10** (1985), 309
[7] Dittmar T, Phd. Thesis, Ruhr-University Bochum, 2008
[8] Azzam R M A and Bashara N M, "Ellipsometry and polarized light", North Holland, 1987.
[9] Dumelow T et al., Rev. Sci. Instrum., **Vol. 73**, 2001, 1225
[10] Azzam R M A et al., J. Opt. Soc. Am. A, **5**, No 5 (1988) 681

AUTHOR INDEX

AUTHOR INDEX

AUTHOR INDEX

Institute of Physics
Dirac House, Temple Back
Bristol BS1 6BE UK

proceedings.com